BIOMAT 2007

International Symposium on
Mathematical and Computational Biology

BIOMAT 2007

International Symposium on
Mathematical and Computational Biology

Armação dos Búzios, Rio de Janeiro, Brazil
24 – 29 November 2007

edited by

Rubem P Mondaini
Universidade Federal do Rio de Janeiro, Brazil

Rui Dilão
Instituto Superior Técnico, Portugal

 **World Scientific**

NEW JERSEY · LONDON · SINGAPORE · BEIJING · SHANGHAI · HONG KONG · TAIPEI · CHENNAI

Published by

World Scientific Publishing Co. Pte. Ltd.

5 Toh Tuck Link, Singapore 596224

USA office: 27 Warren Street, Suite 401-402, Hackensack, NJ 07601

UK office: 57 Shelton Street, Covent Garden, London WC2H 9HE

British Library Cataloguing-in-Publication Data
A catalogue record for this book is available from the British Library.

BIOMAT 2007
International Symposium on Mathematical and Computational Biology

Copyright © 2008 by World Scientific Publishing Co. Pte. Ltd.

ISBN-13 978-981-281-232-2
ISBN-10 981-281-232-6

Printed in Singapore by B & JO Enterprise

PREFACE

This is the volume of selected works of the BIOMAT 2007 International Symposium on Mathematical and Computational Biology. The BIOMAT Symposia is known all over the world as a series of international interdisciplinary conferences which are held annually since the year 2001. The series is also the oldest in Latin America to assemble together professionals from diverse educations to exchange their expertise in the topics of biomathematics and the general study of biosystems. The two days of pre-conference tutorials are aimed to motivate young candidates to scientific research in these interdisciplinary areas and are lectured by selected teachers among the best representatives of the tutorial research topic. Every year the structure of symposia, including tutorials and technical sessions, has been proved to be efficient for motivation of future scientific careers and the feedback obtained by senior scientists from their peers.

The BIOMAT 2007 Symposium was held in the city of Armação dos Búzios, an internationally famous beach resort in the State of Rio de Janeiro, Brazil. There were fourteen Keynote Speakers coming from Europe, Asia, Africa and Americas. We have been most honoured with the presence and collaboration of a scientist with the scientific status of Prof. Kerson Huang, a very famous master of Statistical Mechanics. The author of many important contributions to the scientific literature and famous textbooks used by research students of five continents. We had also about hundred and forty submitted works and an 18 % – 22 % level of acceptance. The accepted papers in full version have been presented in oral and poster continuous sessions. It is worth while to register the deep impact that these sessions have promoted in joint publications of some participants.

As a rule in the BIOMAT Symposia, the set of selected papers is a fruitful combination of state of art research and some review approaches. They can be grouped in the areas of Mathematical Biology, Biological Physics, Biophysics and Bioinformatics. There are new results on some aspects of Lotka-Volterra equations, the proposal of using Differential Geometry to model neurosurgical tools, recent data on Epidemiological Modelling, Pattern Recognition and comprehensive reviews to the Structure of Proteins, the Folding Problem and the influence of Allee effects on Population Dynamics.

The book contains some original results on the growth of gliomas. The role played by membranes channels on activity-dependent modulation of spike transmission. A proposal for reconsidering the concept of gene and the understanding of the mechanism responsible for gene expression. Some useful ideas for the modelling of ordered sets of atom sites. The comparison of agent-based models with the approach of differential equations on the study of selection mechanisms in Germinal Centres. The synchronization phenomenon for Protocell systems driven by linear kinetic equations.

We are indebted to the Board of Trustees of two Brazilian sponsoring agencies: Coordination for the Improvement of Higher Education Personnel CAPES and Foundation for Research Support of Rio de Janeiro State. We thank PETROBRAS, the Brazilian Oil Company and the world leader of oil research on deep sea waters and the PETROBRAS-CENPES Research Centre. The financial contribution from PETROBRAS-CENPES has been specially helpful in this edition of the BIOMAT Symposia series. We thank the directors and representatives of these institutions: Prof. Emídio Cantídio de Oliveira Filho, Prof. Zena Martins, Prof. Paulo Sérgio Parro, from CAPES; Prof. Jerson Lima from FAPERJ; Dr. Gina Vasquez, Dr.Humberto Magalhães from PETROBRAS-CENPES.

On behalf of the Editorial Board of the BIOMAT Consortium, we thank all the authors, participants, sponsors and collaborators of the BIOMAT 2007 Symposium. The success of this conference series has been also framed by their special consideration.

Rubem P. Mondaini
President of the BIOMAT Consortium

Rio de Janeiro, December 2007

CONTENTS

Epidemiological Modeling

Biological and Chemical Pattern Recognition

Modeling of Biosystems

PROTEIN FOLDING AS A PHYSICAL STOCHASTIC PROCESS

KERSON HUANG

Physics Department
Massachusetts Institute of Technology
Cambridge, MA, USA 02139
E-mail: `kerson@mit.edu`

We model protein folding as a physical stochastic process as follows. The unfolded protein chain is treated as a random coil described by SAW (self-avoiding walk). Folding is induced by hydrophobic forces and other interactions, such as hydrogen bonding, which can be taken into account by imposing conditions on SAW. The resulting model is termed CSAW (conditioned self-avoiding walk. Conceptually, the mathematical basis is a generalized Langevin equation. In practice, the model is implemented on a computer by combining SAW and Monte Carlo. To illustrate the flexibility and capabilities of the model, we consider a number of examples, including folding pathways, elastic properties, helix formation, and collective modes.

1. Introduction

One of the outstanding unsolved problems in molecular biology is protein folding[1,2]. The principle through which the amino acid sequence determines the native structure, as wells as the dynamics of the process, remain open questions. Generally speaking, there have been two types of approaches to the problem: bioinformatics[3] and molecular dynamics (MD)[4].

Bioinformatics is purely data analysis, and does not involve dynamics at all. It massages the data base of known proteins in different ways, using very sophisticated computer programs, in order to discover correlations between sequence and structure. By its very nature, it cannot provide any physical understanding.

On the other hand, MD solves the Newtonian equations of motion of all the atoms in the protein on a computer, using appropriate inter-atomic potentials. To describe the solvent, one includes thousands of water molecules explicitly, treating all the atoms in the water on same footing as those on the protein chain. Not surprisingly, such an extravagant use of computing power is so inefficient that one can follow the folding process only to about

1

a microsecond, whereas the folding of a real protein takes from one second to ten minutes.

We shall try an approach from the point of view of statistical mechanics[5]. After all, the protein is a chain molecule immersed in water, and, like all physical systems, will tend towards thermodynamic equilibrium with the environment. Our goal is to design a model that embodies physical principles, and at the same time amenable to computer simulation in reasonable time.

We treat the protein as a chain performing Brownian motion in water, regarded as a medium exerting random forces on the chain, with the concomitant energy dissipation. In addition, we include regular (non-random) interactions within the chain, as well as between the chain and the medium.

The unfolded chain is assumed to be a random coil described by SAW (self-avoiding walk), as suggested by Flory[6] some time ago. That is, each link in the chain corresponds to successive random walks, in which the chain is prohibited from revisiting an occupied position. Two types of interactions are included in our initial formulation:

- the hydrophobic action due to the medium, which causes the chain to fold;
- the hydrogen-bonding within the chain, which leads to helical structure.

Other interactions can be added later.

We model the protein chain in 3D space, keeping only degrees of freedom relevant to folding, which we take to be the torsional angles between successive links. In the computer simulation, we first generate an ensemble of SAW's, and then choose a subensemble through a Monte Carlo method, which generates a canonical ensemble with respect to a Hamiltonian that specifies the interactions. We call the model CSAW[7,8] (conditioned self-avoiding walk). Mathematically speaking, it is based on a Langevin equation[5] describing the Brownian motion of a chain with interaction. There seems little doubt that such an equation does describe a protein molecule in water, for It is just Newton's equation with the environment treated as a stochastic medium.. The model can be implemented efficiently on a computer, and is flexible enough to be used as a theoretical laboratory.

Both CSAW and MD are based on Newtonian mechanics, and differ only in the idealization of the system. In CSAW we replace the thousands of water molecules used in MD by a stochastic medium — the heat reservoir of statistical mechanics. We ignore inessential degrees of freedom, such as small fluctuations in the lengths and angles of the chemical bonds that link the protein chain. The advantages of these idealizations are that

- we avoid squandering computer power on irrelevant calculations;
- we gain a better physical understanding of the folding process.

One often hears a debate on whether the folding process is "thermodynamic" or "kinetic". There is also an oft cited Levinthal "paradox", to the effect that the folding time should be much larger than the age of the universe, since the protein (presumably) had to search through an astronomically large number of states before finding the right one. From our point of view, these are not real issues.

The question of thermal equilibrium merely hangs on whether the protein can reach equilibrium in realistic time, instead being trapped in some intermediate state. For any particular protein, simulation of the Langevin equation will answer the question.

As to Levinthal's "paradox", the protein is blithely unaware of that. It just follows pathways guided by the Langevin equation.

After a brief review of the basics of protein folding and stochastic processes, we shall describe the model in more detail, and illustrate its use through examples involving realistic protein fragments. We will demonstrate folding pathways, elastic properties, helix formation, and protein collective modes.

The results indicate that the model has been successful in describing qualitative features of folding in simple proteins.

2. Protein Basics

2.1. *The protein chain*

The protein chain consists of a sequence of units or "residues", which are amino acids chosen from a pool of 20. This sequence is called the *primary structure*. The center of each amino acid is a carbon atom called C_α. Along the protein chain, the C_α's are connected by covalent chemical bonds in the shape of a "crank" that lies in one plane. Two cranks join at a C_α with a fixed angle between them, the tetrahedral angle $\theta_{\text{tet}} = -\arccos(1/3) \approx 110°$. The amino acids differ from each other only in the side chains connect to the C_α's. There are 20 possible choices for side chains.

The relative orientation of successive cranks is determined by two torsional angles ϕ and ψ, as schematically illustrated in Fig.1. These torsional angles are the only degrees of freedom relevant to protein folding, and small oscillations in bond lengths and bond angles can be ignored. For our purpose, therefore, a protein of N residues has $2(N - 1)$ degrees of freedom.

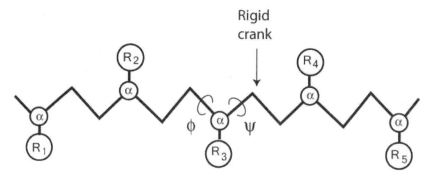

Figure 1. Schematic representation of the protein chain. Centers of residues are carbon atoms labeled α. They are connected by rigid chemical bonds in the shape of a planar crank. The only degrees of freedom we consider are the torsional angles ϕ, ψ that specify the relative orientations of successive cranks. Residues can differ only in the side chains labeled R_i, chosen from a pool of twenty. Atoms connected to the cranks are omitted for clarity.

2.2. Secondary and tertiary structures

At high temperatures, or in an acidic solution, the protein exists in an unfolded state that can be represented by a random coil[6]. When the temperature is lowered, or when the solution becomes aqueous, it folds into a "native state" of definite shape. Fig.2 shows the native state of myoglobin with different levels of detail. Local structures, such as helices, are called *secondary structures*. When these are blurred over, one sees a skeleton called the *tertiary structure*.

Secondary structures are of two main types, the alpha helix and the beta sheet, as shown in Fig.3. The former is stabilized by hydrogens bonds that connect residues 1 to 4, 2 to 5, *etc*. The beta sheet is a global mat sewn together by hydrogen bonds.

2.3. Hydrophobic effect

The molecules of liquid water form hydrogen bonds with each other, resulting in a dense fluctuating network, in which bonding partners change on a

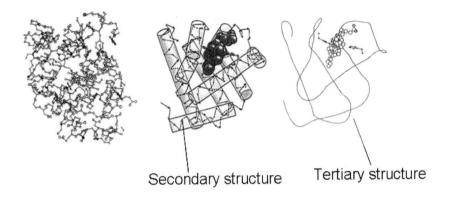

Secondary structure   Tertiary structure

Figure 2. Native state of Myoglobin showing different degrees of detail.

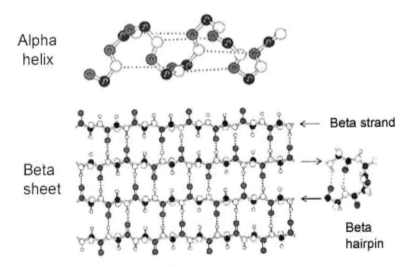

Figure 3. Secondary structures. Dotted lines in the alpha helix denote hydrogen bonds. The beta sheet is composed of "beta strands" matted together by
hydrogen bonds. two ajacent strands are connect by a "beta hairpin".

time scale of 10^{-12}s. A computer simulation of such a network is shown in
Fig.4a[12]. A foreign molecule introduced into water disrupts the network,
unless it can participate in hydrogen bonding. If it can hydrogen-bond with
water, it is said to be "soluble", or "hydrophilic", and will be received by

6

water molecules as one of their kind. Otherwise it is unwelcome, and said to be "insoluble", or "hydrophobic". Protein side chains can be hydrophilic or hydrophobic.

When immerse in water, the protein chain folds in order to shield the hydrophobic residues from water. In effect, the water network squeezes the protein into shape. This is called the "hydrophobic effect". However, a "frustration" arises in this process, because the skeleton is hydrophilic, and likes to be in contact with water, as indicated in Fig.4b. The frustration is resolved by the formation of secondary structures, which use up hydrogen bonds internally. The folded chain reverts to a random coil when the temperature becomes too high, or when the pH of the solution becomes acidic.

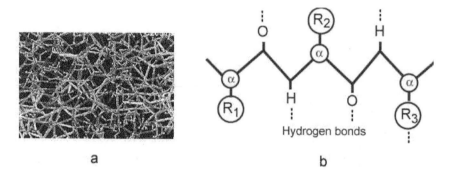

a   b

Figure 4.   (a) Computer simulation of network of hydrogen bonds in liquid water. (b) The hydrophobic side chains R_1 and R_2 cannot form hydrogen bonds, and prefer to be shielded from water. However, the atoms O and H on the main chain need to form hydrogen bonds. A "frustration" thereby arises, and is resolved by formation of secondary structures that use up hydrogen bonds internally.

2.4. Folding stages

As depicted schematically in Fig.5, a typical folding process consists of a very rapid collapse into an intermediate state called the "molten globule". The latter takes a relatively long time to undergo fine adjustments to reach the native state. The collapse time is generally less than 200 μs, while the molten globule can last as long as 10 minutes.

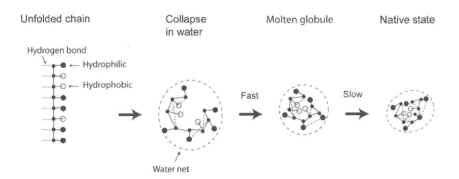

Figure 5. Being squeezed by a water net, the protein chain rapidly collapses into the molten globule state, which slowly adjusts itself into the native state.

2.5. *Statistical nature of the folding process*

We have to distinguish between protein assembly inside a living cell (*in vivo*), and folding in a test tube (*in vitro*). In the former the process takes place within factory molecules called ribosomes, and need the assistance of "chaperon" molecules to prevent premature folding. In the latter, the molecules freely fold or unfold, reversibly, depending on the pH and the temperature.

We deal only with folding *in vitro*, in which ten of thousands of protein molecules undergo the folding process independently, and they do not fold in unison. We are thus dealing with an ensemble of protein molecules, in which definite fractions exist in various stages of folding at any given time. The Langevin equation naturally describes the time evolution of such an ensemble. Behavior of individual molecules fluctuate from the average, even after the ensemble has reached equilibrium. In macroscopic systems containing the order of 10^{23} atoms, such fluctuations are unobservably small. For a protein with no more than a few thousand atoms, however, these fluctuations are expected to be pronounced.

3. Stochastic Process

3.1. *Stochastic variable*

A stochastic process is one involving random forces, and is described through a so-called stochastic variable (or random variable), which does not have a definite value, but is characterized instead by a probability distribution of values. Practically everything we deal with in the macroscopic world involve random variables, from the position of a billiard ball to the value of a stock.

Einstein pointed out the essence of a stochastic variable in his theory of Brownian motion. He emphasized that every Brownian step we can observe is the result of a very large number of smaller random steps, which in turn are the result of a very number of even smaller steps, and so on, until we reach the cutoff imposed by atomic structure. This self-similarity leads to a Gaussian distribution, regardless of the underlying mechanism — a result known as the *central limit theorem*[9].

3.2. *Brownian motion*

The simplest stochastic process is the Brownian motion of a single particle suspended in a medium. Its position $x(t)$ is a stochastic variable described by the Langevin equation

$$m\ddot{x} = F(t) - \gamma\dot{x} \qquad (1)$$

Here, the force exerted by the medium on the particle is split into two parts: a randomly fluctuating force $F(t)$ and a friction $-\gamma\dot{x}$. The random force is a member of a statistical ensemble with the properties

$$\langle F(t) \rangle = 0$$
$$\langle F(t)F(t') \rangle = c_0\delta(t - t') \qquad (2)$$

where the brackets $\langle\rangle$ denote ensemble average. The two forces are not independent, but related through the fluctuation-dissipation theorem:

$$\frac{c_0}{2\gamma} = k_B T \qquad (3)$$

where k_B is Boltzmann's constant and T is the absolute temperature, a property of the medium.

The Langevin equation can be solved exactly, and also be simulated by *random walk*. Both methods lead to diffusion, in which the position has a Gaussian distribution with variance $\sqrt{2Dt}$, where t is the time, and $D = c_0/(2\gamma^2)$ is called the diffusion constant. An equivalent expression is *Einstein's relation*

$$D = \frac{k_B T}{\gamma} \qquad (4)$$

Thus, a random force must generate energy dissipation, and the dissipation constant γ can be deduced from the variance of the distribution of positions.

3.3. *Monte Carlo*

If a particle undergoes Brownian motion in the presence of a regular (non-random) external force $G(x)$, we may not be able to solve the Langevin equation exactly, but we can still simulate it on a computer by *conditioned random walk,* as follows. We first generate a random trial step, but accept it only according to the Monte Carlo algorithm. Let E be the potential energy corresponding to the external force G. Let ΔE be the energy change in the proposed update. The algorithm is as follows:

- if $\Delta E \leq 0$, accept it;
- if $\Delta E > 0$, accept it with probability $\exp\left(-\Delta E/k_B T\right)$.

The last condition simulates thermal fluctuations, which may drive the system to a higher energy. After a sufficiently large number of updates, the sequence of state generated will yield a canonical ensemble with temperature T. That is, the Monte Carlo procedure tends to minimize not the energy, but the free energy.

Mathematically speaking, conditioned random walk simulates a generalized Langevin equation, as indicated in the following:

$$m\ddot{x} = \underbrace{[F(t) - \gamma\dot{x}]}_{\text{Treat via random walk}} + \underbrace{G(x).}_{\text{Treat via Monte Carlo}} \qquad (5)$$

Of course, we could integrate the whole equation as a stochastic differential equation, as an alternative to Monte Carlo. The equivalence of these two methods is illustrated by example in the appendix of Ref.[15]

4. CSAW

In protein folding, we are dealing with the Brownian motion of a chain with interactions. All we need to do, in principle, is to generalize conditioned random walk to conditioned SAW (self-avoiding walk). The resulting model is called CSAW (conditioned self-avoiding walk).

We can generate a SAW representing an unfolded protein chain by the *pivot algorithm*[10,11], as follows. Choose an initial chain in 3D continuous space, and hold one end of the chain fixed.

- Choose an arbitrary point on the chain as pivot.
- Rotate the end portion of the chain rigidly about the pivot (by changing the torsional angles at the pivot point).
- If this does not result in any overlap, accept the configuration, otherwise repeat the procedure.

By this method, we can generate a uniform ergodic ensemble of SAW's, which simulates a Langevin equation of the form

$$m_k \ddot{\mathbf{x}}_k = \mathbf{F}_k(t) - \gamma_k \dot{\mathbf{x}} + \mathbf{U}_k, \quad (k = 1, \cdots, N) \tag{6}$$

where the subscripts k label the residues along the chain. The terms $\mathbf{U}_k$ denote the regular (non-random) forces that maintain the rigid bonds between successive residues, and that prohibit the residues from overlapping one another.

We now add other regular forces G_k, which include the hydrophobic interaction and hydrogen-bonding. Treating this force via Monte Carlo results in CSAW, which simulates a generalized Langevin equation as indicated in the following:

$$m_k \ddot{\mathbf{x}}_k = \underbrace{(\mathbf{F}_k - \gamma_k \dot{\mathbf{x}} + \mathbf{U}_k)}_{\text{Treat via SAW}} + \underbrace{\mathbf{G}_k}_{\text{Treat via Monte-Carlo}}. \quad (k = 1 \cdots N) \tag{7}$$

Now we shall specify the forces G_k explicitly.

5. Implementation of CSAW

To reiterate, the system under consideration is a sequence of centers corresponding to C_α atoms, connected by planar "cranks". The degrees of freedom of the system are the pairs of torsional angles $\{\phi_i, \psi_i\}$ specifying the relative orientation of two successive cranks. There are O and H atoms attached to each crank, through rigid bonds lying in the same plane as the crank. The residues can differ from one another only through the side chains attach to C_α, and there are 20 of them to choose from. As indicated in Fig.6, the center of the side chain is located at an apex of a tetrahedron with C_α at the center.

We can start with a chain of bare cranks, and then add other components one by one, as desired. We can first represent the side chains by hard spheres, and put in the atoms in a more elaborate version. In this manner, we can tinker with different degrees of buildup, and investigate the relative importance of each element.

For Monte Carlo, we take the energy E to be

$$E = -g_1 K_1 - g_2 K_2 \tag{8}$$

$K_1 = $ Total contact number of all hydrophobic residues

$K_2 = $ Number of hydrogen bonds

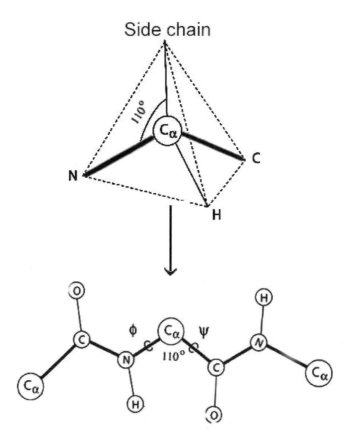

Figure 6. The side chain is at the apex of a tetrahedron with C_α at the center.

The first term in E expresses the hydrophobic effect. The contact number of a residue is the number of atoms touching its side chain. In the simplest version, in which we do not explicitly put in the side chain, the contact number is simply the number of atoms in contact with C_α, *not counting the other C_α's lying next to it along the chain.* This is illustrated in Fig.7a.

The contact number measures how well a residues is being shielded from the medium. When two hydrophobic residues are in contact, the total contact number increases by 2, an this induces an effective attraction between hydrophobic residues. The unfolded chain corresponds to $g_1 = 0$.

The second term in E describes hydrogen bonding. As illustrated in

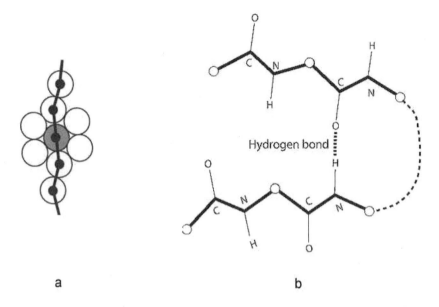

a   b

Figure 7. (a) The shaded hydrophobic residue illustrated here has four contact neighbors. The permanent neighbors along the chain are not counted. (b) Hydrogen bonding occurs between O and H on the main chain, from different residues.

Fig.7b, a hydrogen bond is deemed to have formed between O and H from different cranks when

- the distance between O and H is 2.5 A, within given tolerance;
- The bonds $C = O$ and $N - H$ are antiparallel, within given tolerance.

Only the combinations $g_1/k_B T$ and $g_2/k_B T$ appear in the Monte Carlo procedure. They are treated as adjustable parameters.

Note that E only includes the potential energy. We can leave out the kinetic energy because it contributes only a constant factor to the configurational probability of the ensemble.

6. Exploratory Runs

It is instructive to run the program with minimal components, as described in Refs.[7,8]. For a chain of 30 residues, the main findings are the following:

- Under hydrophobic forces alone, without hydrogen-bonding, the chain folds into a reproducible shape. This shows that the hydrophobic effect alone can produce tertiary structure. There is no secondary structure in this case, and the chain rapidly collapses to the final structure without passing through an intermediate state..

- When there is no hydrophobic force and the interaction consists purely of hydrogen-bonding, the chain rapidly folds into one long alpha helix.
- When both hydrophobic force and hydrogen bonding are taken into account, secondary structure emerges. The folding process exhibits two-stage behavior, with a fast collapse followed by slow "annealing", in qualitative agreement with experiments.

We now recount some simulations of realistic protein fragments.

7. Folding Pathways and Energy Landscape

Chignolin is a synthetic peptide of 10 residues [13], in the shape of a "beta hairpin" – a turn in a beta sheet as depicted in Fig.3. Jinzhi Lei [14] of Tsinghua University modeled it in CSAW, with side chains modeled as hard spheres. The native state emerges after about 70000 trial steps, as shown in Fig.8. The computation took less than 5 minutes on a work station. In contrast, an MD simulation on the same work station did not reach the native state in one month's computation. The run was repeated 100 times independently, to obtain an ensemble of folding paths.

To display the folding pathways, we project them onto a two-dimensional subspace of the configuration space, chosen as follows. Define a 10×10 distance matrix $D_{ij} = |\mathbf{R}_i - \mathbf{R}_j|$, where $\mathbf{R}_i$ is the vector position of the ith C_α. Let its eignevalues be $\lambda_1, \ldots, \lambda_{10}$ in ascending order. Through experimentation, we find that it is best to project the pathways onto the λ_1-λ_{10} plane, and we rotate the viewpoint to obtain the clearest representation. This is achieved by using λ_1 and $\lambda_1 + \lambda_{10}$ as axes. Fig.9 shows the evolution of 100 folding paths. We can see that the ensemble of 100 points, identified by given shading, migrates towards an attractor as time goes on. The energy landscape is shown below the migration map.

In Fig.10 we show 4 individual paths. They get trapped in various local pockets, and breakout after long searches for outlets. In this respect, the paths are similar to Levy flights.

Finally, in Fig.11, we exhibit the elastic property of the protein chain by plotting the energy as a function of molecular radius, in a semilog plot. The behavior is consistent with an exponential force law. The flat portion

14

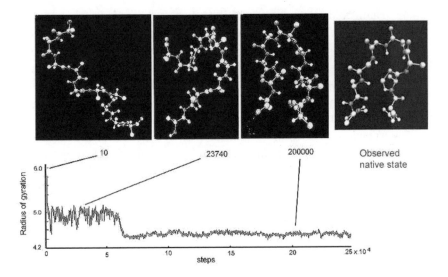

Figure 8.   Folding of Chignolin, a beta hairpin with ten residues.

in the middle corresponds to the breaking of hydrogen bonds that held the
beta hairpin together.

8. Nucleation and Growth of an Alpha Helix

Next we report on Polyalanine (Ala$_{20}$), a protein fragment of 20 identical
amino acids alanine, which is hydrophobic[15]. The native state is known to
be a single alpha helix. We tune g_1/k_BT and g_2/k_BT to maximize helical
content.

An ensemble of 100 folding paths was generated. Fig.12 shows the fractions of unfolded, intermediate, and folded molecules, as functions of time.
The solid curves are fits made according to a specific model, in which the
molecular radius reaches equilibrium first, while the helical content continues to grow. The helical growth is described by a set of rate equations, while
the relaxation of the radius is akin to that of an elastic solid. This shows
that the tertiary structure was established before the secondary structure,
and their evolutions are governed by different mechanisms.

Fig.13 shows a contour plot of the ensemble average of helicity, with
time on the horizontal axis, and residue number along the vertical. We can
see that the alpha helix grew from two specific nucleation points.

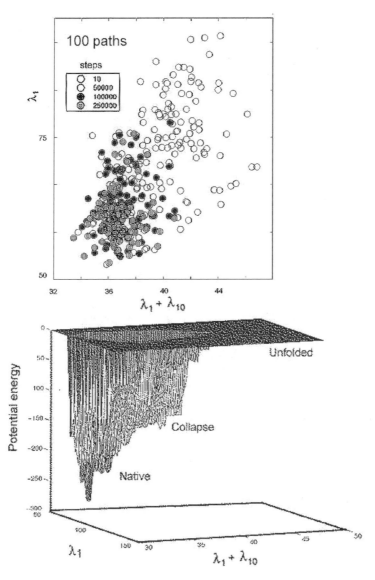

Figure 9. Evolution of 100 folding paths of Chignolin. The ensemble evolves towards an attractor. Lower panel shows the energy landscape. See text for explanation of the axes.

9. All-atom Model

Finally we show some preliminary results of Weitao Sun [16] of Tsinghua University on the histone 1A7W, which has 68 residues. This is a test of

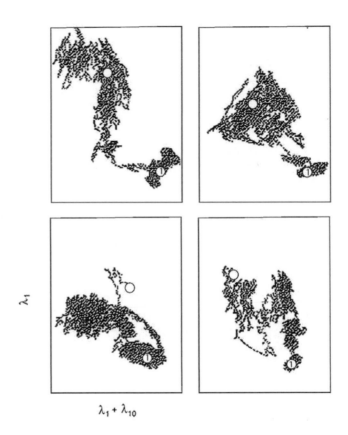

$\lambda_1 + \lambda_{10}$

Figure 10. Invidual pathways in the folding of Chignolin. Starting point are marked with an open circle, and endpoints are marked 1.

an all-atom CSAW model, in which atoms on the side chains are explicitly included. The model also includes the electrostatic interactions among all atoms. Fig.14 compares the simulated shape of the protein with the native state. It was found that inclusion of electrostatic interactions makes a noticeable improvement.

The main purpose of this calculation is to study the evolution of the dynamical structure function

$$S(k, \omega) = \left\langle \left| n(\mathbf{k}, \omega) \right|^2 \right\rangle$$

where $n(\mathbf{k}, \omega)$ is the space-time Fourier transform of the particle density,

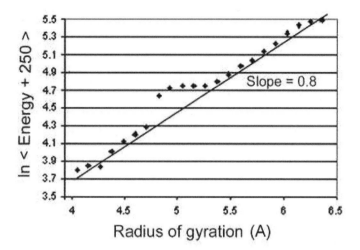

Figure 11. Elastic property of Chignolin: semilog plot of potential energy vs. radius, averaged over an ensemble of 50 samples. The flat part corresponding to the breaking of hydrogen bonds. The general shape of the curve is consistent with an exponential force law. Energy unit is not calibrated.

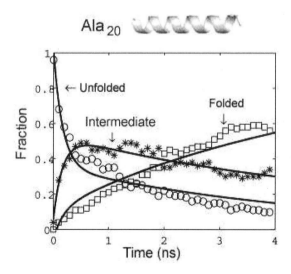

Figure 12. Fractions of Polyalanine at various stages of folding, as functions of time. Picuture at top shows the native state of the protein fragment.

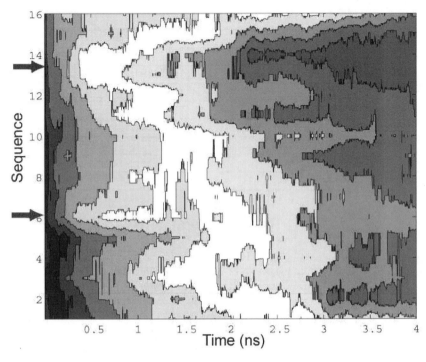

Figure 13. Contour map of ensemble average of helicity as a function of time and residue sequence, in the folding of Polyalanine. Nuclearion occured near the two positions marked by arrows.

and $\langle\rangle$ denotes ensemble average. In principle, this function can be experimentally measured via inelastic x-ray scattering. A peak occurring at particular k, ω will indicate the existence of an excitation mode. The integral of $S(k, \omega)$ over k will yield the normal mode spectrum, and the integral over ω will yield the static structure factor. Preliminary results are shown in Fig.15. We see that at update=3000 the collective modes of the final structure have not yet formed, but they emerge at update=5000. The lower panel of Fig.15 shows details of a sound-wave mode with constant velocity.

10. Discussion and Outlook

In treating protein folding as a physical process, the CSAW model differs from MD in two important aspects, namely

- irrelevant degrees of freedom are ignored;

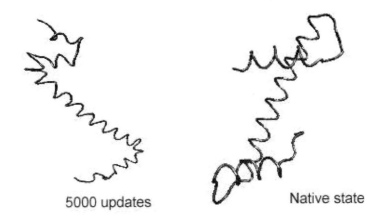

5000 updates   **Native state**

Figure 14. Folding the histone 1A7W (68 residues) with an all-atom CSAW model including electrostatic interactions.

- the environment is treated as a stochastic medium.

These, together with simplifying treatment of interactions, enable the model to produce qualitatively correct results with minimal demands on computer time.

An important simplification is separating the hydrophobic effect and hydrogen bonding, as expressed by the separate terms in the potential energy (8). Since both effects arise physically from hydrogen bonding, it is not obvious that we can make such a separation. The implicit assumption is that hydrogen bonding with water involves only the side chains, while internal hydrogen bonding involves only atoms along the main chain. This property is supported by statistical data, but should be a result rather an assumption of the model. We should try to remedy this in an improved version of the model.

The successful examples discussed here deal either with the alpha helix or the beta hairpin. Our next goal is to study the formation of a beta sheet. This is a much more difficult problem, for it involves global instead of local properties of the protein chain. Not knowing which elements are crucial for the project, we have made the following enhancements to-date:

- All-atom side chains can now be installed, with fractional hydrophobicity.
- Electrostatic interactions among all atoms can be included.
- Hard-sphere repulsions between atoms can be replaced by Lennard-

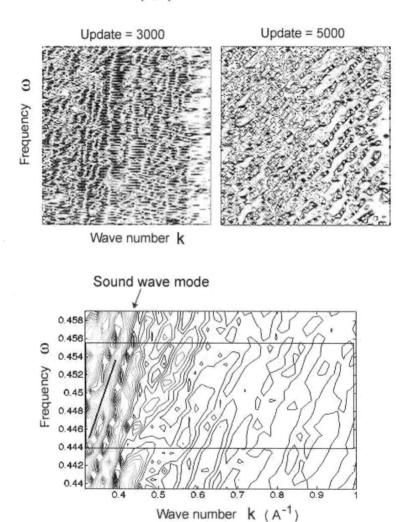

Figure 15. Evolution of the dynamical structure function reveals the emergence of collective modes at update=5000. Lower panel shows details of a sound-wave mode with velocity correponding to the slope of the straight line. The frequency is in inverse units of dt, a computational time increment not yet calibrated to real time.

Jones potentials.

- Hydrogen bonds can switch among qualifying partners, with given probability.

We hope to make progress on this problem.

References

1. C. Branden and J. Tooze, *Introduction to Protein Structure*, 2nd ed., (Garland Publishing, New York, 1999).
2. V. Daggett and A.R. Fersht, *Nat. Rev.: Mol. Cell Biol.* **4**, 497 (2003).
3. M. Kanehisa, *Post-Genome Informatics*, (Oxford University Press, Oxford, 2000).
4. H.A. Scheraga, M. Khalili, and A. Liwo, *Annu. Rev. Phys. Chem.*, **58**, 57 (2007).
5. K. Huang, *Lectures on Statistical Physics and Protein Folding* (World Scientific Publishing, Singapore, 2005).
6. P. Flory, *Principles of polymer chemistry* (Cornell University Press, London, 1953).
7. K. Huang, "CSAW: Dynamical model of protein folding", arXiv:cond-mat/0601244 v1 12 Jan 2006.
8. K. Huang, *Biophys. Rev. Lett.*, **2**, 139 (2007).
9. K. Huang, *Introduction to Statistical Physics* (Taylor & Francis, London, 2001) Chaps.16,17.
10. B. Li, N. Madras, and A.D. Sokal, *J. Stat. Phys.* **80**, 661 (1995).
11. T. Kennedy, *J. Stat. Phys.* **106**, 407 (2002).
12. M. Matsumoto, S. Saito, and I. Ohmine, *Nature*, **416**, 409 (2002).
13. A. Suenga *et. al.* Chem. Asian J. **2**, 591 (2007).
14. L.Z. Lei (unpublished).
15. L.Z. Lei and K. Huang, "Dynamics of alpha-helix formation in the CSAW model", arXiv 0706 3256 v1 [cond-mat.soft] 22 Jun 2007.
16. W.T. Sun (unpublished).

OPTIMAL METHODS FOR RE-ORDERING DATA MATRICES IN SYSTEMS BIOLOGY AND DRUG DISCOVERY APPLICATIONS

PETER A. DIMAGGIO JR., SCOTT R. MCALLISTER,
CHRISTODOULOS A. FLOUDAS*

Princeton University,
Department of Chemical Engineering,
Engineering Quadrangle,
Princeton, NJ 08544, USA
E-mail: floudas@titan.princeton.edu

XIAO-JIANG FENG, JOSHUA D. RABINOWITZ,
HERSCHEL A. RABITZ

Princeton University,
Department of Chemistry,
Princeton, NJ, 08544, USA

The analysis of large-scale data sets via clustering techniques is utilized in a number of applications. Many of the methods developed employ local search or heuristic strategies for identifying the "best" arrangement of features according to some metric. In this article, we present rigorous clustering methods based on the optimal re-ordering of data matrices. Distinct mixed-integer linear programming (MILP) models are utilized for the clustering of (a) dense data matrices, such as gene expression data, and (b) sparse data matrices, which are commonly encountered in the field of drug discovery. Both methods can be used in an iterative framework to bicluster data and assist in the synthesis of drug compounds, respectively. We demonstrate the capability of the proposed optimal re-ordering methods on several data sets from both systems biology and molecular discovery studies and compare our results to other clustering techniques when applicable.

1. Introduction

Problems of data organization and data clustering are prevalent across a number of different disciplines. The goal of data clustering, regardless of

---

*Author to whom all correspondence should be addressed; Tel: (609) 258-4595.

the application, is to organize data in such a way that objects which exhibit "similar" attributes are grouped together. The definition of similarity depends on the application and may correspond to the direct comparison of values or the degree of correlation among trends or patterns of values. Several methods have been proposed for the clustering of large-scale, dense data. The most common approaches to the data clustering problem are typically categorized as hierarchical[1] or partitioning [2] clustering. Although algorithms to identify the optimal solutions to these categories of problems do exist[3,4,5], they are frequently solved using heuristic search techniques that result in suboptimal clusters because the comparisons between terms are evaluated locally. Various other frameworks for data clustering have been proposed, including model-based clustering [6,7], neural networks [8], simulated annealing[9], genetic algorithms[10,11], information-based clustering[12], decomposition based approaches[13,14], and data classification [15].

Rearrangement clustering has emerged as an effective technique for optimally minimizing the sum of the pair-wise distances between rearranged rows and columns for dense data matrices with few missing elements. The bond energy algorithm (BEA) was originally proposed as a method for finding "good" solutions to this problem[16] and it was subsequently discovered that this problem could be formulated as a traveling salesman problem (TSP) which can be solved to optimality[17,18] using existing methods. The concept of biclustering[19,20,21,22,23,24,25,26,27,28,29] has emerged in the context of microarray experiments where the rearrangement of the rows and columns is needed to identify a bicluster, which corresponds to a subset of genes related under a certain set of conditions. This is particularly useful when a gene is involved in more than one biological process or belongs to a group of genes that are co-expressed under limited conditions[23]. An assortment of techniques and formulations has been proposed, many of which utilize heuristics to identify biclusters and often require normalization to elucidate patterns of interest, reviewed in Madeira and Olivera, 2004[30].

Another challenging problem in data clustering is the rearrangement of data matrices that are very sparse. These types of data matrices are common in drug discovery where the x- and y-axis of a data matrix can correspond to different functional groups for two distinct substituent sites on a molecular scaffold. Each possible x and y pair corresponds to a single molecule which can be synthesized and tested for a certain property, such as the percent inhibition of a protein function. This procedure is not limited to two dimensions (or scaffold sites) and can be generalized to address N scaffold sites. For even moderate size matrices, synthesizing and testing only half of the molecules is labor intensive and not economically feasible. Thus, it is of paramount importance to have a reliable method for guiding the synthesis process to select molecules that have a high probability of success. Finding such lead molecules can be highly difficult, even with the assistance of combinatorial chemistry and high-throughput screening given the enormous number of potential small molecule structures. A common practice is to employ quantitative structure-activity relationship (QSAR) methods[31] to computationally predict the biological properties of the library compounds. Existing methods for constructing predictive QSAR models involve three basic steps: (1) assaying the properties of a training set of molecules, (2) selecting physical/chemical descriptors to relate the small molecule structures to the measured properties, (3) identifying mathematical functions that quantitatively relate the molecular descriptor to the properties.

To address the clustering of both dense and sparse data matrices, we introduce rigorous deterministic optimization algorithms which utilize optimal re-ordering to cluster the rows and columns of data matrices from systems biology and drug discovery studies. Various objective functions that are dependent on the sparsity of the data can be used to quantify the degree of similarity among rows and columns in the final arrangement. We introduce a network flow model and a traveling salesman problem (TSP) representation which consider only nearest neighbor comparisons in the final arrangement to model the permutations of the rows and columns for dense data sets. An alternative MILP model is presented for the optimal re-ordering of sparse data matrices and takes into consideration long-range effects among rearranged rows and columns. To demonstrate the capabilities of the proposed methods, we applied our rigorous approaches to (a) metabolite concentration data[32], (b) colon cancer gene expression data[33], and (c) drug inhibition data provided by Pfizer Inc. We compared the results of our method for the dense data matrices in (a) and (b) with other clustering techniques. For these data sets, our method provides a denser grouping of related metabolites and genes (i.e., rows of the data matrix) than other clustering methods, which suggests that optimal ordering has distinct advantages over local ordering. We also demonstrate that our global optimization method can reconstruct underlying fundamental patterns in the columns of these dense data matrices. The optimal re-ordering of the sparse drug inhibition data in (c) is shown to cluster the compounds with desirable properties into an easily identifiable submatrix of the rearranged matrix. This indicates that the re-ordering technique can be utilized in an iterative framework where a local interpolation-based synthesis strategy can be used to predict the expected inhibition values for the unsampled compounds. We illustrate the utility of such a framework by performing an "in silico" synthesis on a uniformly-sampled subset of the original data matrix. It is shown that this method is very effective in guiding the synthesis procedure towards molecules with desirable inhibition values.

2. Mathematical Models

In this section, we present the components of the mathematical model: (1) the variables, (2) the objective functions used to quantify the similarity among elements in the final arrangement, and (3) corresponding problem formulations which provide the optimal rearrangement of rows and columns for dense and sparse data matrices. We also present a method for identifying cluster boundaries in dense matrices which is used to iteratively bicluster the resulting submatrices via optimal re-ordering.

2.1. *Variable Definitions for Re-ordering*

The index pair (i,j) corresponds to a specific row i and column j of a matrix, where the value of this pair is denoted as $a_{i,j}$. The cardinality (or in this case, the dimension) of the rows and columns of the matrix will be represented as $|I|$ and $|J|$, respectively. For the sake of brevity in this section and the remainder of the article, we present the terminology and mathematical model only for the rows of the matrix, but an analogous representation follows for the columns.

For the re-ordering of dense data matrices, we are concerned with the assignment of adjacent or neighboring elements in the final ordering, which is represented using the following binary 0-1 variables.

$$
y_{i,i'}^{row} = \begin{cases} 1, & \text{if row i is adjacent to row i'} \\ & \text{in the final arrangement} \\ 0, & \text{otherwise} \end{cases}
$$

For instance, if the binary variable $y_{8,3}^{row}$ is equal to one then row 8 is immediately above row 3 in the final arrangement of the matrix. The assignment of $y_{8,3}^{row} = 0$ implies that row 8 is *not* immediately above row 3 in the final arrangement, but does not provide any additional information regarding the final positions of rows 8 and 3 in the matrix.

The re-ordering of sparse data matrices should take into consideration long-range effects between elements in the final ordering. Thus, we model these final positions as an assignment problem, where the binary 0-1 variables $w_{i,k}^{row}$ indicate the assignment of a particular row, i, to a position in the final ordering, $1 \leq k \leq |I|$, as defined below.

$$w_{i,k}^{row} = \begin{cases} 1, & \text{if row } i \text{ is assigned to position } k \text{ in the final ordering} \\ 0, & \text{otherwise} \end{cases}$$

An assignment of $w_{8,3}^{row} = 1$ indicates that row 8 from the initial ordering is assigned to position 3 in the final ordering. It is important to note that the index k in $w_{i,k}^{row}$ represents the final positions in the rearrangement and thus allows us to compute *distances*, $d_{i,i'}$, between re-ordered elements and will be defined subsequently. The binary variables $y_{i,i'}^{row}$ and $w_{i,k}^{row}$ define two separate conventions for modeling the rearrangement of dense and sparse data matrices, respectively.

2.2. Objective Functions

The proposed rigorous method optimally rearranges the rows and columns of a data matrix according to a given metric of similarity, which is left to the user to specify and depends on the sparsity of the data. We first present common expressions that can be used for quantifying the pair-wise similarity between two adjacent rows of a dense matrix. An intuitive metric of similarity is to minimize the relative difference in value for adjacent rows of a matrix, as presented in Equation 1.

$$\sum_i \sum_{i'} \sum_j y_{i,i'}^{row} \cdot |a_{i,j} - a_{i',j}| \tag{1}$$

The emphasis can be placed on penalizing specifically large differences in value by squaring the difference in value between two adjacent rows and columns, as shown in Equation 2.

$$\sum_i \sum_{i'} \sum_j y_{i,i'}^{row} \cdot (a_{i,j} - a_{i',j})^2 \tag{2}$$

A metric similar to the root-mean squared deviation of values can also be used to guide the rearrangement of the matrix, as shown in Equation 3.

$$\sum_i \sum_{i'} y_{i,i'}^{row} \cdot \sqrt{\frac{\sum_j (a_{i,j} - a_{i',j})^2}{|J|}} \tag{3}$$

The aforementioned objective functions can be also tailored to exploit physical trends in the data set. For instance, suppose it is known a priori that the values of the data are monotonic when arranged in a particular order and that this final configuration is desirable. Then the terms in any of these objective functions could be easily restricted to include only those rows that violate such a monotonicity trend (i.e., $a_{i,j} > a_{i',j}$). It should be noted that the objective functions defined in Equations 1 through 3 are symmetric, whereas incorporating monotonicity into these expressions introduces asymmetry. Each of these proposed metrics can result in distinctly different permutations of the final rearranged matrix. The above objective functions are typically used for dense data matrices, however the model is not limited to these forms.

In the rearrangement of sparse data matrices, it is desirable to extend the similarity comparisons beyond the nearest neighboring elements since there are many missing values. For this purpose, we present a pairwise similarity measure that depends both on the compound properties and the distance between them in the reordered dimension.

$$\sum_i \sum_{i'} \left(\frac{1}{R} \cdot \frac{|I| - d_{i,i'}}{|I| - 1} \right) \cdot \sum_j (a_{i,j} - a_{i',j})^2 \tag{4}$$

Here $d_{i,i'}$ is the distance between the two compounds i and i' after rearrangement. This distance value value will be 1 if the two compounds are adjacent to each other in the final arrangement and $|I| - 1$ if they are on opposite ends of the data matrix. The contribution of the pair-wise comparisons is normalized by R, which is the number of compound pairs where both $a_{i,j}$ and $a_{i',j}$ are available from synthesis and property assaying for all i and i'. Note that while we have presented the squared difference between elements in Equation 4, we could also use any of the other forms as used in Equations 1 and 3. The goal of the re-ordering algorithms is to minimize the objective functions presented in this section.

2.3. Models for Optimal Re-ordering

In this section, we present various mathematical models for performing the physical permutations of the rows and columns for both dense and sparse data matrices. Note that the objective functions introduced in the previous section are independent of how the rows are physically permuted.

2.3.1. Network Flow Model: Dense Data Matrices

A network flow model[34] can be used to perform the physical permutations for the rows and columns of dense data matrices. The final ordering of the row permutations can be represented as a directed acyclic graph, where an edge connects two rows if these rows are *adjacent* in the final ordering[28]. As previously mentioned, the binary variables $y_{i,i'}^{row}$ represent the assignment of a neighboring row i' directly below row i in the final arrangement. In network flow terminology, we say that the binary variable $y_{i,i'}^{row}$ represents the existence of the edge between rows i and i'. We introduce another set of binary variables, $y\_source_i^{row}$ and $y\_sink_i^{row}$, to indicate which rows are assigned at the top and bottom of the final rearranged matrix, respectively.

$$y\_source_i^{row} = \begin{cases} 1, & \text{if row i is the top-most row} \\ & \quad \text{in the final arrangement} \\ 0, & \text{otherwise} \end{cases}$$

$$y\_sink_i^{row} = \begin{cases} 1, & \text{if row i is the bottom-most row} \\ & \quad \text{in the final arrangement} \\ 0, & \text{otherwise} \end{cases}$$

The flows values assigned to the edges connecting the rows are continuous variables denoted by $f_{i,i'}^{row}$. These flows start from a fictitious source row and end at a fictitious sink row.

$$f_{i,i'}^{row} \equiv \text{the flow from row } i \text{ to row } i'$$
$$f\_source_i^{row} \equiv \text{the flow entering the source row } i$$
$$f\_sink_i^{row} \equiv \text{the flow leaving the sink row } i$$

It should be noted that the variables $y_{i,i}^{row}$ and $f_{i,i}^{row}$ are zero since row i can never be adjacent to *itself*. The physical act of connecting two rows by an edge (i.e., putting two rows adjacent to one another in the final arrangement) is modeled via the following constraint equations.

$$\sum_{i' \neq i} y_{i',i}^{row} + y\_source_i^{row} = 1 \qquad \forall i \tag{5}$$

$$\sum_{i' \neq i} y_{i,i'}^{row} + y\_sink_i^{row} = 1 \qquad \forall i \tag{6}$$

These constraints enforce that each row, i, has only one neighboring row above it (or is the top-most row) and only one neighboring row below it (or is the bottom-most row) in the final arrangement, respectively. The next two constraints ensure that only one top-most (source) row and only one bottom-most (sink) row should be assigned in the final matrix.

$$\sum_i y\_source_i^{row} = 1 \tag{7}$$

$$\sum_i y\_sink_i^{row} = 1 \tag{8}$$

The set of constraints defined by Equations 5 through 8 are sufficient for assigning unique neighbors to every row. However, *cyclic* arrangements of the rows also satisfy these constraint equations (i.e., it is possible to have $y_{i,i'}^{row} = y_{i',i''}^{row} = y_{i'',i}^{row} = 1$, which results in a cyclic final ordering of i, i', i'', $i, \dots$). To ensure that the final arrangement of the rows is acyclic, unique flow values are assigned to each edge, $y_{i,i'}^{row}$, that connects rows i and i'. The value for the flow entering the source row (or top-most row) is defined to be the total number of rows ($|I|$) to indicate that this is the top-most row in the final arrangement.

$$f\_source_i^{row} = |I| \cdot y\_source_i^{row} \qquad \forall i \qquad (9)$$

Note that the above constraints, in conjunction with Equation 7, ensure that only one source flow is assigned to an edge. Starting from this source row, each subsequent row in the final arrangement will have an entering flow value of $|I| - 1$, $|I| - 2$, and so on. This cascading property of the flow values will ensure a unique final ordering of the rows and eliminate cyclic arrangements. A flow conservation equation is used to model this cascading of the flows by requiring that the flow entering a row is exactly one unit greater than the flow leaving that row.

$$\sum_{i'} (f_{i',i}^{row} - f_{i,i'}^{row}) + f\_source_i^{row} - f\_sink_i^{row} = 1 \qquad \forall i \qquad (10)$$

Since we have defined the convention that $f\_source_i^{row}$ starts at $|I|$, then $f\_sink_i^{row}$ has a flow value of *zero* and thus can be eliminated from the above constraints.

Lastly, we can assign general upper and lower bounds for all flow values since a flow connecting two rows i and i' (i.e., $y_{i,i'}^{row} = 1$) can never be greater than $|I| - 1$ nor less than 1.

$$f_{i,i'}^{row} \leq (|I| - 1) \cdot y_{i,i'}^{row} \qquad \forall (i, i') \qquad (11)$$

$$f_{i,i'}^{row} \geq y_{i,i'}^{row} \qquad \forall (i, i') \qquad (12)$$

These constraints also ensure that if rows i and i' are not connected by an edge (i.e., $y_{i,i'}^{row} = 0$) then no flow is assigned ($f_{i,i'}^{row} = 0$). The set of constraints (5)-(12) comprise the entire mathematical model necessary for performing the row and column permutations for dense data matrices, which are guided by any of the aforementioned objective functions in Equations 1 through 3.

2.3.2. *TSP Model: Dense Data Matrices*

The re-ordering of the rows and columns of dense matrices can also be modeled as a traveling salesman problem (TSP), which is one of the most well-studied problems in the area of combinatorial optimization. The problem objective is to visit a list of N cities and return to the starting city via the minimum cost route (often referred to as the optimal tour). Despite the conceptual simplicity of this representation, finding the best tour and guaranteeing its optimality remains challenging for large-scale problems. It has been pointed out that the row and column re-ordering problems can be solved as two independent traveling salesman problems[35].

In the TSP formulation, each row in the matrix is a vertex, $i \in |I|$. The existence of an edge between rows i and i' is again represented by the binary variable $y_{i,i'}^{row}$. For each edge there is an associated cost, $c_{i,i'}$, of "traveling" from row i to i'. Thus, the objective of the problem is to visit each row in the matrix only once via these edges while incurring the minimum total cost and the order in which these rows are visited denotes their final positions in the matrix. The problem definition requires that the tour start and end at the same row, so we introduce a dummy city to connect the top-most and bottom-most row in the final arrangement with edges that have zero cost. A formal definition of the problem is provided below.

$$\min \sum_{i,i'} c_{i,i'} \cdot y_{i,i'}^{row} \tag{13}$$

$$\sum_{i'} y_{i,i'}^{row} = 1 \quad \forall i \tag{14}$$

$$\sum_{i'} y_{i',i}^{row} = 1 \quad \forall i \tag{15}$$

The cost associated with traversing an edge, $c_{i,i'}$, is computed using the aforementioned objective functions. As in the network flow model, cyclic tours satisfy Equations 14 and 15, thus additional constraints are required to eliminate these *subtours*. These constraints are efficiently incorporated into TSP solvers, such as Concorde[36], via cutting plane methods and are beyond the scope of this paper so will not be discussed here.

Although the idea of traveling implies moving from one row to the next, if the cost of traveling in either direction is the same for any row, then the problem is symmetric and only *undirected* edges between rows need to be considered. However, the objective functions that incorporate monotonicity violations are by definition asymmetric and require an asymmetric TSP formulation. The asymmetric traveling salesman problem can be recast as a symmetric traveling salesman problem by introducing a duplicate set of N rows and restricting the overall connectivity of edges. The details of how to perform such a transformation have been described elsewhere [37,38] and will not be presented here.

2.3.3. *Assignment Problem Model: Sparse Data Matrices*

The sparse data matrix problem is difficult to formulate as an efficient deterministic optimization problem. For simplicity of model presentation, we assume that the library has two substitution sites. In this case, the library becomes a two-dimensional matrix, where a row represents a substituent on site 1 being fixed and a column corresponds to a substituent on site 2 being fixed. The basic model can be formulated as an assignment problem, where binary variables, $w_{i,k}^{row}$, indicate the assignment of a particular row, i, to a position k in the final ordering.

Intuitive constraints to impose on the row assignments are that a final position can contain only one row and a row can be assigned to at most one final position. This is given by Equations 16 and 17.

$$\sum_k w_{i,k}^{row} = 1 \quad \forall i \tag{16}$$

$$\sum_i w_{i,k}^{row} = 1 \quad \forall k \tag{17}$$

We can represent the final position of row i in the matrix as a positive variable, p_i, and relate $w_{i,k}^{row}$ to p_i using simple equality constraints as shown in Equation 18.

$$p_i = \sum_k k \cdot w_{i,k}^{row} \quad \forall i \tag{18}$$

These positions can range from 1 to the total number of rows, $|I|$, which is represented by the bounds given in Equation 19.

$$1 \le p_i \le |I| \quad \forall i > 1 \tag{19}$$

Issues related to problem symmetry can be alleviated by restricting the final assignment of row 1 to occupy only one half of the final matrix, as shown in Equation 20.

$$1 \leq p_1 \leq \lfloor (|I| + 1)/2 \rfloor \tag{20}$$

From these final positions, we can also define the distance between any two rows i and i' in the final arrangement by the positive variable $d_{i,i'}$. The distance between two final positions i and i' is given by the *nonlinear* equation $d_{i,i'} = |p_i - p_{i'}|$. However, exact lower bounds on this distance can be represented by two sets of linear inequality constraints, as shown in Equations 21 and 22.

$$d_{i,i'} \geq p_i - p_{i'} \quad \forall i < i' \tag{21}$$

$$d_{i,i'} \geq p_{i'} - p_i \quad \forall i < i' \tag{22}$$

Any distance must be at least 1 and cannot be greater than the total number of rows minus 1 ($|I| - 1$), which results in the bounds in Equation 23.

$$1 \leq d_{i,i'} \leq |I| - 1 \quad \forall i, i' > i \tag{23}$$

Generating tight upper bounds on the distance variables is not as straight-forward and as a result the mixed-integer linear problem can be extremely difficult to solve. Specifically, the resulting linear programming relaxations may not be tight and significant computational effort is required for the branch and bound tree to find and/or prove the optimal solution. However, several additional constrains can be derived to help bound and restrict the distance variables to reasonable values. One such constraint utilizes the fact that the summation of all the final distances is equal to a known constant, C, as shown in Equation 24.

$$\sum_i \sum_{i' > i} d_{i,i'} = C \tag{24}$$

For instance, if there are only four rows, then $C = 3 \cdot 1 + 2 \cdot 2 + 1 \cdot 3 = 10$. Another constraint provides a valid upper bound on the distance by incorporating information about the final position of a row assignment. If row i is assigned to final position k, then the maximum distance between row i and any other row i' must be less than $MAX(|I| - k, k - 1)$, which is the maximum distance to either end of the matrix from position k. We represent $MAX(|I| - k, k - 1)$ as the function $F(k)$. This constraint is expressed in generalized form in Equations 25 and 26.

$$d_{i,i'} \leq \sum_k F(k) \cdot w_{i,k}^{row} \quad \forall i < i' \tag{25}$$

$$d_{i,i'} \leq \sum_k F(k) \cdot w_{i',k}^{row} \quad \forall i < i' \tag{26}$$

Another constraint on the distance variables can be derived from the fact that once row i has been assigned to some final position k, then the sum of the distances from position k to all other positions is known, which we represent as a general function $G(k)$. For instance, if a row i is assigned to final position 2 in a 4-row problem, then the sum of its distances to all other rows i' in the final arrangement is $(2-1) + (3-2) + (4-2) = 4 = G(2)$. The general form for this constraint is presented in Equation 27.

$$\sum_{i'>i} d_{i,i'} + \sum_{i'<i} d_{i,'i} = \sum_k G(k) \cdot w_{i,k}^{row} \quad \forall i \tag{27}$$

Since the distances are Euclidean we can impose the restriction that they satisfy the triangle inequality constraint in Equation 28.

$$d_{i,i'} \leq d_{i,i''} + d_{i'',i} \quad \forall i, i', i'' \tag{28}$$

However, this results in $O(|I|^3)$ additional constraints and leads to memory issues for standard commercial solvers for even moderately sized problems. To circumvent this memory issue, we introduce these constraints dynamically during the program execution as *cuts*. In other words, we include only those triangle inequality constraints for which Equation 28 is violated and this ensures that only the most necessary constraints are added to the problem. Another set of constraints which bounds the distances with respect to a fixed point, $1 \leq p^* \leq |I|$, are introduced as cuts into the problem, where p* is chosen during program execution.

$$d_{i,i'} \leq \sum_k |k - p^*| \cdot \left(w_{i,k}^{row} + w_{i',k}^{row}\right) \quad \forall i, i' > i, p^* \tag{29}$$

The proposed mixed-integer linear programming formulation, which is the minimization of Equation 4 subject to Equations 16 through 29, when solved to optimality can guarantee that it has found the best ordering for the given objective function in Equation 4. The largest reordering problem we have solved to optimality has contained about 90 rows. However, even though this method may not close the gap between the linear programming relaxation and the best integer solution, it is a very effective technique for establishing a lower bound on the value of the objective function. It should be mentioned here that heuristic approaches can be integrated into the deterministic optimization framework for generating good integer solutions.

2.4. *Iterative Framework for Biclustering Dense Data*

We can bicluster dense data matrices by optimally re-ordering in an iterative fashion[28]. The framework begins by optimally re-ordering a single dimension of the data matrix. Let us denote the dimension that is re-ordered as the columns and the dimension that is not re-ordered as the rows of the data matrix. For instance, in gene expression data the columns would correspond to the time series or set of conditions over which the expression for the genes of interest (i.e., the rows) were measured. The objective function value for each pair-wise term between neighboring columns in the final ordering is evaluated and the median of these values is computed. The median was selected as the evaluating metric since it is statistically less biased to outliers than the average. Cluster boundaries are defined to lie between those columns which have the *largest* median (since the objective function is being minimized). The cluster boundaries are used to partition the original matrix into several submatrices. The rows of each submatrix are then optimally re-ordered and clusters in this dimension are defined using the median value of the objective function between neighboring rows in the final ordering. The number of cluster boundaries used to partition the matrix into smaller submatrices should exploit information regarding the dimension that is re-ordered and can be specified by the user.

3. Results and Discussion

In this section, we present the results for our proposed re-ordering methods on three distinct data matrices (two dense and one sparse) and draw comparisons against other clustering techniques when applicable.

3.1. *Metabolite Concentration Data: Dense Data Matrix*

The proposed method for dense data matrices[28] was first tested on data comprised of concentration profiles for 68 metabolites (the rows of the data matrix) recorded over time (columns of the data matrix) for the organisms *E. coli* and *S. cerevisiae* under the conditions of nitrogen and carbon starvation for both organisms[32]. We applied our biclustering algorithm to this data using the objective function defined in Equation 2. The re-ordering problem for the columns was solved to global optimality using CPLEX[39] in

168 seconds on Intel 3.0 GHz Pentium 4 processor using the network flow model. The optimal ordering for the columns using the objective function in Equation 2 is shown in Figure 1 where the top four cluster partitions for the columns are denoted by the solid vertical lines.

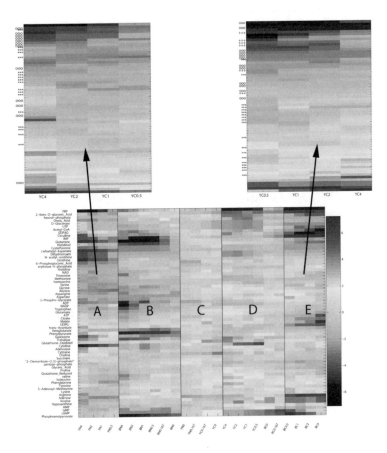

Figure 1. Partitioning of columns into regions A, B, C, D, and E using cluster boundaries and optimal re-ordering results for metabolites for regions A and E. The relative groupings of metabolites are illustrated using the labels "***" for amino acid metabolites, "ooo" for biosynthetic intermediates, and "+++" for TCA compounds.

It is interesting to note that the two most significant cluster boundaries perfectly partition subsets of the *E. coli* and *S. cerevisiae* conditions. The four boundaries that separate either *E. coli* from *S. cerevisiae* conditions or nitrogen from carbon starvation conditions in the re-ordered columns are within the top nine most significant cluster boundaries. Another interesting feature of the column rearragment is that the starvation conditions for nitrogen and carbon cluster exactly together. That is, all the nitrogen starvation conditions occupy one half of the matrix and the carbon starvation conditions occupy the remaining half of the matrix. This suggests that the algorithm has the ability to reconstruct underlying fundamental patterns. The regions between these cluster boundaries, labeled A, B, C, D, and E in Figure 1, are also optimally re-ordered using the proposed method. For the sake of brevity, let us consider the results obtained from optimally re-ordering the submatrices for regions A and E, as shown in the enlarged regions of Figure 1. The submatrices for regions A and E were optimally re-ordered in 4085 and 4587 CPU seconds, respectively, using the network flow model.

The optimally re-ordered metabolites for region A result in a dense grouping of the amino acids and biosynthetic intermediates under the conditions of nitrogen starvation in *S. cerevisiae*. One should note the almost monotonic behavior of the concentration profile, which groups the decreasing concentrations at the top and the increasing concentrations at the bottom of the matrix. The twelve amino acids glycine, asparigine, serine, alanine, methionine, threonine, histidine, aspartate, tryptophan, phenylalanine, isoleucine, and valine are found in a cluster of 26 metabolites. There is also a strong aggregation of the biosynthetic intermediates carbamoyl-aspartate, ornithine, dihydrooroate, N-acetyl-ornithine, IMP, cystathionine, and orotic acid in the first nine rows of the data matrix. This supports the observation that most biosynthetic intermediates decrease in concentration over all starvation conditions based on the hypothesis that the cells turn off de novo biosynthesis as an early, strong, and consistent response to nutrient deprivation[32]. One should also note that carbamoyl-aspartate and dihydrooroate, separated by only two positions in the final ordering, are both on the pyrimidine pathway [32].

The optimally re-ordered metabolites for region E over the conditions of carbon starvation in *E. coli* yields an excellent grouping of amino acid and TCA metabolites. In a cluster of 27 metabolites, 16 are amino acids (out of a total of 19 amino acids in the data) and 8 are ordered consecutively: serine, glycine, valine, glutamate, tryptophan, alanine, threonine, and methionine (see the "***" symbols in Figure 1). This richness of amino acid metabolites is consistent with the observation that amino acids tend to accumulate during carbon starvation[32]. Another interesting feature is that four out of the six TCA metabolites (trans-aconitate, citrate, malate, and acetly-coa, represented by the "+++" symbols in Figure 1) are within six positions of each other in the optimal ordering. The biosynthetic intermediates also order well for this submatrix (as shown by the "ooo" symbols in Figure 1), where all twelve are placed in the top half of the re-ordered matrix, which is rich in decreased concentration metabolites. An interesting observation is the final position of FBP relative to phospoenolpyruvate (PEP), which are exactly opposite each other in the re-ordered matrix. PEP is a positive regulator of pyruvate kinase, which is the major enzyme consuming PEP [32]. Since carbon-starvation resulted in a decrease of FBP, this presumably down-regulates the activity of pyruvate kinase, which in turn results in PEP accumulation.

As a basis for comparison with traditional clustering techniques, we examined the results for hierarchical clustering applied to the metabolite concentration data[32]. For the hierarchical clustering results, the majority of amino acids are positioned in the top half of the arranged matrix, with the largest consecutive ordering of amino acids being alanine, glutamate, threonine, methionine, and serine. The results for region E using the proposed biclustering method provides a denser grouping of the amino acids than hierarchical clustering does and a consecutive ordering of eight amino acids was identified. In the hierarchical results, four TCA cycle compounds (aconitate, malate, citrate, and succinate) were assigned to a cluster of ten metabolites [32]. The re-orderings provided for region E arranged four TCA cycle compounds (trans-aconitate, citrate, malate, and acetly-coa) almost consecutively, with only two metabolites separating three consecutive ones. Overall, when compared to hierarchical clustering, our method arranges the metabolites in an order which more closely reflects their known metabolic functions.

The objective function values for Equation 2 were evaluated for the final ordering as provided by the hierarchical clustering results and then compared to the optimal values that were determined using our method over all columns and rows (shown in Table 1). The "Gap" column in Table 1 is a standard measure for quantifying the deviation of an ordering from optimality. Based on Table 1, the final ordering provided by the hierarchical results is suboptimal with respect to the squared difference objective function in Equation 2.

Table 1. Comparison Between Optimal Objective and Hierarchical Objective Value for Squared Difference Objective (see Equation 2)

Data Set	Dimension	Optimal Objf	Hierarchical Objf	Gap
Metabolite	Rows	4,415.8	6,377.6	30.8 %
Concentration	Columns	1,753.0	2,677.9	34.5 %
Colon	Rows	26,602.6	40,878.5	34.9 %
Cancer	Columns	32,174.0	43,627.2	26.3 %

3.1.1. Results for Other Biclustering Algorithms

Since the rearranged data appears to naturally form biclusters, we compared the results for our method with the biclustering algorithms ISA, Cheng and Church's, OPSM, BiMax, and SAMBA on the metabolite concentration data set. Each algorithm was run using the default parameter values, which were adjusted in the event that no biclusters were found. The bicluster from results were visualized using the BiVoc algorithm[40].

The results for Cheng and Church's algorithm[19] reported a total of 10 biclusters. A bicluster of 30 metabolites contained 15 of the 18 metabolites assigned to the amino acid category [32]. The longest consecutive ordering of amino acids within this bicluster are serine, methionine, threonine, glutamate, and alanine, which is exactly the same as that reported in the hierarchical clustering results. The majority of the other metabolites in this bicluster are biosynthetic intermediates. The remaining nine biclusters do not result in any consistent grouping of related metabolites. Only one bicluster was reported from the ISA algorithm[20]. The four metabolites in this bicluster are acetyl-coa, aspartate, ADP, and cAMP, under the conditions of carbon starvation in *E. coli*. Acetyl-coa is a TCA cycle compound and aspartate is categorized as a amino acid metabolite[32], so there is no direct relationship between the metabolites identified in this bicluster. The OPSM algorithm[41] produced a bicluster that contained the amino acid metabolites valine, isoleucine, alanine, phosphoenolpyruvate, ATP, proline, asparigine, and glutamate under the conditions of nitrogen and carbon starvation in *S. cerevisiae* and *E. coli*. The other OPSM biclusters revealed little correlation among the metabolites grouped together. The BiMax algorithm[25] resulted in several biclusters of increased concentration and many of the metabolites were assigned to at least three biclusters. However, little relation among the metabolites in the BiMax biclusters was found. The SAMBA algorithm [21] was also applied to the metabolite concentration data and resulted in a total of three biclusters with little overlap of the metabolites. No apparent grouping of related metabolites was observed for these biclusters.

3.2. *Colon Cancer Data: Dense Data Matrix*

We also tested the proposed method for dense data matrices on gene expression data for 62 colon tissue samples, 22 of which were normal and 40 of which were tumor tissues[33]. In the original work by Alon et al.[33], the 2000 genes with the highest minimal intensity across the samples were examined using a deterministic-annealing algorithm[42]. Two-way clustering was performed on both the genes and the tissue samples and it was found that the algorithm was able to approximately separate the tissues into a normal-rich cluster and a tumor-rich cluster. Table 2 reports separation of the tissues into tumor-rich and normal-rich groups, where the letter classifies the tissue as normal (N) or tumor (T) and the number corresponds to the patient from which the sample was extracted. Only three normal tissues (N8, N12, N34) were assigned in the tumor-rich tissue region and a total of five tumor tissues (T30, T36, T33, T37, T2) were placed in the normal-rich tissue region. The clustering of the genes revealed a strong correlation among the ribosomal proteins, where a cluster consisting of 22 ribosomal proteins was discovered.

We applied our method to the same set of 2000 genes of highest minimal intensity and 62 tissue samples using the traveling salesman representation[29] and the objective function defined in Equation 2. The original data was normalized by performing Z-normalization over all genes and all tissues. The optimal re-ordering for the tissues (or columns) was achieved a CPU time of 0.17 seconds using the TSP model. The normal and tumor tissue samples were partitioned into normal-rich and tumor-rich regions based on the largest two cluster boundaries. In Table 2, we report the separation achieved using our method, where three normal tissues (N34, N8, N7) were incorrectly assigned to the tumor-rich tissue region and six tumor tissues (T2, T37, T40, T36, T30, T33) were found in the normal-rich tissue region. These results are consistent with the findings of Alon et al. in that N8 and N34 were incorrectly grouped with the tumor tissues and T30, T36, T33, T37, T2 were incorrectly grouped with the normal tissues. The genes (or rows) of the data matrix were then optimally re-ordered over both the corresponding tumor and normal-rich submatrices. For the tumor-rich submatrix, our method organized 30 out of the 48 ESTs homologous to ribosomal proteins into one dense cluster and these results are similar to the findings of Alon et al. [33]. Interdispersed among the ribosomal protein cluster are ESTs homologous to genes that are related to cell growth, such as elongation factors, which is consistent with previous findings [33,43].

Table 2. Partitioning of Tumor-Rich and Normal-Rich Tissues for Colon Cancer Data

Method	Tumor Rich Tissues	Normal Rich Tissues
Alon et al.	T16, T28, T13, T9, T21, T35, T10, T27, T8, T5, T4, T1, T15, T26, T39, T11, T6, T19, T12, T22, T34, T7, **N8**, T3, **N12**, T17, T25, T18, T23, T31, T20, **N34**, T24, T29, T38, T14, T40, T32	N9, **T30**, **T36**, N6, **T33**, N11, N1, N39, N28, N35, N32, N4, N33, N7, **T37**, N5, N27, N3, N2, N40, N36, **T2**, N29, N10
Proposed Method	T2, T37, T25, T17, T14, T15, T18, T4, T1, T11, T19, T23, T16, T5, T10, T34, T24, T26, T12, T9, T21, T6, T20, T29, T31, T32, T8, T13, T22, T38, T28, T35, **N34**, T27, T39, **N8**, **N7**	**T7**, **T3**, N36, **T40**, N40 N33, N27, N35, **T36**, N28, N29, **T30**, N32, N39, **T33**, N1, N6, N2, N11, N5, N4, N3, N10, N9, N12
Hierarchical Clustering	T37, T2, **N36**, **N4**, T12, T6, T9, **N8**, T19, **N10**, **N3**, **N5**, **N12**, **N9**, T29, T20, T40, T38, T28, **N34**, T31, T26, T24, T21, T34, T39, T27, T13, T8, T3, T22, T7, T11, T1, T14, T17, T15, T16, T23, T4, T35, T10	N11, N33, N40, N35, N27, **T36**, N32, **T30**, N29, **T33**, N29, N28, N1, N6, **T18**, **T32**, **T25**, N2, **T5**, N7

The colon cancer data was also examined using hierarchical clustering with the Euclidean objective function. The genes related to ribosomal proteins were clustered together as well, but only 24 out of the 48 were grouped into a larger cluster. Table 2 shows that the separation of normal and tumor tissues was not as consistent for hierarchical clustering, where nine normal tissues (N36, N4, N8, N10, N3, N5, N12, N9, N34) were found in the tumor-rich tissue region and seven tumor tissues (T36, T30, T33, T18, T32, T25, T5) were placed in the normal-rich tissue region. In Table 1, we present the deviation from optimality for the ordering reported from hierarchical clustering with reference to the optimal ordering over all columns and rows as determined by our method.

3.3. Percent Inhibition Data: Sparse Data Matrix

We applied our method for the optimal ordering of sparse matrices to a data matrix of 151 rows and 93 columns containing percent inhibition data for an unknown set of compounds[44]. The initial ordering of the compounds was provided by Pfizer Inc and is shown in Figure 2. The most desirable compounds are the strongest inhibitors of an unknown target, which correspond to the highest percentage inhibition values in this set. The original matrix has a total of 4110 known data values out of a possible 14043 (29%).

This data matrix was reordered using the assignment model for all 4110 known values and a uniformly sampled subset of known values to test the rigorousness of the approach. The best identified reordering of this data matrix using all of the 4110 known values is presented in Figure 3. The best found reordering for a sampling of 50% of the available data values (or 15% of the whole library space) is presented in Figure 4 and the corresponding placement of the *unsampled* compounds based on this reordering is revealed in Figure 5. Note how the proposed re-ordering method is able to group many of the desired compounds in an easily identifiable subset of the matrix (i.e., the upper left corner of the matrix).

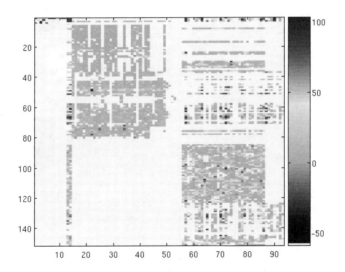

Figure 2.   Original sparse data matrix using all known values

3.3.1. *Iterative Synthesis Strategy*

The dense grouping of the high inhibition compounds implies that this re-ordering method can be used to guide additional compound synthesis. Once the optimal (or best) substituent ordering has been established, it is necessary to develop an appropriate strategy for molecular discovery. If the ordered compound property landscape is indeed smooth and regular, a simple local interpolation measure can be applied to adequately represent

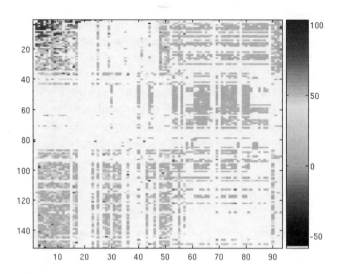

Figure 3. Optimal rearrangement of the sparse data matrix using all known values

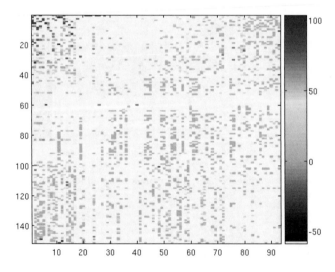

Figure 4. Optimal rearrangement of the sparse data matrix sampled at 15%

the *expected* property value of a compound that is missing[45]. The most basic local interpolation measure would assume that the property value of a compound is similar to the average property values of its neighbors.

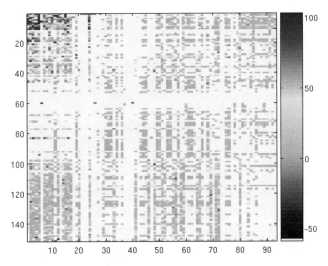

Figure 5. Optimal rearrangement of the sparse data matrix sampled at 15%, all known values included

Two main issues arise if a property value is predicted by the average of neighboring compounds. If we define a set of compounds to be neighbors of specific compound, $a_{i,j}$, we may want to weight the contributions of these neighboring values by their distance from the specified compound. The distance between two points (i, j) and (i', j') in the ordered compound property landscape is defined as the Euclidian distance, as shown in Equation 30.

$$d_{i',j'}^{i,j} = \sqrt{(i - i')^2 + (j - j')^2} \tag{30}$$

Furthermore, we must account for the frequency of known points around a point (i, j). Let $R_{i'j'}^{ij}$ represent the set of compounds, (i', j'), that will be used to determine the estimated property value of the compound at position (i, j) as defined by Equation 31.

$$R_{i'j'}^{ij} = \left\{ (i', j') : d_{i',j'}^{i,j} \leq d^{thresh} \text{ and } (a_{i',j'} \text{ is known or } (i = i' \text{ and } j = j')) \right\} \tag{31}$$

The threshold distance, d^{thresh}, is the maximum distance a neighboring point can be from the specified compound (in this article we used the relation $d^{thresh} = (|I| \cdot 0.1 + |J| \cdot 0.1)/2$. Once we have defined $R_{i'j'}^{ij}$, Equation 32 is used to define a normalization factor, $\Omega_{i,j}$. This normalization factor gives a higher weight to the closer neighbors and always retains the weight for the compound (i, j) to avoid long-range effects for missing values in sparsely populated regions.

$$\Omega_{i,j} = \sum_{(i',j') \in R_{i'j'}^{ij}} \frac{1}{d_{i',j'}^{i,j} + 1} \tag{32}$$

The estimated property value at (i, j), $\rho_{i,j}$, is then defined by Equation 33, which is a normalized average of the weighted neighboring property values.

$$\rho_{i,j} = \sum_{(i',j') : a_{i',j'} \text{ is known}} \frac{a_{i',j'}}{\left(d_{i',j'}^{i,j} + 1 \right) \cdot \Omega_{i,j}} \tag{33}$$

One strategy for synthesis, given an initial sampling of compound property values and its associated optimal substituent ordering, is to sort the estimated property values, $\rho_{i,j}$, and synthesize some number of compounds with the highest predicted property values. As more property values are determined, the process of reordering the substituents and estimating the property values can be repeated in an iterative fashion. An important feature to examine for such an iterative strategy is how much initial data is required to synthesize the important compounds while simultaneously minimizing the number of total compounds synthesized. To test the utility of such a protocol, we have developed an algorithm which begins with a small population of known compounds and uses the reordering results to perform an "in silico" synthesis. After each reordering, Equation 33 is used to estimate the inhibition values for the unknown compounds. We then select the top 50 unknown compounds of highest predicted inhibition for in silico synthesis (i.e., reveal their actual values) and subsequently reorder this new library.

Our iterative procedure was applied to four sparse data matrices: samplings of 50% (2050 compounds or 15% of the whole library space), 25% (1025 compounds or 7.3% of the whole library space), 10% (411 compounds or 3% of the whole library space), and 5% (206 compounds or 1.5% of the whole library space) at random from the original data matrix (4110 known values)[45]. The synthesis from each initial sampling was carried out until 3000 compounds were revealed. The results for the synthesis of compounds above 80% inhibition are shown in Figure 6.

The horizontal dashed line at the top of Figure 6 denotes the total number of compounds that are above 80 percent inhibition in the original data. The diagonal dashed-dotted line in this figure represents the average gain per synthesis for the original synthesis procedure based on the data provided by Pfizer. In Figure 6, we see that the gain from each of the

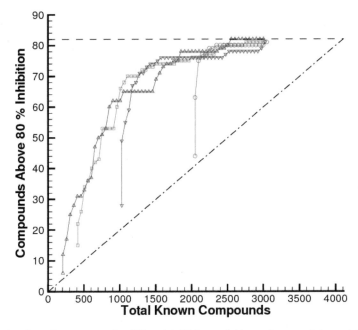

Figure 6. Iterative strategy for different initial populations of sparse data matrix for a threshold of 80 percent inhibition. The ○, ▽, □, and △ curves represent starting with 50, 25, 10, and 5 percent of the available data, respectively.

synthesis curves is fairly consistent among the initial samplings. Each of the curves exhibits a sharp slope in the beginning, indicating a good yield of higher inhibition compounds per synthesis iteration of 50 compounds. It is interesting to note that the synthesis curve starting from only 5% of the original data (206 known values) achieves a better yield of high inhibition compounds than all the other curves when 3000 known compounds are revealed. In fact, this synthesis curve finds all of the available compounds above the inhibition thresholds of 70%, 80% and 90% in less than 3000 compounds (the results for 70% and 90% inhibition are not shown here).

4. Conclusions

Several rigorous methods for the optimal re-ordering of dense and sparse data matrices were presented in this article. The re-ordering of the rows and columns for dense data matrices can be accomplished via either a network flow model or a traveling salesman problem representation. Sparse data matrices can be rearranged using an assignment problem model, which allows for long-range comparisons among rows and columns in the final ordering to accommodate missing values. Several different objective functions can be used to quantify the degree of similarity between adjacent rows and columns in the final arrangement and the selection of the appropriate metric is dependent on the density of the data. We applied the proposed re-ordering methods to (a) metabolite concentration data[32], (b) colon cancer gene expression data[33], and (c) drug inhibition data provided by Pfizer Inc. For each of these data sets in (a) and (b), our method provides a denser grouping of related metabolites and genes than other clustering methods do, which suggests that optimal re-ordering has distinct advantages over a local re-ordering. It was also shown that the proposed method for dense data matrices has the ability to separate objects into distinct groups, as was illustrated with the separation of the starvation conditions in the metabolite concentration data and the separation of normal and tumor tissues in the colon cancer data set. We also demonstrated using the sparse data set in (c) that the assignment problem model can be utilized in an iterative fashion for descriptor-free drug discovery. The iterative synthesis strategy we presented and applied to sparse samplings of the data matrix was able to uncover a significant percentage of the lead molecules while using only a fraction of total compound library, even when starting from a mere 3% of the total library space.

Acknowledgements

The authors gratefully acknowledge financial support from the National Science Foundation and the U.S. Environmental Protection Agency. Although the research described in the article has been funded in part by the U.S. EPA, it has not been subjected to any EPA review and therefore does not necessarily reflect the views of the Agency, and no official endorsement should be inferred. The authors are also grateful to Dr. T.-C. Hua of Pfizer for providing the data analyzed in this work.

References

1. M. B. Eisen, P. T. Spellman, P. O. Brown, and D. Botstein. Cluster analysis and display of genome-wide expression patterns. *Proc. Natl. Acad. Sci.*, 95:14863–14868, 1998.
2. J. A. Hartigan and M. A. Wong. Algorithm AS 136: a K-means clustering algorithm. *Appl. Stat.*, 28:100–108, 1979.
3. H. D. Sherali and J. Desai. A global optimization RLT-based approach for solving the hard clustering problem. *J. Glo. Opt.*, 32:281–306, 2005a.

4. H. D. Sherali and J. Desai. A global optimization RLT-based approach for solving the fuzzy clustering problem. *J. Glo. Opt.*, 33:597–615, 2005b.

5. Z. Bar-Joseph, E.D. Demaine, D.K. Gifford, N. Srebro, A.M. Hamel, and T.S. Jaakola. K-ary clustering with optimal leaf ordering for gene expression data. *Bioinfo.*, 19(9):1070–1078, 2003.

6. A. W. F. Edwards and L. L. Cavalli-Sforza. A method for cluster analysis. *Biometrics*, 21:362–375, 1965.

7. J. H. Wolfe. Pattern clustering by multivariate mixture analysis. *Multivariate Behavioral Research*, 5:329–350, 1970.

8. A. K. Jain and J. Mao. Artificial neural networks: a tutorial. *IEEE Computer*, 29:31–44, 1996.

9. R . W. Klein and R. C. Dubes. Experiments in projection and clustering by simulated annealing. *Pattern Recognition*, 22:213–220, 1989.

10. V. V. Raghavan and K. Birchand. A clustering strategy based on a formalism of the reproductive process in a natural system. In *Proceedings of the Second International Conference on Information Storage and Retrieval*, pages 10–22, 1979.

11. J. N. Bhuyan, V. V. Raghavan, and K. E. Venkatesh. Genetic algorithm for clustering with an ordered representation. In *Proceedings of the Fourth International Conference on Genetic Algorithms*, pages 408–415, 1991.

12. N. Slonim, G.S. Atwal, G. Tkacik, and W. Bialek. Information-based clustering. *Proc. Natl. Acad. Sci.*, 102(51):18297–18302, 2005.

13. M.P. Tan, J.R. Broach, and C.A. Floudas. A novel clustering approach and prediction of optimal number of clusters: Global optimum search with enhanced positioning. *J. Glo. Opt.*, 2006. in press.

14. M.P. Tan, J.R. Broach, and C.A. Floudas. Evaluation of normalization and pre-clustering issues in a novel clustering approach: Global optimum search with enhanced positioning. *J. Bioin. Comp. Bio*, 2007. in press.

15. S. Busygin, O.A. Prokopyev, and P.M. Pardalos. An optimization based approach for data classification. *Opt. Meth. Soft.*, 22(1):3–9, 2007.

16. W.T. McCormick Jr, P.J. Schweitzer, and T.W. White. Problem decomposition and data reorganization by a clustering technique. *Oper. Res.*, 20(5):993–1009, 1972.

17. J.K Lenstra. Clustering a data array and the traveling-salesman problem. *Oper. Res.*, 22(2):413–414, 1974.

18. J.K Lenstra and A.H.G. Rinnooy Kan. Some simple applications of the traveling-salesman problem. *Oper. Res. Quart.*, 26(4):717–733, 1975.

19. Y. Cheng and G.M. Church. Biclustering of expression data. *Proc. ISMB 2000*, pages 93–103, 2000.

20. J. Ihmels, G. Friedlander, S. Bergmann, O. Sarig, Y. Ziv, and N. Barkai. Revealing modular organization in the yeast transcriptional network. *Nat. Genet.*, 31:370–377, 2002.

21. A. Tanay, R. Sharan, and R. Shamir. Discovering statistically significant biclusters in gene expression data. *Bioinfo.*, 18:S136–S144, 2002.

22. S. Busygin, O.A. Prokopyev, and P.M. Pardalos. Feature selection for consistent biclustering via fractional 0-1 programming. *J. Comb. Opt.*, 10:7–21,

2005.

23. H.L. Turner, T.C. Bailey, W.J. Krzanowski, and C.A. Hemingway. Biclustering models for structured microarray data. *IEEE/ACM Transactions on Computational Biology and Bioinformatics*, 2(4):316–329, 2005.

24. S. Yoon, C. Nardini, L. Benini, and G. De Micheli. Discovering coherent biclusters from gene expression data using zero-suppressed binary decision diagrams. *IEEE/ACM Transactions on Computational Biology and Bioinformatics*, 2(4):339–354, 2005.

25. A. Prelic, S. Bleuler, P. Zimmermann, A. Wille, P. Buhlmann, W. Gruissem, L. Hennig, L. Thiele, and E. Zitzler. A systematic comparison and evaluation of biclustering methods for gene expression data. *Bioinfo.*, 22(9):1122–1129, 2006.

26. D.J. Reiss, N.S. Baliga, and R. Bonneau. Integrated biclustering of heterogeneous genome-wide datasets for the inference of global regulatory networks. *BMC Bioinfo.*, 7:280–302, 2006.

27. F. Divina and J. Aguilar. Biclustering of expression data with evolutionary computation. *IEEE Trans. Knowl. Data Eng.*, 18(5):590–602, 2006.

28. P. DiMaggio, S. McAllister, C.A. Floudas, X.J. Feng, J. Rabinowitz, and H. Rabitz. Biclustering via optimal re-ordering of data matrices. submitted to J. Glo. Opt.

29. P. DiMaggio, S. McAllister, C.A. Floudas, X.J. Feng, J. Rabinowitz, and H. Rabitz. Optimal re-ordering of data matrices in systems biology; a biclustering approach. in prep.

30. S. C. Madeira and A.L. Oliveira. Biclustering algorithms for biological data analysis: A survey. *IEE-ACM Trans. Comp. Bio.*, 1(1):24–45, 2004.

31. R. Perkins, H. Fang, W. Tong, and W. Welsh. Quantitative structure-activity relationship methods: perspectives on drug discovery and toxicology. *Environ. Toxicol. and Chem.*, 22:1666–1679, 2003.

32. M. J. Brauer, J. Yuan, B. Bennett, W. Lu, E. Kimball, D. Bostein, and J.D. Rabinowitz. Conservation of the metabolomic response to starvation across two divergent microbes. *Proc. Natl. Acad. Sci.*, 103:19302–19307, 2006.

33. U. Alon, N. Barkai, D.A. Notterman, K. Gish, S. Ybarra, D. Mack, and A.J. Levine. Broad patterns of gene expression revealed by clustering analysis of tumor and normal colon tissues probed by oligonucleotide arrays. *Proc. Natl. Acad. Sci.*, 96:6745–6750, 1999.

34. L.R. Ford and D.R. Fulkerson. *Flows in Networks*. Princeton University Press, 1962.

35. S. Climer and W. Zhang. Rearrangement clustering: Pitfalls, remedies, and applications. *J. Machine Learning Res.*, 7:919–943, 2006.

36. D. Applegate, R. Bixby, V. Chvatal, and W. Cook. Finding cuts in the tsp (a preliminary report). Technical Report 95-05, DIMACS, 1995.

37. R. Jonker and T. Volgenant. Transforming asymmetric into symmetric traveling salesman problems. *Oper. Res. Let.*, 2:161–163, 1983.

38. R. Jonker and T. Volgenant. Transforming asymmetric into symmetric traveling salesman problems: erratum. *Oper. Res. Let.*, 5:215–216, 1986.

39. CPLEX. *ILOG CPLEX 9.0 User's Manual*. 2005.

40. G.A. Grothaus, A. Mufti, and T.M. Murali. Automatic layout and visualization of biclusters. *Alg. Mol. Bio.*, 1:1–15, 2006.

41. A. Ben-Dor, B. Chor, R.Karp, and Z.Yakhini. Discovering local structure in gene expression data: The order-preserving submatrix problem. In *RE-COMB*, 2002.

42. K. Rose. Deterministic annealing for clustering, compression, classification, regression, and related optimization. *Proc. IEEE*, 11:2210–2239, 1998.

43. G. Getz, E. Levine, and E. Domany. Coupled two-way clustering analysis of gene microarray data. *Proc. Natl. Acad. Sci.*, 97(22):12079–12084, 2000.

44. S. McAllister, X.J. Feng, P. DiMaggio, C.A. Floudas, J. Rabinowitz, and H. Rabitz. Accelerating molecular discovery by optimal substituent reordering. submitted to Angewandte Chemie.

45. S. McAllister, P. DiMaggio, X.J. Feng, C.A. Floudas, J. Rabinowitz, and H. Rabitz. Enhancing molecular discovery using descriptor-free rearrangement clustering techniques and sparse data sets. in prep.

THE SOLUTION OF THE DISTANCE GEOMETRY PROBLEM IN PROTEIN MODELING VIA GEOMETRIC BUILDUP

DI WU

Department of Mathematics, Western Kentucky University, USA

ZHIJUN WU

Department of Mathematics, Iowa State University, USA

· YAXIANG YUAN

Institute of Computational Mathematics,
Chinese Academy of Science, China

A well-known problem in protein modeling is the determination of the structure of a protein with a given set of inter-atomic or inter-residue distances obtained from either physical experiments or theoretical estimates. A general form of the problem is known as the distance geometry problem in mathematics, the graph embedding problem in computer science, and the multidimensional scaling problem in statistics. The problem has applications in many other scientific and engineering fields as well such as sensor network localization, image recognition, and protein classification. We describe the formulations and complexities of the problem in its various forms, and introduce a geometric buildup approach to the problem. Central to this approach is the idea that the coordinates of the atoms in a protein can be determined one atom at a time, with the distances from the determined atoms to the undetermined ones. It can determine a structure more efficiently than other conventional approaches, yet without requiring more distance constraints than necessary. We present the general algorithm and its theory and review the recent development of the algorithm for controlling the propagation of the numerical errors in the buildup process, for determining rigid vs. unique structures, and for handling problems with inexact distances (distances with errors). We show the results from applying the algorithm to some of the model problems and justify the potential use of the algorithm in protein modeling.

1. Distance Based Protein Modeling

Proteins are an important class of biological molecules. They are encoded in genes and produced in cells through genetic translation. They are life supporting (or sometimes, destructing) ingredients and are indispensable for almost all biological processes. For example, humans have hundreds of thousands of different proteins and would not be able to maintain normal life even if short of a singe type of protein (Figure 1a). On the other hand, with the help of some proteins, viruses are able to grow, translate, integrate, and replicate, causing diseases (Figure 1b). Some proteins themselves are toxic and even infectious such as the proteins in poisonous plants and in beef causing the Mad Cow Disease (Figure 1c)[1].

A protein consists of a linear chain of amino acids connected with strong chemical bonds. The amino acids and their order in the chain are fixed for each different protein, and they are specified by the gene (a sequence of DNA molecules) from which the protein is generated. Once the chain of amino acids for a protein is produced, it immediately folds into a unique and stable 3D structure, which is crucial for the protein to function. Since the function of the protein depends on its structure, the determination of the structure becomes a necessary step for the understanding of the biological properties of every protein[1].

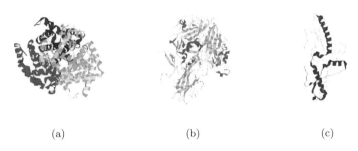

(a)   (b)   (c)

Figure 1. **Example proteins** (a) hemoglobin protein, 1BUW, in blood; (b) protein 2PLV, supporting poliovirus; (c) prion protein 1I4M-D, causing the Mad Cow Disease in human.

Unfortunately, there is no direct physical means to observe a protein structure at an atomic level. There are only techniques that can be used to measure certain physical properties of the protein upon which the structure can be deduced. X-ray crystallography and nuclear magnetic resonance spectroscopy (NMR) are major experimental techniques of such in practice. They are responsible for the determination of 80% and 15% of the protein structures (total about 30,000) so far deposited in the Protein Data Bank (PDB), respectively[2]. The experimental techniques have many limitations, though. X-ray crystallography requires purifying and crystallizing proteins, which may take months or years to finish, if not failed. The results often vary with varying experiments for reasons not fully understood[3]. NMR can only be applied to small proteins for otherwise the spectral data would become too difficult to clarify[4]. The structures determined by NMR are not as accurate and detailed as well[1]. Theoretical or computational approaches such as homology modeling, structural alignment, threading, energy minimization, dynamic simulation, etc., have been developed[5,6], but they are more successful in building theoretical models or refining experimental structures than determining the structures completely independently, although recent progress as shown in the CASP competitions[7] and in utilizing more powerful computing resources is indeed exciting and encouraging[8].

In this paper, we discuss a well-known problem in protein modeling, for the determination of the structure of a protein with a given set of inter-atomic or inter-residue distances obtained from either physical experiments or theoretical estimates (Figure 2). A more general and abstract form of the problem is known as the distance geometry problem in mathematics[9], the graph embedding problem in computer science[10], and the multidimensional scaling problem in statistics[11]. In general, the problem can be stated as to find the coordinates for a set of points in some topological space given the distances for certain pairs of points. Therefore, in addition to protein modeling where everything is discussed only in three-dimensional Euclidean space, the problem has applications in many other scientific and engineering fields as well, such as sensor network localization[12], image recognition[13], and protein classification[14], to name a few. In any case, the problem may or may not have a solution in a given topological space, and even if it does have a solution, the solution may not be easy to find, depending on the given distances. For example, in any k-dimensional Euclidean space, the problem is polynomial time solvable if the distances for all the pairs of points are provided, and is NP-complete otherwise in general[10].

In protein modeling, the distances or their ranges for certain pairs of atoms or residues in a given protein may be obtained from either physical experiments such as NOE (Nuclear Overhauser Effects), J-coupling, and dipolar coupling in NMR[4,15,16], or theoretical estimates such as the bond lengths and bond angles known from general organic chemistry[1], or statistical estimates on certain inter-atomic or inter-residue distances based on their distributions in databases of known protein structures[17,18,19]. Then, a structure may be determined for the protein by using the available distances. However, the given distances may not necessarily be sufficient for determining the structure uniquely, or even just rigidly. Here, by uniquely we mean that the structure is unique under translation and rotation, and by rigidly we mean that any part of the structure cannot be changed continuously without violating the given distance restraints. Sometimes, the distances may contain errors and may be inconsistent in the sense that they may have violated some basic geometric conditions such as the triangle inequality for the distances among any three atoms. In that case, a structure that fits the given distances will not even exist. After all, even if a structure does exist, it is still not trivial to determine based on the given distances. A distance geometry problem needs to be solved, which is computationally intractable in general[10].

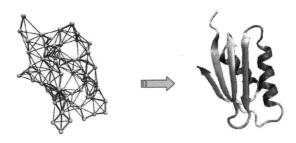

Figure 2. Distance based protein modeling Given a set of inter-atomic distances or their ranges, find the coordinates of the atoms in the protein.

Crippen and Havel and several other research groups[20,21] pioneered the work on using the solution of a distance geometry problem for protein structure determination, especially for NMR structure modeling, where the distances for certain pairs of atoms and in particular, the pairs of hydrogen atoms that are within say, 5 Å distance, can be estimated through J-couplings and NOE, with additional ones that can be derived from known bond lengths and bond angles. However, in NMR modeling, the distances obtained are restricted to a small subset of all pairs of atoms in the protein. Otherwise, if the distances for all pairs of atoms are available, a structure would be much easier to build upon. The NMR distances also contain experimental errors and are not necessarily always consistent. A structure that can fit the distances approximately rather than exactly may be the best we can hope for in practice. Moreover, in NMR, instead of exact distances, the ranges or lower and upper bounds of the distances are usually provided, due to the fact that the structures are flexible in solution and the distances are not fixed. An ensemble of structures rather than a single one that can fit in the distance ranges are therefore sought in real practice to show the dynamic nature of the structure[22,23]. For these various reasons, the focus on NMR modeling has been more on developing methods for extracting the bounds on the missing distances (bound smoothing), removing the inconsistencies in the distances (distance metrication), and fitting the structures in the distance ranges (optimization), as described in the embed algorithm[20,21] and implemented in NMR modeling software such as the CNS[24,25]. Therefore, the solution of an exact distance geometry problem has not been improved much since the embed algorithm was first developed, and its impact in NMR modeling has been rather limited. On the other hand, important theoretical and algorithmic issues related to the solution of the problem still remain to be resolved, while its applications in more general areas of distance-based protein modeling are expanding[26,27,28,29].

Existing approaches to the solution of the distance geometry problem include, for example, the embedding algorithm by Crippen and Havel[20,21], the alternating projection method by Glunt and Hayden[32,33], the graph reduction approach by Hendrickson[30,31], the global optimization method by Moré and Wu[34,35], the stochastic/perturbation method by Zou, Byrd, and Schnabel[36], the multidimensional scaling method by Kearsly, Tapia, and Trosset[37,38], the dc programming method by Le Thi Hoai and Pham Dinh[39], the semi-definite programming approach by Biswas, Liang, Toh, and Ye[40], and the stochastic search method by Grosso, Locatelli, and Schoen[41].

We investigate the solution of the distance geometry problem within a so-called geometric buildup framework. Dong and Wu[42,43] first implemented a geometric buildup algorithm for the solution of the distance geometry problem with exact distances and justified the linear computation time for the case when the distances required in every buildup step are always available. Central to the geometric buildup approach is the idea to determine only a small group of atoms at the beginning and then complete the whole molecule by repeatedly determining one or more atoms every time using the available distances between the determined and undetermined atoms. The advantage of using a geometric buildup approach is that it works directly on the given distances and exploits the special structure of a given problem, and hence may be able to solve the problem more efficiently than a general approach. We present the general algorithm of this approach, and discuss related computational issues including control of numerical errors, determination of rigid vs. unique structures, and tolerance of distance errors, based on the recent development of the algorithm[44,45,46]. The theoretical basis of the approach is established based on the theory of distance geometry. A group of necessary and sufficient conditions for the determination of a structure with a given set of distances using a geometric buildup algorithm are justified. The applications of the algorithm to model protein problems are demonstrated.

2. The Distance Geometry Problem

Let n be the number of atoms in a given protein and $x_1, \ldots, x_n$ be the coordinate vectors for the atoms, where $x_i = (x_{i,1}, x_{i,2}, x_{i,3})^T$ and $x_{i,1}$, $x_{i,2}$, and $x_{i,3}$ are the first, second, and third coordinates of atom i. If the coordinates $x_1, \ldots, x_n$ are known, the distances $d_{i,j}$ between atoms i and j can be computed with $d_{i,j} = \|x_i - x_j\|$, where $\| \cdot \|$ is the Euclidean norm. Conversely, if the distances $d_{i,j}$ are given, the coordinates $x_1, \ldots, x_n$ for the atoms can also be obtained based on the distances $d_{i,j}$, but the computation is not as straightforward. The solution of a system of equations as can be stated in the following for $x_1, \ldots, x_n$ is required.

$$\|x_i - x_j\| = d_{i,j}, \quad (i,j) \in S, \tag{1}$$

where S is a subset of all atom pairs. The latter problem is known as a distance geometry problem in mathematics[9], a graph embedding problem in computer science[10], and a multidimensional scaling problem in statistics[11]. In practice, the distances may have errors, and therefore, a more general yet practical form of the problem would be to find the coordinates of the atoms $x_1, \ldots, x_n$, given only a set of lower and upper bounds, $l_{i,j}$ and $u_{i,j}$, of the distances $d_{i,j}$ such that

$$l_{i,j} \leq \|x_i - x_j\| \leq u_{i,j}, \quad (i,j) \in S. \tag{2}$$

The distance geometry problem is polynomial time solvable if the distances for all pairs of atoms are available. However, it has been proved to be NP-hard in general. Even if errors are allowed for the distances, the problem is still hard, if only small errors are allowed.

2.1. Problems with Exact Distances

We first consider the simple case when a complete set of exact distances is given. By exact distances we mean the distances are given in exact values, not in ranges, and by a complete set of distances we mean the distances for all pairs of atoms are included. A solution to the distance geometry problem with such a set of distance data can be obtained efficiently by using for example an algorithm that requires the singular value decomposition (SVD) of an induced distance matrix.

Assume that a set of coordinates $x_1, \ldots, x_n$ can be found for a given set of distances $d_{i,j}$, where $i, j = 1, \ldots, n$. Then, $\|x_i - x_j\| = d_{i,j}$ for all i, $j = 1, \ldots, n$, and

$$\|x_i\|^2 - 2x_i^T x_j + \|x_j\|^2 = d_{i,j}^2, \quad i, j = 1, \ldots, n. \tag{3}$$

Since the molecular structure is invariant under any translation or rotation, we set a reference system so that the origin is located at the last atom or in other words, $x_n = (0, 0, 0)^T$. It follows that

$$d_{i,n}^2 - 2x_i^T x_j + d_{j,n}^2 = d_{i,j}^2, \quad i, j = 1, \ldots, n-1. \tag{4}$$

Define a coordinate matrix X and an induced distance matrix D,

$$X = \{x_{i,j} : \ i = 1, \ldots, n-1, \ j = 1, 2, 3\} \quad \text{and}$$
$$D = \{(d_{i,n}^2 - d_{i,j}^2 + d_{j,n}^2)/2 : \ i, j = 1, \ldots, n-1\}. \tag{5}$$

Then, $XX^T = D$ and D must be of maximum rank 3.

The distance geometry problem can be defined in a general space R^k with $x_1, \ldots, x_n$ in R^k and $d_{i,j}$ the Euclidean distances between atoms i and j. Then, the equation $XX^T = D$ still holds, and D must be of maximum rank k, where $X = \{x_{i,j} : i = 1, \ldots, n, j = 1, \ldots, k\}$.

Theorem 2.1. [9] *Let $\{d_{i,j} : i, j = 1, \ldots, n\}$ be a set of distances in R^k, for some $k \leq n$. Then, the induced matrix D as defined in (5) is of maximum rank k.*

Proof. It follows from the fact that $D = XX^T$ for a coordinate matrix X in $R^{n-1} \times R^k$ and X is of maximum rank k. □

The equation $XX^T = D$ can be solved using the singular value decomposition of D. Let $D = U\Sigma U^T$ be the singular value decomposition of D, where U is an orthogonal matrix and Σ a diagonal matrix with the singular values of D along the diagonal. If D is a matrix of rank less than or equal to k, the decomposition can be obtained with U being $(n-1) \times k$ and Σ being $k \times k$. Then, $X = U\Sigma^{1/2}$ solves the equation $XX^T = D$. Here the singular value decomposition of D requires $O(kn^2)$ floating-point operations[47], and therefore, the distance geometry problem with a complete set of exact distances can be solved in polynomial time.

Note that although in practice, the distances may not be available for all the pairs of atoms, the solution of the problem with all exact distances can still be important for the solution of the general problem with a sparse set of distances. For example, in the embed algorithm, a complete set of distances among all the atoms is generated after bound smoothing, and the solution of a distance geometry problem with all exact distances is always required afterwards[20,21]. Also, if a subset of atoms has all the distances among the atoms, but the whole set of atoms does not, the coordinates of the subset of atoms can still be determined efficiently by solving a distance geometry problem with all exact distances for the subset of atoms. The procedure may also be applied repeatedly as some of the atoms are determined and the availability of the distances among them is changed, until no such subsets of atoms can be found[48,49].

2.2. Problems with Sparse Distances

We now consider the problem with an incomplete set of exact distances. Let S be a subset of all pairs of atoms such that (i, j) is in S if the distance $d_{i,j}$ between atoms i and j is given. Then, the problem is to find the coordinates $x_1, \ldots, x_n$ for the atoms so that

$$\|x_i - x_j\| = d_{i,j}, \quad (i, j) \in S. \tag{6}$$

Let $G = (V, E, W)$ be a weighted graph, where $V = \{v_1, \ldots, v_n\}$ is the set of vertices, $E = \{e_{i,j} : (i, j) \text{ in } S\}$ the set of edges, and $W = \{w_{i,j} = d_{i,j} : (i, j) \text{ in } S\}$ the weights on the edges. Then, the distance geometry problem for molecular structure determination can be considered as a graph embedding problem for G in R^3, i.e., to find a mapping from the vertices $v_1, \ldots, v_n$ in V to a set of points $x_1, \ldots, x_n$ in R^3 so that the distances between points i and j for all (i, j) in S are equal to the weights $d_{i,j}$ on the corresponding edges $e_{i,j}$.

The graph embedding problem can be considered in a Euclidean space of any dimension. In any case, it has been proved that the graph embedding problem is an NP-hard problem even for the one-dimensional case[10]. The proof can be demonstrated via the solution of a special class of one-dimensional graph embedding problem, the problem of folding a closed chain in a line (in one-dimensional space, Figure 3). Let $G = (V, E, W)$, with $V = \{v_1, \ldots, v_{n+1}\}$, $E = \{e_{i,i+1} : i = 1, \ldots, n\} \cup \{e_{1,n+1}\}$, and $W = \{w_{i,i+1} = l_i : i = 1, \ldots, n\} \cup \{w_{1,n+1} = 0\}$, where l_i is the length of the link between node i and node $i + 1$ in the chain. Then, the problem

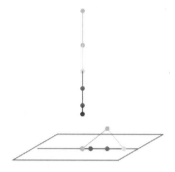

Figure 3. **Folding a closed chain** The integer set partition problem can be reduced to the problem of folding a closed chain in a line, a one-dimensional graph embedding problem, thereby proving that the one-dimensional graph embedding problem must be NP-hard, for the integer set partition problem has proved to be.

can be stated formally as to find a mapping from the nodes $\{v_1, \ldots, v_{n+1}\}$ of G to a set of points $\{x_1, \ldots, x_{n+1}\}$ in R so that

$$|x_{i+1} - x_i| = l_i, \quad i = 1, \ldots, n, \quad |x_{n+1} - x_1| = 0. \tag{7}$$

Theorem 2.2. *The integer set partition problem can be reduced to the problem of folding a closed chain in a line.*

Proof. Let $A = \{a_1, \ldots, a_n\}$ be a given set of positive integers. Define a graph $G = (V, E, W)$, with $V = \{v_1, \ldots, v_{n+1}\}$, $E = \{e_{i,i+1} : i = 1, \ldots, n\} \cup \{e_{1,n+1}\}$, and $W = \{w_{i,i+1} = a_i : i = 1, \ldots, n\} \cup \{w_{1,n+1} = 0\}$. The graph defines a closed chain. Suppose that the chain can be folded in a line or in other words, the graph can be embedded in R. Then, v_i is placed at x_i in R for $i = 1, \ldots, n+1$, and

$$|x_{i+1} - x_i| = a_i, \quad i = 1, \ldots, n, \quad |x_{n+1} - x_1| = 0.$$

Let $A_1 = \{a_i = |x_{i+1} - x_i| = x_{i+1} - x_i\}$ and $A_2 = \{a_i = |x_{i+1} - x_i| = x_i - x_{i+1}\}$. Then,

$$\sum_{i=1}^{n} (x_{i+1} - x_i) = \sum_{a_i \in A_1} (x_{i+1} - x_i) - \sum_{a_i \in A_2} (x_i - x_{i+1}).$$

However,

$$\sum_{i=1}^{n} (x_{i+1} - x_i) = x_{n+1} - x_1 = 0.$$

It follows that

$$\sum_{a_i \in A_1} (x_{i+1} - x_i) - \sum_{a_i \in A_2} (x_i - x_{i+1}) = \sum_{a_i \in A_1} a_i - \sum_{a_i \in A_2} a_i = 0 \,,$$

and A_1 and A_2 solves the set partition problem for A. $\qquad\square$

It follows from the above theorem that the problem of folding a closed chain in a line cannot be in P, for otherwise, the set partition problem would be solvable in P via the solution of an equivalent chain folding problem, which is contradictory to the fact that the set partition problem is in NP[50].

2.3. Problems with Inexact Distances

In protein modeling practice, the distances are often provided with estimated ranges only. The related distance geometry problem then becomes to find the coordinates $x_1, \ldots, x_n$ of the atoms, so that the distances between atoms i and j, for all (i, j) in a subset S of all pairs of atoms, are within their estimated ranges, i.e.,

$$l_{i,j} \le \|x_i - x_j\| \le u_{i,j} \,, \quad (i, j) \in S \,. \tag{8}$$

where $l_{i,j}$ and $u_{i,j}$ are the lower and upper bounds of the distances between atoms i and j. Let $d_{i,j} = (l_{i,j} + u_{i,j})/2$ and $\varepsilon_{i,j} = (u_{i,j} - l_{i,j})/2$. The above problem can be written as

$$|\|x_i - x_j\| - d_{i,j}| \le \varepsilon_{i,j} \,, \quad (i, j) \in S \,, \tag{9}$$

and be viewed as to find an approximate solution to the distance geometry problem for a set of exact distances $d_{i,j}$ with each distance $\|x_i - x_j\|$ allowed to have an error $\varepsilon_{i,j}$ from $d_{i,j}$. We call such a solution an ε-approximate solution.

If large errors are allowed, an approximate solution is certainly easier to obtain than an exact solution. However, if only small errors are allowed, the problem for finding an approximate solution can be as hard as for finding an exact solution. To see this, again, we can consider the simple case of folding a closed chain in a line, but this time, we allow the links to be connected loosely. Let $G = (V, E, W)$, with $V = \{v_1, \ldots, v_{n+1}\}$, $E = \{e_{i,i+1} : i = 1, \ldots, n\} \cup \{e_{1,n+1}\}$, and $W = \{w_{i,i+1} = l_i : i = 1, \ldots, n\} \cup \{w_{1,n+1} = 0\}$, where l_i is the length of the link between node i and node $i+1$ in the chain. Then, the problem can be stated formally as to find a mapping from the nodes $\{v_1, \ldots, v_{n+1}\}$ of G to a set of points $\{x_1, \ldots, x_{n+1}\}$ in R so that

$$|\|x_{i+1} - x_i\| - l_i| \le \varepsilon_i \,, \quad i = 1, \ldots, n, \quad |x_{n+1} - x_1| \le \varepsilon_{n+1} \,, \tag{10}$$

for a set of errors $\{\varepsilon_1, \ldots, \varepsilon_{n+1}\}$.

Moré and Wu[51] showed that the above problem is also NP-hard when the allowed errors are small. In fact, the set partition problem can again be reduced to this problem with $\varepsilon_i < 1/(2n)$ for $i = 1, \ldots, n+1$. Here, we give another proof that requires only $\Sigma_i \varepsilon_i < 1$, removing the dependence of the required bound of the errors on the problem size n explicitly.

Theorem 2.3. *The integer set partition problem can be reduced to the problem of folding a closed chain with total allowed error $\Sigma_i \varepsilon_i < 1$.*

Proof. Let $A = \{a_1, \ldots, a_n\}$ be a given set of positive integers. Define a graph $G = (V, E, W)$, with $V = \{v_1, \ldots, v_{n+1}\}$, $E = \{e_{i,i+1} : i = 1, \ldots, n\} \cup \{e_{1,n+1}\}$, and $W = \{w_{i,i+1} = a_i : i = 1, \ldots, n\} \cup \{w_{1,n+1} = 0\}$. The graph defines a closed chain. Suppose that the chain can be folded in a line with an error ε_i allowed on each length a_i and $\Sigma_i \varepsilon_i < 1$. Then, v_i is placed at x_i in R for $i = 1, \ldots, n+1$, and

$$\big|\,|x_{i+1} - x_i| - a_i\,\big| \le \varepsilon_i, \quad i = 1, \ldots, n, \quad |x_{n+1} - x_1| \le \varepsilon_{n+1}.$$

Let $A_1 = \{a_i = |x_{i+1} - x_i| = x_{i+1} - x_i\}$ and $A_2 = \{a_i = |x_{i+1} - x_i| = x_i - x_{i+1}\}$. Then,

$$
\begin{aligned}
\sum_{i=1}^{n}(x_{i+1} - x_i) &= \sum_{a_i \in A_1}(x_{i+1} - x_i) - \sum_{a_i \in A_2}(x_i - x_{i+1}) \\
&\ge \sum_{a_i \in A_1}(a_i - \varepsilon_i) - \sum_{a_i \in A_2}(a_i + \varepsilon_i) \\
&= \sum_{a_i \in A_1} a_i - \sum_{a_i \in A_2} a_i - \sum_{i=1}^{n} \varepsilon_i,
\end{aligned}
$$

and

$$
\begin{aligned}
\sum_{i=1}^{n}(x_{i+1} - x_i) &= \sum_{a_i \in A_1}(x_{i+1} - x_i) - \sum_{a_i \in A_2}(x_i - x_{i+1}) \\
&\le \sum_{a_i \in A_1}(a_i + \varepsilon_i) - \sum_{a_i \in A_2}(a_i - \varepsilon_i) \\
&= \sum_{a_i \in A_1} a_i - \sum_{a_i \in A_2} a_i + \sum_{i=1}^{n} \varepsilon_i.
\end{aligned}
$$

Therefore,

$$\sum_{i=1}^{n}(x_{i+1} - x_i) - \sum_{i=1}^{n}\varepsilon_i \leq \sum_{a_i \in A_1} a_i - \sum_{a_i \in A_2} a_i \leq \sum_{i=1}^{n}(x_{i+1} - x_i) + \sum_{i=1}^{n}\varepsilon_i .$$

However,

$$-\varepsilon_{n+1} \leq \sum_{i=1}^{n}(x_{i+1} - x_i) = x_{n+1} - x_1 \leq \varepsilon_{n+1} .$$

It follows that

$$-1 \leq -\sum_{i=1}^{n+1}\varepsilon_i \leq \sum_{a_i \in A_1} a_i - \sum_{a_i \in A_2} a_i \leq \sum_{i=1}^{n+1}\varepsilon_i \leq 1 .$$

Note that the two sums in the middle are over the integers and their difference cannot be a fraction. Therefore,

$$\sum_{a_i \in A_1} a_i - \sum_{a_i \in A_2} a_i = 0 ,$$

and A_1 and A_2 solves the set partition problem for A. $\qquad\square$

3. The Geometric Buildup Approach

Central to the geometric buildup approach to the distance geometry problem is the idea to determine only a small group of atoms at the beginning and then complete the whole molecule by repeatedly determining one or more atoms every time using the available distances between the determined and undetermined atoms. The advantage of using a geometric buildup approach is that it works directly on the given distances and exploits the special structure of a given problem, and hence may be able to solve the problem more efficiently than a general approach. Dong and Wu[42] first applied a geometric buildup algorithm to the solution of the distance geometry problem, and showed that the algorithm can find a solution to the problem in $O(n)$ floating-point operations if the distances for all the pairs of atoms are available. The work was later extended to sparse distances[43] with an updating scheme to control the propagation of numerical errors in the buildup process[44]. The recent development on the algorithm includes the enhancement of the algorithm on rigid vs. unique structure determination[45] and the extension of the algorithm to handling inexact or inconsistent distance data[46].

3.1. *The General Algorithm*

Given an arbitrary set of distances, the algorithm first finds four atoms that are not in the same plane and determines the coordinates for the four atoms, using for example the singular value decomposition algorithm as described in Section 2.1, with all the distances among them (assuming available). Then, for any undetermined atom j, the algorithm repeatedly performs a procedure as follows: Find four determined atoms that are not in the same plane and have distances available to atom j, and determine the coordinates for atom j. Let $x_i = (x_{i,1}, x_{i,2}, x_{i,3})^T$, $i = 1, 2, 3, 4$, be the coordinate vectors of the four atoms. Then, the coordinates $x_j = (x_{j,1}, x_{j,2}, x_{j,3})^T$ for atom j can be determined by using the distances $d_{i,j}$ from atoms $i = 1, 2, 3, 4$ to atom j (Figure 4). Indeed, x_j can be obtained from the solution of the following system of equations,

$$\|x_i\|^2 - 2x_i^T x_j + \|x_j\|^2 = d_{i,j}^2, \quad i = 1, 2, 3, 4. \tag{11}$$

By subtracting equation i from equation $i+1$ for $i = 1, 2, 3$, we can eliminate the quadratic terms for x_j to obtain

$$-2(x_{i+1} - x_i)^T x_j = (d_{i+1,j}^2 - d_{i,j}^2) - (\|x_{i+1}\|^2 - \|x_i\|^2), \quad i = 1, 2, 3. \tag{12}$$

Let A be a matrix and b a vector, and

$$A = -2 \begin{bmatrix} (x_2 - x_1)^T \\ (x_3 - x_2)^T \\ (x_4 - x_3)^T \end{bmatrix}, \quad b = \begin{bmatrix} (d_{2,j}^2 - d_{1,j}^2) - (\|x_2\|^2 - \|x_1\|^2) \\ (d_{3,j}^2 - d_{2,j}^2) - (\|x_3\|^2 - \|x_2\|^2) \\ (d_{4,j}^2 - d_{3,j}^2) - (\|x_4\|^2 - \|x_3\|^2) \end{bmatrix} \tag{13}$$

We then have $Ax_j = b$. Since x_1, x_2, x_3, x_4 are not in the same plane, A must be nonsingular, and we can therefore solve the linear system to obtain a unique solution for x_j. Here, solving the linear system requires only constant time. Since we only need to solve $n-4$ such systems for $n-4$ coordinate vectors x_j, the total computation time is proportional to n, if in every step, the required coordinates x_i and distances $d_{i,j}$, $i = 1, 2, 3, 4$ are always available.

Figure 5 shows an example protein structure determined by using the general geometric buildup algorithm, with the distances for all the pairs of atoms in the protein, as demonstrated in Dong and Wu[42]. The structure is determined accurately and uniquely. The RMSD value of the structure compared with its X-ray reference structure is 1.0e-04 Å. The computation time is much more efficient than the conventional singular value decomposition algorithm as described in Section 2.1.

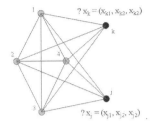  

$$\|x_k - x_1\| = d_{k,1}$$
$$\|x_k - x_2\| = d_{k,2}$$
$$\|x_k - x_3\| = d_{k,3}$$
$$\|x_k - x_4\| = d_{k,4}$$

$$\|x_j - x_1\| = d_{j,1}$$
$$\|x_j - x_2\| = d_{j,2}$$
$$\|x_j - x_3\| = d_{j,3}$$
$$\|x_j - x_4\| = d_{j,4}$$

Three dimensional case:
Four distances suffice to determine an atom.

Two dimensional case:
Three distances suffice to determine an atom.

Figure 4. **Geometric buildup** In two-dimensional space, if there are three determined atoms that are not in the same line and there are distances from these atoms to an undetermined atom, the undetermined atom can be determined uniquely using the three distances. In three-dimensional space, if there are four determined atoms that are not in the same plane and there are distances from these atoms to an undetermined atom, the undetermined atom can be determined uniquely using the four distances.

The theoretical basis of the general geometric buildup algorithm can be traced back in the theory of distance geometry[9]. Several authors had discussions on the theoretical issues related to such an approach as well, including Saxe[10], Sippl and Scheraga[48,49], and Huang, Liang, and Pardalos[52]. Based on the distance geometry theory, any point in a Euclidean space can be determined in terms of the distances from this point to a special set of points.

The General Geometric Buildup Algorithm
1. Determine an initial set of atoms.
2. Repeat:
   For each undetermined atom j,
      If atom j has distances to four independent and determined atoms,
      Determine atom j with these distances.
         End
   End
   If no atoms are determined in the loop, unsuccessfully stop.
3. All atoms are successfully determined.

Definition 3.1. A set of points B in a space S is a metric basis of S provided any point in S can be uniquely determined by its distances to the points in B.

Definition 3.2. A set of $k+1$ points in R^k is called an independent set of

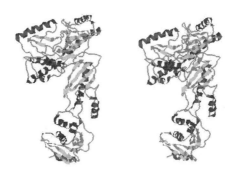

Figure 5. (RMSD = 1.0e-04 Å) **Geometric buildup** The X-ray crystal structure (left) of the HIV-1 RT p66 protein (4200 atoms) and the structure (right) determined by the geometric buildup algorithm using the distances for all pairs of atoms in the protein. The algorithm took only 188,859 floating-point operations, while a conventional singular-value decomposition algorithm required 1,268,200,000 floating-point operations.

points if it is not a set of points in R^{k-1}.

Theorem 3.1. *A set of k+1 independent points in R^k form a metric basis for R^k.*

Proof. It follows directly by generalizing the basic geometric buildup step to the k-dimensional Euclidean space. Let $x_i = (x_{i,1}, \ldots, x_{i,k})^T$ be the coordinate vectors of an independent set of points $i = 1, \ldots, k+1$ in R^k. Let $x_j = (x_{j,1}, \ldots, x_{j,k})^T$ be the coordinate vector for any point j in R^k with distances $d_{i,j}$ from points $i = 1, \ldots, k+1$ to point j. Then,

$$\|x_i\|^2 - 2x_i^T x_j + \|x_j\|^2 = d_{i,j}^2, \quad i = 1, \ldots, k+1, \qquad (14)$$

and $Ax_j = b$, where

$$A = -2 \begin{bmatrix} (x_2 - x_1)^T \\ (x_3 - x_2)^T \\ \cdots \\ (x_{k+1} - x_k)^T \end{bmatrix}, \quad b = \begin{bmatrix} (d_{2,j}^2 - d_{1,j}^2) - (\|x_2\|^2 - \|x_1\|^2) \\ (d_{3,j}^2 - d_{2,j}^2) - (\|x_3\|^2 - \|x_2\|^2) \\ \cdots \\ (d_{k+1,j}^2 - d_{k,j}^2) - (\|x_{k+1}\|^2 - \|x_k\|^2) \end{bmatrix} \qquad (15)$$

Since the points $i = 1, \ldots, k+1$ are not in R^{k-1}, the matrix A must be nonsingular and x_j is determined uniquely. $\qquad \square$

Given the above properties, we can easily see that a necessary condition for uniquely determining the coordinates of the atoms with a given set of distances is that each atom must have at least four distances to other atoms, and a sufficient condition is that in every step of the geometric buildup algorithm, there is an undetermined atom and the atom has four distances from four determined atoms who are not in the same plane. In general, we have

Theorem 3.2. *A necessary condition for the unique determination of the coordinates of a group of points $x_1, \ldots, x_n$ in R^k with a given set of distances among the points is that each point must have at least $k+1$ distances from other $k+1$ points, assuming that this point is not in R^{k-1} with any k of the $k+1$ points.*

Proof. It follows immediately from the fact that in R^k, a point can be defined uniquely only if it has $k+1$ distances from $k+1$ independent points, assuming it is not in R^{k-1} with any k of the $k+1$ points. If it has only k distances from k points, the point will have at least two reflective positions. $\square$

Theorem 3.3. *A sufficient condition for the unique determination of the coordinates of a group of points $x_1, \ldots, x_n$ in R^k with a given set of distances among the points is that in every step of the geometric buildup algorithm, there is an undetermined point with $k+1$ distances from $k+1$ independent and determined points.*

Proof. The geometric buildup algorithm gives a constructive proof for the theorem, because if the condition holds in every step of the algorithm, the algorithm will be able to determine the coordinates of all the points uniquely. $\square$

3.2. *Control of Numerical Errors*

The general geometric buildup algorithm can be sensitive to the numerical errors generated during the calculation of the coordinates of the atoms. With this algorithm, the coordinates of many atoms are determined by using the coordinates of previously determined atoms, and therefore, the errors in the previously determined atoms are passed to and accumulated in later determined atoms. As a result, the coordinates for later determined atoms may become completely incorrect, especially if there is a long sequence of atoms to be determined.

Wu and Wu[44] proposed an updating scheme to prevent the accumulation of the numerical errors. The idea of the scheme is based on the fact that the coordinates of any four atoms can be determined without any other information if all the distances among them are given. Therefore, the coordinates of any four determined atoms should be recalculated whenever possible using the distances among them, before they are used as a basis set of atoms for the determination of other atoms. The recalculated coordinates do not depend on the coordinates of previously determined atoms and therefore do not inherit any errors from them. They are determined from "scratch" and will not pass previous errors to later atoms as well. In this way, the coordinates of many atoms can be "corrected", and the errors in the calculated coordinates can be prevented from growing into incorrect structural results.

The recalculation of the coordinates of the four atoms in the above algorithm usually is done in an independent coordinate system, which is not related to the overall structure already constructed by the algorithm. However, they can be moved back to the original structure by aligning them to their original locations with an appropriate translation and rotation (Figure 6). In other words, the new coordinates of the four atoms can be translated and rotated so that the root-mean-square-deviation (RMSD) between the new coordinates and the old ones is minimized.

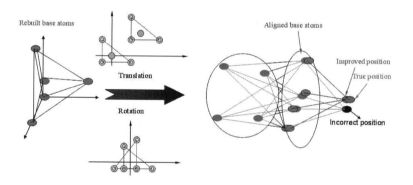

Figure 6. **Re-determination of base atoms** The four base atoms are re-determined if the distances among them are given. The atoms are then moved to and aligned with their original positions, and used to determine other atoms.

Let $y_1, \ldots, y_4$ be the coordinate vectors of the four atoms calculated in the regular geometric buildup process, and $x_1, \ldots, x_4$ the recalculated coordinate vectors. Let Y and X be the corresponding coordinate matrices. If the distances among all the four atoms are available, X can be obtained for example using the singular value decomposition algorithm described in Section 2.1. In order to move X to the position where Y is located in the molecule, the geometric centers of X and Y are calculated first:

$$x_c^T = \sum_{i=1}^{4} X(i,:)/4\,, \quad y_c^T = \sum_{i=1}^{4} Y(i,:)/4\,. \tag{16}$$

Then, X is translated so that the geometric centers of X and Y are at the same location,

$$X \Leftarrow X + e(y_c - x_c)^T\,, \tag{17}$$

where $e = (1,1,1,1)^T$. After the translation, a rotation for X is selected so that the root-mean-square-deviation of X and Y is minimized. In fact, the calculation of such a deviation can be done by solving an optimization problem,

$$\min_{Q} \|Y - XQ\|_F\,, \quad QQ^T = I\,, \tag{18}$$

where $\|\ \|_F$ is the matrix Frobenius norm and Q the rotation matrix. Let $C = X^T Y$, and let $C = U\Sigma V^T$ be the singular-value decomposition of C. Then, it is not difficult to verify that $Q = UV^T$ solves the above optimization problem[47].

The Updated Geometric Buildup Algorithm
1. Determine an initial set of atoms.
2. Repeat:
   For each undetermined atom j,
       If atom j has distances to four independent and determined atoms,
           If the distances among the determined atoms are given in the original data,
           Recalculate their coordinates with these distances.
           End
           Determine atom j with these distances.
       End
   End
   If no atoms are determined in the loop, unsuccessfully stop.
3. All atoms are successfully determined.

Figure 7 demonstrates in some scenarios for how the structure determined by a geometric buildup algorithm can be affected by the accumulated numerical errors and how they can be corrected by using the updating scheme, as given in Wu and Wu[44]. The figure shows the structures (red lines) of protein 4MBA (1086 atoms) determined using ≤ 5 Å distances, first by the general geometric buildup algorithm (Figure 7a) and then by the updating algorithm (Figure 7b). The graphs show that the general algorithm results in a structure that disagrees with the X-ray reference structure (blue lines) in many regions, while the updating algorithm generates a structure that agrees with the X-ray reference structure (blue lines) almost completely.

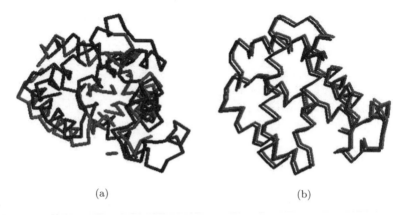

(a)   (b)

Figure 7.   **Control of rounding errors** (a) The structure (gray lines) of 4MBA determined by using a general geometric buildup algorithm and compared with the original structure of 4MBA (black lines). (b) The structure (gray lines) of 4MBA determined by using an updating geometric buildup algorithm and compared with the original structure of 4MBA (black lines).

3.3. *Rigid vs. Unique Buildup*

For the unique determination of a structure, it is necessary that every atom has at least four distances from other atoms. Further, the general geometric buildup algorithm requires four distances from four determined atoms to the atom to be determined in every buildup step. These conditions may not be satisfied by a given set of distances in practice. If the first condition is not satisfied, the structure will not be guaranteed unique. If the second condition is not satisfied, the general geometric buildup algorithm will not be able to determine the structure, even if the first condition is satisfied and the structure is unique.

In order to handle more sparse distance data, we can consider determining the structures only rigidly instead of uniquely. The necessary condition to have a rigid structure requires only three distances for each atom. Therefore, in every buildup step, the geometric buildup algorithm can be modified

to require only three distances from three determined atoms to the atom
to be determined. The atom can then be determined rigidly, although with
two possible positions. In the end, the algorithm may produce multiple
structures, due to the multiple choices of the positions of the atoms, but
the structures are rigid and in finite number.

More formally, in any buildup step, let $x_i = (x_{i,1}, x_{i,2}, x_{i,3})^T$, $i = 1, 2, 3$,
be the coordinate vectors of three determined atoms that are not in a line.
Let $x_j = (x_{j,1}, x_{j,2}, x_{j,3})^T$ be the coordinate vector for an undetermined
atom j and $d_{i,j}$ the distances from atoms $i = 1, 2, 3$ to atom j. Then, x_j
can be obtained from the solution of the following system of equations,

$$\|x_i\|^2 - 2x_i^T x_j + \|x_j\|^2 = d_{i,j}^2, \quad i = 1, 2, 3. \tag{19}$$

By subtracting equation i from equation $i+1$ for $i = 1, 2$, we can eliminate
the quadratic terms for x_j to obtain

$$-2(x_{i+1} - x_i)^T x_j = (d_{i+1,j}^2 - d_{i,j}^2) - (\|x_{i+1}\|^2 - \|x_i\|^2), \quad i = 1, 2. \tag{20}$$

Let A be a matrix and b a vector, and

$$A = -2 \begin{bmatrix} (x_2 - x_1)^T \\ (x_3 - x_2)^T \end{bmatrix}, \quad b = \begin{bmatrix} (d_{2,j}^2 - d_{1,j}^2) - (\|x_2\|^2 - \|x_1\|^2) \\ (d_{3,j}^2 - d_{2,j}^2) - (\|x_3\|^2 - \|x_2\|^2) \end{bmatrix}. \tag{21}$$

We then have $A x_j = b$. Let $x_j = A^T y_j$, where $y_j = (y_{j,1}, y_{j,2})^T$. Then,
$A A^T y_j = b$. Since x_1, x_2, x_3 are not in the same line, A must be full
rank and $A A^T$ be nonsingular. We can therefore solve the linear system
$A A^T y_j = b$ to obtain a unique solution for y_j. Let $x_j' = (x_{j,1}, x_{j,2})^T$ and
$A' = A(1 : 2, 1 : 2)$. Then, $x_j' = [A']^T y_j$. By using one of the equations in
(19), we can obtain two possible values for $x_{j,3}$, assuming that the equation
has real solutions. In the end, we obtain two solutions for (19).

The advantage of using the modified buildup algorithm is that the algorithm requires fewer distance constraints than the general buildup algorithm. It can handle even more sparse distance data, yet determine meaningful structures. The modified algorithm may find multiple structures,
but they all are rigid, and in some cases, it can find a unique structure
as well, because the requirement by the general buildup algorithm on the
availability of the special four distances in every buildup step is sufficient
for the determination of a unique structure, but not necessary.

However, a problem with the modified buildup algorithm is that it may
produce too many possible structures: Since in every step, an atom is only
determined rigidly, there may be at least two possible positions for it. We
have to keep both positions unless later on we find that one of them can be

excluded with other distance constraints. Moreover, the three determined atoms may also have multiple positions. Let the ith determined atom have l_i possible positions, $i = 1, 2, 3$. Then, in the worst case, there can be $2 \times l_1 \times l_2 \times l_3$ possible positions for the atom to be determined. Therefore, as the algorithm proceeds, the total number of possible positions for an atom to be determined may grow into exponentially many.

To reduce the number of possible positions for an atom, we can allow the algorithm to determine the atom uniquely first if there are more than three required distances available, and determine it rigidly otherwise. Also, in every buildup step, after the atom is determined, either rigidly or uniquely, we can examine all given distances from this atom to other determined atoms for their possible positions. If some positions have violated their distance constraints, they can be removed for further consideration. In this way, the structures generated in the end are guaranteed to satisfy all available distance constraints among the atoms, and they may be reduced to a unique structure after all infeasible structures are identified and removed.

The Rigid Geometric Buildup Algorithm

1. Determine an initial set of atoms.
2. Repeat:
   For each undetermined atom j,
       If atom j has distances to four independent and determined atoms,
           Determine atom j with these distances.
           Check multiple structures with additional available distances.
       End
       If atom j has distances to three independent and determined atoms,
           Determine atom$j$ with these distances.
           Record multiple structures generated from reflections.
       End
   End
   If no atoms are determined in the loop, unsuccessfully stop.
3. All atoms are successfully determined.

Figure 8 shows how a structure can be determined rigidly and how multiple structures can be generated and also reduced. Figure 8a shows that atom i is first determined with three available distances. There are two positions for atom i due to reflection, which makes two possible structures. Figure 8b shows that atom j again is determined with three available distances, with two positions for each of the possible structures. Total four possible structures are made. In Figure 8c, atom k is determined uniquely with four distances, and therefore, the number of possible structures is not increased. However, there is an additional distance between atoms i and k. By examining all the structures, we find that two of them do not satisfy this distance constraint, and they can be removed from the structure pool, as shown in Figure 8d.

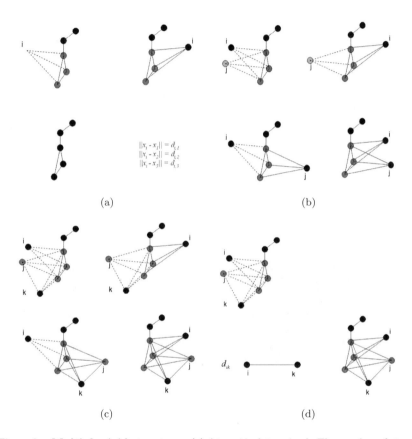

(a)

(b)

(c)

(d)

Figure 8. **Multiple rigid structures** (a) Atom i is determined. The number of structures is two. (b) Atom j is determined. The number of structures is increased to four. (c) Atom k is determined. (d) Two structures are removed because they do not satisfy the distance constraint for atom i and k.

Figure 9 further demonstrates the application of the rigid geometric buildup algorithm to a small protein, 1AKG, and the nature of the multiple structures it can generate, as given along with other examples in[45]. The protein 1AKG is a small polypeptide with 16 amino acids and 110 atoms. The general geometric buildup algorithm is able to determine to the struc-

72

ture for this protein completely, with distances ≤ 4.5 Å, and the RMSD value of the structure is 8.3e-07 Å against the original structure. Here, the number of distances used is 1638, which is about 14% of all the distances. However, with distances ≤ 3.5 Å, the general geometric buildup algorithm fails, but the rigid algorithm is still able to find a reasonable number of rigid structures. Here, the number of distances used is 898, which is only 7.5% of all the distances. There are total 8192 multiple conformations found by the rigid algorithm. The one closest to the original structure has the RMSD value equal to 4.3e-07 Å. Note that $8192 = 2^{13}$, and therefore, the multiple structures are perhaps generated just from a sequence of 13 reflections of the atomic positions. In fact, as can be observed in the figure, most of the reflections happen for the side-chain atoms when they are in the surface of the protein, and they only affect the determination of a small part of the structure. On the other hand, the major parts of the protein with the backbone atoms and the atoms in the interior of the protein are all uniquely determined.

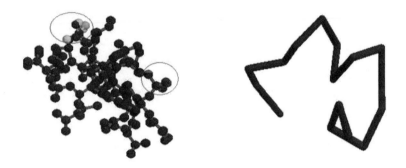

Figure 9. **Rigid structure determination** Shown is the structure of protein 1AKG, with 16 residues, 110 atoms. The distances < 3.5 Å were used. Total 8192 rigid structures were determined. They all were almost identical except for the circled small regions.

Similar to the general geometric buildup algorithm, the theoretical basis for the rigid geometric buildup algorithm can be established and generalized to any k-dimensional Euclidean space. For this purpose, we define a reduced metric basis for a space and k independent points in R^k.

Definition 3.3. A set of points B in a space S is a reduced metric basis of S provided any point in S can be determined rigidly by its distances to the points in B.

Definition 3.4. A set of k points in R^k is said to be an independent set of points if it is not a set of points in R^{k-2}.

Theorem 3.4. *A set of k independent points in R^k form a reduced metric basis for R^k.*

Proof. It follows directly by generalizing the modified geometric buildup step to the k-dimensional Euclidean space. Let $x_i = (x_{i,1}, \ldots, x_{i,k})^T$ be the coordinate vectors of an independent set of points $i = 1, \ldots, k$ in R^k. Let $x_j = (x_{j,1}, \ldots, x_{j,k})^T$ be the coordinate vector for any point j in R^k with distances $d_{i,j}$ from points $i = 1, \ldots, k$ to point j. Then

$$\|x_i\|^2 - 2x_i^T x_j + \|x_j\|^2 = d_{i,j}^2, \quad i = 1, \ldots, k \tag{22}$$

and $Ax_j = b$, where

$$A = -2 \begin{bmatrix} (x_2 - x_1)^T \\ (x_3 - x_2)^T \\ \cdots \\ (x_k - x_{k-1})^T \end{bmatrix}, \quad b = \begin{bmatrix} (d_{2,j}^2 - d_{1,j}^2) - (\|x_2\|^2 - \|x_1\|^2) \\ (d_{3,j}^2 - d_{2,j}^2) - (\|x_3\|^2 - \|x_2\|^2) \\ \cdots \\ (d_{k,j}^2 - d_{k-1,j}^2) - (\|x_k\|^2 - \|x_{k-1}\|^2) \end{bmatrix}. \tag{23}$$

Let $x_j = A^T y_j$, where $y_j = (y_{j,1}, \ldots, y_{j,k-1})^T$. Then, $AA^T y_j = b$. Since $x_1, \ldots, x_k$ are not in R^{k-2}, A must be full rank and AA^T be nonsingular. We can therefore solve the linear system $AA^T y_j = b$ to obtain a unique solution for y_j. Let $x_j' = (x_{j,1}, \ldots, x_{j,k-1})^T$ and $A' = A(1 : k-1, 1 : k-1)$. Then, $x_j' = [A']^T y_j$. By using one of the equations in (22), we can obtain two possible values for $x_{j,k}$, assuming that the equation has real solutions. In the end, we obtain two solutions for (22), and the positions for point j are determined rigidly. $\qquad \square$

Given the above properties, we can easily see that a necessary condition for rigidly determining the coordinates of the atoms with a given set of distances is that each atom must have at least three distances to other atoms, and a sufficient condition is that in every step of the geometric buildup algorithm, there is an undetermined atom and the atom has three distances from three determined atoms who are not in the same line. In general, we have

Theorem 3.5. *A necessary condition for the rigid determination of the coordinates of a group of points $x_1, \ldots, x_n$ in R^k with a given set of distances among the points is that each point must have at least k distances*

from other k points, assuming that this point is not in R^{k-2} with any $k-1$ of the k points.

Proof. It follows immediately from the fact that in R^k, a point can be defined rigidly only if it has k distances to k independent points, assuming it is not in R^{k-2} with any $k-1$ of the k points. If it has only $k-1$ distances from $k-1$ points, the position of the point will be flexible. $\square$

Theorem 3.6. *A sufficient condition for the rigid determination of the coordinates of a group of points $x_1, \ldots, x_n$ in R^k with a given set of distances among the points is that in every step of the geometric buildup algorithm, there is an undetermined point with k distances from k independent and determined points.*

Proof. The modified geometric buildup algorithm gives a constructive proof for the theorem, because if the condition holds in every step of the algorithm, the algorithm will be able to determine the coordinates of all the points rigidly. $\square$

3.4. *Tolerance of Inexact Distances*

In practice, the distance data often contains errors. As a result, the distances may become inconsistent or in other words, may have violated some basic geometric rules such as the triangle inequality for the distances among any three atoms. The general geometric buildup algorithm usually assumes that the distances are consistent and therefore, in every step, only four (or three) distances are required for the determination of the coordinates of an atom uniquely (or rigidly), although there may be more available. However, this will not be the case if the distances are not consistent. In order for the algorithm to handle inexact distances (distances with errors), the general buildup procedure has to be modified. First, in every buildup step, if l distances are found from an undetermined atom to l determined atoms, $l \geq 4$, all l distances should be used for the determination of the unknown atom. Second, if $l \geq 4$, an over-determined system of equations is obtained for the determination of the position of the unknown atom. If the distances have errors, the system may not be consistent. Therefore, we can only solve the system approximately by using for example a least-squares method. Third, a new updating scheme may be necessary to prevent the accumulation of the rounding errors. The updating scheme described in Section 3.2 may not be practical any more for $l \gg 4$ because it requires all the distances available among l determined atoms.

A simple way to extend the geometric buildup algorithm to handle the possible errors from the distance data is as follows. In every buildup step, in addition to the four required distances, we can include all the available distances, say l distances, from the determined atoms to the one to be determined (see Figure 10). Let $x_i = (x_{i,1}, x_{i,2}, x_{i,3})^T$, $i = 1, \ldots, l$, be the coordinate vectors of the l determined atoms and $d_{i,j}$ the distances from atoms $i = 1, \ldots, l$ to the undetermined atom j. Then, the coordinates $x_j = (x_{j,1}, x_{j,2}, x_{j,3})^T$ for atom j can be obtained from the solution of the following system of equations,

$$\|x_i\|^2 - 2x_i^T x_j + \|x_j\|^2 = d_{i,j}^2, \quad i = 1, \ldots, l. \tag{24}$$

By subtracting equation i from equation $i + 1$ for $i = 1, \ldots, l - 1$, we can eliminate the quadratic terms for x_j to obtain

$$-2(x_{i+1} - x_i)^T x_j = (d_{i+1,j}^2 - d_{i,j}^2) - (\|x_{i+1}\|^2 - \|x_i\|^2), \quad i = 1, \ldots, l - 1. \tag{25}$$

Let A be a matrix and b a vector, and

$$A = -2 \begin{bmatrix} (x_2 - x_1)^T \\ (x_3 - x_2)^T \\ \ldots \\ (x_l - x_{l-1})^T \end{bmatrix}, \quad b = \begin{bmatrix} (d_{2,j}^2 - d_{1,j}^2) - (\|x_2\|^2 - \|x_1\|^2) \\ (d_{3,j}^2 - d_{2,j}^2) - (\|x_3\|^2 - \|x_2\|^2) \\ \ldots \\ (d_{l,j}^2 - d_{l-1,j}^2) - (\|x_l\|^2 - \|x_{l-1}\|^2) \end{bmatrix}. \tag{26}$$

We then have $Ax_j = b$. This system is certainly over-determined if $l > k + 1$. However, it can be solved by using a standard linear least-squares method. For example, we can compute the QR factorization of A to obtain an equation $QRx_j = b$, where Q is $(l-1) \times 3$ and R is 3×3. If at least four of the l determined atoms are not in the same plane, A must be full rank and R be nonsingular. We can then solve the linear system $QRx_j = b$ to obtain a unique solution $x_j = R^{-1}Q^T b$, which minimizes $\|b - Ax_j\|$. Here, solving the linear system $QRx_j = b$ requires $O(l)$ computing time, but QR factorization may take $O(l^2)$ time. Since we only need to solve $\sim n$ such linear least-squares problems for $\sim n$ coordinate vectors x_j, the total computation time must be in order of $l_m^2 n$, if in every step, the required coordinates x_i and distances $d_{i,j}$ are always available, where $l_m = \max_j\{|S_j|\}$, $S_j = \{i : (i, j) \text{ in } S\}$.

Again, the theory for the extended geometric buildup algorithm can be established and generalized to any k-dimensional Euclidean space in a

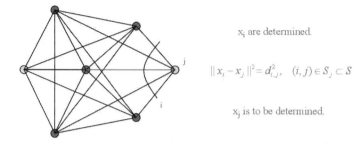

x_i are determined.

$$\| x_i - x_j \|^2 = d_{i,j}^2, \quad (i,j) \in S_j \subset S$$

x_j is to be determined.

Figure 10. **Tolerance of distance errors** The extended algorithm tries to determine the coordinates of each atom by taking all available distance constraints into account and by minimizing the errors for all the constraints. In this way, all the constraints are intended to be satisfied, and the algorithm is also more stable with possible errors in the distance data.

similar fashion as that for the general geometric buildup algorithm. For this purpose, we define an extended metric basis for a space and an extended set of independent points in R^k.

Definition 3.5. A set of points B in a space S is an extended metric basis of S provided any point in S can be determined uniquely by its distances from the points in B.

Definition 3.6. A set of l points is said to be an extended set of independent points in R^k if it contains $k + 1$ independent points.

Theorem 3.7. *An extended set of l independent points in R^k form an extended metric basis for R^k.*

Proof. It follows directly by generalizing the extended geometric buildup step to the k-dimensional Euclidean space. Let $x_i = (x_{i,1}, \ldots, x_{i,k})^T$ be the coordinate vectors for an extended set of independent points $i = 1, \ldots, l$ in R^k. Let $x_j = (x_{j,1}, \ldots, x_{j,k})^T$ be the coordinate vector for any point j in R^k with distances $d_{i,j}$ from points $i = 1, \ldots, l$ to point j. Then

$$\| x_i \|^2 - 2x_i^T x_j + \| x_j \|^2 = d_{i,j}^2, \quad i = 1, \ldots, l, \tag{27}$$

and $Ax_j = b$, where

$$A = -2 \begin{bmatrix} (x_2 - x_1)^T \\ (x_3 - x_2)^T \\ \cdots \\ (x_l - x_{l-1})^T \end{bmatrix}, \quad b = \begin{bmatrix} (d_{2,j}^2 - d_{1,j}^2) - (\| x_2 \|^2 - \| x_1 \|^2) \\ (d_{3,j}^2 - d_{2,j}^2) - (\| x_3 \|^2 - \| x_2 \|^2) \\ \cdots \\ (d_{l,j}^2 - d_{l-1,j}^2) - (\| x_l \|^2 - \| x_{l-1} \|^2) \end{bmatrix}.$$
$$\tag{28}$$

Multiply the equation by A^T to obtain $A^T A x_j = A^T b$. Since $k + 1$ of the l determined points are independent, A must be full rank and $A^T A$ be nonsingular. We can then solve the linear system $A^T A x_j = A^T b$ to obtain a unique solution $x_j = [A^T A]^{-1} A^T b$. □

The above algorithm may not necessarily be stable for preventing rounding errors from growing, because in every step, the coordinates of the unknown atom must have rounding errors, which can still be propagated and accumulated into later calculations. On the other hand, different from the general geometric buildup algorithm, it is difficult to employ an updating scheme as described in Section 3.2 for the extended algorithm, because the scheme requires the availability of the distances among all l determined atoms, which is not so realistic when l is large. In order to control the rounding errors as well as tolerate the distance errors, a nonlinear instead of linear least-squares approximation can in fact be used in the buildup procedure instead. The idea is to determine the unknown atom in each buildup step by using not only the l distances from l determined atoms to the unknown atom, but also the distances among all the l determined atoms. The l distances from l determined atoms to the unknown atom must be given. The distances among the l determined atoms may not necessarily be provided, but they can be calculated. In any case, once all these distances become available, the coordinates for the unknown atom and the l known atoms can all be calculated (or recalculated) using these distances.

Let $x_1, \ldots, x_l$ and x_{l+1} be the coordinate vectors of atoms $1, \ldots, l + 1$. If the distances among all these atoms, $d_{i,j}$, $i, j = 1, \ldots, l+1$, are available, then, $\|x_i - x_j\| = d_{i,j}$ for all $i, j = 1, \ldots, l + 1$, and

$$\|x_i\|^2 - 2x_i^T x_j + \|x_j\|^2 = d_{i,j}^2, \quad i, j = 1, \ldots, l + 1. \tag{29}$$

Since the structure formed by these atoms is invariant under any translation or rotation, we can set a reference system so that the origin is located at the last atom or in other words, $x_{l+1} = (0, 0, 0)^T$. It follows that $\|x_i\| = d_{i,l+1}$, $\|x_j\| = d_{j,l+1}$, and

$$d_{i,l+1}^2 - 2x_i^T x_j + d_{j,l+1}^2 = d_{i,j}^2, \quad i, j = 1, \ldots, l. \tag{30}$$

We now have a system of equations similar to the one discussed in Section 2.1. Define a coordinate matrix X and an induced distance matrix D,

$$X = \{x_{i,k} : i = 1, \ldots, l, \, k = 1, 2, 3\} \quad \text{and}$$
$$D = \{(d_{i,l+1}^2 - d_{i,j}^2 + d_{j,l+1}^2)/2 : i, j = 1, \ldots, l\}. \tag{31}$$

Then, $XX^T = D$. Let $D = U\Sigma U^T$ be the singular value decomposition of D, where U is an orthogonal matrix and Σ a diagonal matrix with the singular values of D along the diagonal. If D is a matrix of rank less than or equal to 3, $X = V\Lambda^{1/2}$ solves the equation $XX^T = D$, where $V = U(:, l : 3)$ and $\Lambda = \Sigma(1 : 3, 1 : 3)$. In other words, if the distances $d_{i,j}$ are available for all $i, j = 1, \ldots, l + 1$, we can always construct an induced matrix D for the distances and then, based on the singular value decomposition of D, obtain the coordinates for all the atoms $1, \ldots, l$ as given in X with atom $l + 1$ fixed at $(0, 0, 0)^T$.

Note that the distances may have errors. Then, the matrix D may in fact have a higher rank than k or in other words, the equation $XX^T = D$ may not have an exact solution. However, $X = V\Lambda^{1/2}$ as defined above is still a good approximation to the solution of the equation (30) in the following least-squares sense.

Theorem 3.8. *Let $D = U\Sigma U^T$ be the singular value decomposition of D. Let $V = U(:, 1 : 3)$ and $\Lambda = \Sigma(1 : 3, 1 : 3)$. Then, $X = V\Lambda^{1/2}$ minimizes $\|D - XX^T\|_F$, where $\|\quad\|_F$ is the matrix Frobenius norm.*

Proof. [30] Let $f(X) = \|D - XX^T\|^2$. Then $(D - XX^T)X = 0$ for any stationary point X of f. It follows that $(D - XX^T)X = (D - XX^T)XX^T = 0$ and

$$f(X) = \text{trace}\,(D^2) - \text{trace}\,(2DXX^T - XX^TXX^T)$$
$$= \text{trace}\,(D^2) - \text{trace}\,(XX^TXX^T).$$

Let $\sigma_1 \geq \ldots \geq \sigma_l \geq 0$ be the singular values of D and $\lambda_1 \geq \lambda_2 \geq \lambda_3 > 0$ be the singular values of XX^T. Then,

$$f(X) = \text{trace}\,(D^2) - \text{trace}\,(XX^TXX^T)$$
$$= \sum_{j=1}^{l} \sigma_j^2 - \sum_{j=1}^{k} \lambda_j^2\,.$$

Let $XX^T = V\Lambda V^T$ be the singular value decomposition of XX^T, where V is an $l \times 3$ orthogonal matrix and $\Lambda = \text{diag}\{\lambda_1, \lambda_2, \lambda_3\}$. Since $DXX^T = XX^TXX^T$, $V^TDV = \Lambda$ and therefore, $\{\lambda_j : j = 1, 2, 3\} \subset \{\sigma_j : j = 1, \ldots, n\}$. It follows that $f(X)$ is minimized when $\lambda_j = \sigma_j$ for $j = 1, 2, 3$. $\square$

Geometric Buildup with Linear Least-Squares

1. Find four atoms that are not in the same plane.
2. Determine the coordinates of the atoms with the distances among them.

3. Repeat:
>   For each of the undetermined atoms,
>> If the atom has l distances to l determined atoms that are not in the same plane,
>>> Determine the atom with the least-squares fit to the distances.
>> End
>   End

4. If no atom can be determined in the loop, stop.

5. All atoms are determined.

The extended buildup procedure has the following properties. First, the coordinates of the unknown atom are determined by using l previously determined atoms, to which the unknown atom has distances given. Second, the coordinates are determined by solving a system of distance equations approximately. They are the best possible estimations in a nonlinear least-squares sense as stated in Theorem 3.4.2, and can therefore be evaluated even if the distances have errors. Third, the calculations not only determine the coordinates of the unknown atom, but also recalculate the coordinates of all the involved atoms including the determined ones. Most importantly, these coordinates do not depend completely on the results from previous calculations. Rather, they are determined by using the provided distances among the atoms (determined and undetermined) as much as possible, thereby reducing the risk of large error propagation and accumulation. In this sense, the method should be more stable numerically than the one using linear linear-squares approximation[46].

Geometric Buildup with Nonlinear Least-Squares

1. Find four atoms that are not in the same plane.

2. Determine the coordinates of the atoms with the distances among them.

3. Repeat:
>   For each of the undetermined atoms,
>> If the atom has l distances to l determined atoms that are not in the same plane,
>>> Determine the $l+1$ atoms with the distances among them.
>>> Put the atoms back to their original positions by proper translation and rotation
>> End
>   End

4. If no atom can be determined in the loop, stop.

5. All atoms are determined.

Of course, the calculations of the coordinates are conducted in an independent reference system with its origin at the position of the atom to be determined. In order to recover the coordinates of the atoms in their original structure, we need to make a proper translation and rotation for the coordinates just like we need to do in the updating scheme for the general geometric buildup algorithm. More specifically, let Y be an $l \times 3$ matrix having the original coordinates of the l determined atoms. Let X be an $l \times 3$ matrix with the recalculated coordinates of the determined atoms. First, we translate X to Y with a translation vector $y_c - x_c$, where y_c and x_c are the geometric centers of X and Y, respectively. Then, we can rotate the coordinates of all the atoms by using a rotation matrix $Q = UV^T$, where U and V are obtained from the singular value decomposition, $X^T Y = U\Sigma V^T$. That is, if x_i is the coordinate vector of atom i, $i = 1, \ldots, l + 1$, then, we set x_i to Qx_i.

Table 1 and 2 show some results from applying the extended geometric buildup algorithm with either linear or nonlinear least-squares approximation to the determination of the structures of a set of proteins using the distance data generated from the experimental structures of the proteins with 5 Å and 6 Å cutoff values. Table 1 contains the RMSD (root-mean-square deviation) values of the structures (compared with their original structures) obtained by using the extended buildup algorithm with linear least-squares on the generated data sets. The RMSD values show that the algorithm solved almost all the problems with cutoff distances equal to 6 Å, but failed for those with cutoff distance equal to 5 Å. The last cutoff value is critical because in NMR modeling, usually only less than or equal 5 Å distances can be estimated. In any case, the results show that with linear least-squares, the new buildup algorithm performed well in general if the distance data was not too sparse. The reason that it did not work well for very sparse data was that a long sequence of buildup steps had to be carried out and a large amount of rounding errors was accumulated.

Table 2 contains the RMSD (root-mean-square deviation) values of the structures (compared with their original structures) obtained by using the new buildup algorithm with nonlinear least-squares on the data sets. The RMSD values show that the algorithm solved almost all the problems with cutoff distances equal to 5 Å and 6 Å. Therefore, the results indicated that with nonlinear least-squares, the new buildup algorithm performed well in general. The reason it worked well for very sparse data was that it calculated the coordinates of the undetermined as well as determined atoms in every buildup step using the distances among them (most presumably

Table 1. RMSD Values of Structures Computed with
Linear Least-Squares

ID	TA	≤ 5Å		$\leq 6a$Å	
		DA	RMSD	DA	RMSD
1PTQ	402	402	1.4e-00	402	2.6e-09
1HOE	558	558	5.8e-02	558	3.1e-09
1LFB	641	641	2.0e-02	641	2.1e-10
1PHT	814	809	1.2e+01	814	8.2e-09
1POA	914	914	6.6e-00	914	1.9e-09
1AX8	1003	1003	5.2e-00	1003	1.8e-05
4MBA	1086	1083	4.9e-00	1086	3.8e-06
1F39	1534	1534	1.4e+01	1534	6.3e-08
1RGS	2015	2010	2.0e+01	2015	1.1e-01
1BPM	3672	3669	6.4e+04	3672	3.6e-02
1HMV	7398	7389	1.2e+03	7398	3.5e+01

Note: *ID – Protein ID, TA – Total number of atoms,
DA – Total number of determided atoms, RMSD –
RMSD values of the computed structure against the
original structures.

given in the original distance data) and therefore, stopped the propagation
of the rounding errors.

4. Concluding Remarks

In this paper, we have discussed a well-known problem in protein modeling, for the determination of the structure of a protein with a given set of inter-atomic or inter-residue distances obtained from either physical experiments or theoretical estimates. A more general and abstract form of the problem is known as the distance geometry problem in mathematics, the graph embedding problem in computer science, and the multidimensional scaling problem in statistics. In general, the problem can be stated as to find the coordinates for a set of points in some topological space given the distances for certain pairs of points. Therefore, in addition to protein modeling where everything is discussed only in three-dimensional Euclidean space, the problem has applications in many other scientific and engineering fields as well, such as sensor network localization, image recognition, and protein classification, to name a few. In any case, the problem may or may

Table 2. RMSD Values of Structures Computed with Nonlinear Least-Squares

ID	TA	$\leq 5\text{Å}$		$\leq 6a\text{Å}$	
		DA	RMSD	DA	RMSD
1PTQ	402	402	5.5e-14	402	5.0e-14
1HOE	558	558	1.6e-13	558	2.7e-13
1LFB	641	641	9.5e-14	641	5.5e-14
1PHT	814	809	1.1e-13	814	1.8e-13
1POA	914	914	3.2e-13	914	1.5e-13
1AX8	1003	976	4.0e-13	1003	4.6e-12
4MBA	1086	1083	1.8e-13	1086	2.6e-13
1F39	1534	1534	7.9e-13	1534	1.9e-13
1RGS	2015	2010	8.3e-12	2015	2.4e-12
1BPM	3672	3669	8.1e-11	3672	1.0e-11
1HMV	7398	7389	1.1e-08	7398	5.5e-07

Note: *ID – Protein ID, TA – Total number of atoms, DA – Total number of determided atoms, RMSD – RMSD values of the computed structure against the original structures.

not have a solution in a given topological space, and even if it does have a solution, the solution may not be easy to find, depending on the given distances. For example, in any k-dimensional Euclidean space, the problem is polynomial time solvable if the distances for all the pairs of points are provided, and is NP-complete otherwise in general.

We have investigated the solution of the distance geometry problem within a so-called geometric buildup framework. Central to the geometric buildup approach is the idea to determine only a small group of atoms at the beginning and then complete the whole molecule by repeatedly determining one or more atoms every time using the available distances between the determined and undetermined atoms. The advantage of using a geometric buildup approach is that it works directly on the given distances and exploits the special structure of a given problem, and hence may be able to solve the problem more efficiently than a general approach. We have discussed the formulations and complexities of the distance geometry problem in its various forms, and described the general geometric buildup algorithm and its theoretical basis. We have also discussed the issues of the general algorithm for controlling rounding errors, determining rigid vs.

unique structures, and handing inexact distances, and reviewed various versions of the algorithm that can address these issues and showed their test results.

A basic principle for the general geometric buildup algorithm is that whenever there are four determined atoms that are not in the same plane and there are distances from these atoms to an undetermined atom, the undetermined atom can immediately be determined uniquely by solving a system of four distance equations using the available distances. If for every atom, the required atoms and the distances can be found, the whole structure can be determined uniquely. The distance equations can in fact be reduced to a set of linear equations and hence solved in constant time. Therefore, as we have detailed in the paper, in ideal cases, a geometric buildup algorithm can solve a distance geometry problem with only $4n$ distances in $O(n)$ computing time, while the conventional singular value decomposition algorithm requires all $n(n-1)/2$ distances and $O(n^2)$ computing time, where n is the number of atoms to be determined.

However, the requirement for four determined atoms and hence four corresponding distances in every step of the buildup procedure is sufficient but not necessary for the unique determination of a structure. Therefore, the general geometric buildup algorithm can in fact be modified so that in every buildup step, only three determined atoms and hence three corresponding distances are required. There may be multiple structures that can be determined in this way, but they are still rigid and can possibly end up unique as well. Indeed, as we have reviewed in the paper, a modified geometric buildup algorithm has been developed and tested successfully on a set of proteins. The results showed that the modified algorithm was able to produce meaningful structures rigidly with very sparse distance data, although they may be multiple in many cases.

The geometric buildup algorithm, either rigid or unique, can be sensitive to the numerical errors though, for the coordinates of the atoms are determined using the coordinates of previously determined atoms and the rounding errors in the previously determined atoms can be passed to and accumulated in later determined atoms, resulting in incorrect structural results. An updating scheme has been developed to prevent the accumulation of the numerical errors, as we have described in the paper. The idea of the scheme is based on the fact that the coordinates of any four atoms can be determined without any other information if all the distances among them are given. Therefore, the coordinates of any four determined atoms can be recalculated whenever possible using the distances among them, before they are used as a basis set of atoms for the determination of other atoms. The recalculated coordinates do not depend on the coordinates of previously determined atoms and therefore do not inherit any errors from them.

The general geometric buildup algorithm cannot tolerate errors in given distances either, for the distances then may not be consistent and the systems of distance equations may not be solvable. However, in practice, the distances must have errors because they come from either experimental measures or theoretical estimates. We have demonstrated how an extended geometric buildup algorithm can be developed to prevent the accumulation of the rounding errors in the buildup calculations successfully and also tolerate the errors in the given distances. In this algorithm, in every buildup step, all (instead of a subset of) the distances available for each unknown atom are taken into account for the determination of the position of the atom by using a least-squares approximation (instead of solving a system of equations exactly). We have shown that the least-squares approximation could actually be obtained by using a special singular value decomposition method, which could not only provide an approximate solution to the original system of distance equations, but also prevent the accumulation of the rounding errors in the buildup procedure effectively.

As we have discussed in the introduction section of the paper, a further complicated yet practical case of the distance geometry problem is when the distances are given with only their lower and upper bounds. The problem then becomes to find the coordinates $x_1, \ldots, x_n$ for the atoms for a given set of lower and upper bounds, $l_{i,j}$ and $u_{i,j}$, of the distances $d_{i,j}$ such that

$$l_{i,j} \le \|x_i - x_j\| \le u_{i,j}, \quad (i,j) \in S.$$

The general geometric buildup algorithm and its modifications or extensions presented in this paper have not been developed to deal with distance bounds yet. However, the general buildup procedure should be extendable for the solution of such a problem as well. Here, different from other implementations, in every buildup step, an atom should be determined by satisfying a set of distance bounds instead of exact distances. The computation will certainly be more involved and subject to even more arbitrary errors. The solution to such a problem will not be unique, either. In fact, there can be an ensemble of solutions all satisfying the given distance inequalities. On the other hand, in practice, it is actually preferred to obtain the entire ensemble of solutions instead of a few samples. How to implement a buildup algorithm to achieve that can be challenging and will be the topic of our future investigation.

Acknowledgments

The authors would like to acknowledge the support from the Ogden College of Science and Engineering, Western Kentucky University, the Department of Mathematics and the Baker Center for Bioinformatics and Biological Statistics, Iowa State University, and the Laboratory for Scientific and Engineering Computing, Chinese Academy of Sciences. The work has been funded partially by the NIH/NIGMS grant R01GM081680 and by the NSF of China.

References

1. T. E. Creighton, Proteins: Structures and Molecular Properties, 2nd Edition, Freeman and Company, 1993.
2. H. M. Berman, J. Westbrook, Z. Feng, G. Gilliland, T. N. Bhat, H. Weissig, L. N. Shindyalov, and P. E. Bourne, The Protein Data Bank, Nuc. Acid. Res., 28, 2000, 235-242.
3. J. Drenth, Principles of Protein X-ray Crystallography, Springer-Verlag, 1994.
4. H. Gunther, NMR Spectroscopy: Basic Principles, Concepts, and Applications in Chemistry, John Wiley & Sons, 1995.
5. T. Schlick, Molecular Modeling and Simulation: An Interdisciplinary Guide, Springer, 2003.
6. P. E. Bourne and H. Weissig, Structural Bioinformatics. John Wiley & Sons, Inc., 2003.
7. J. Moult, K. Fidelis, B. Rost, T. Hubbard, and A. Tramontano, Critical assessment of methods for protein structure prediction (CASP), Proteins: Structure, Function, Bioinformatics, 61, 2005, 3-7.
8. V. S. Pande, I Baker, J. Chapman, S. P. Elmer, S. Khaliq, S. M. Larson, Y. M. Rhee, M. R. Shirts, C. D. Snow, E. J. Sorin, and B. Zagrovic, Atomistic protein folding simulations on submillisecond time scale using worldwide distributed computing, Biopolymers, 68, 2003, 91-109.
9. L. M. Blumenthal, Theory and Applications of Distance Geometry, Oxford Clarendon Press, 1953.
10. J. B. Saxe, Embeddability of weighted graphs in k-space is strongly NP-hard, in Proc. 17th Allerton Conference in Communications, Control and Computing, 1979, 480-489.
11. W. S. Torgerson, Theory and Method of Scaling, John Wiley & Sons, 1958.
12. P. Biswas, T. Liang, T. Wang, and Y. Ye, Semidefinite programming based algorithms for sensor network localization. ACM J on Transactions on Sensor Networks, 2, 2006, 188-220.
13. H. Klock and J. M. Buhmann, Multidimensional scaling with deterministic annealing, in Lecture Notes in Computer Science 1223: Energy Minimization

Methods in Computer Vision and Patter Recognition, M Pilillo and E. R. Hancock, eds., Springer-Verlag, 1997, 246-260.

14. J. T. Hou, G. E. Sims, C. Zhang, and S. H. Kim, A global representation of the protein fold space, Proc. Natl. Acad. Sci. USA, 100, 2003, 2386-2390.

15. K. Wuthrich, NMR of Proteins and Nucleic Acids, John Wiley & Sons, 1986.

16. G. M. Clore and A. M. Gronenborn, New methods of structure refinement for macromolecular structure determination by NMR, Proc. Natl. Acad. Sci. USA, 95, 1998, 5891-5898.

17. F. Cui, R. Jernigan, and Z. Wu, Refinement of NMR-determined protein structures with database derived distance constraints, J Bioinformatics and Computational Biology, 3, 2005, 1315-1329.

18. D. Wu, F. Cui, R. Jernigan, and Z. Wu, PIDD: A database for protein inter-atomic distance distributions, Nucleic Acids Research, 35, 2007, D202-D207.

19. D. Wu, R. Jernigan, and Z. Wu, Refinement of NMR-determined protein structures with database derived mean-force potentials, Proteins: Structure, Function, Bioinformatics, 68, 232-242, 2007.

20. G. M. Crippen and T. F. Havel, Distance Geometry and Molecular Conformation, John Wiley & Sons, 1988.

21. T. F. Havel, Distance geometry, in Encyclopedia of Nuclear Magnetic Resonance, D. M. Grant and R. K. Harris, eds., John Wiley & Sons, 1995, 1701-1710.

22. T. Havel, An evaluation of computational strategies for use in the determination of protein structure from distance constraints obtained by nuclear magnetic resonance, Prog. Biophys. Molec. Biol., 56, 1991, 43-78.

23. A. T. Brnger and M. Niles, Computational challenges for macromolecular modeling, in Reviews in Computational Chemistry, 5, K. B. Lipkowitz and D. B. Boyd, eds., VCH Publishers, 1993, 299-335.

24. Kuszewski, M. Niles, and A. T. Brnger, Sampling and efficiency of metric matrix distance geometry: A novel partial metrization algorithm, J. Biomolecular NMR, 2, 1992, 33-56.

25. A. T. Brnger, P. D. Adams, G. M. Clore, W. L. DeLano, P. Gros, R. W. Grosse-Kunstleve, . S. Jiang, J. Kuszewski, N. Nilges, N. S. Pannu, R. J. Read, L. M. Rice, T. Simonson, and G. L. Warren, Crystallography and NMR System: A new software suite for macromolecular structure determination, Acta Cryst., D54, 1998, 905-921.

26. T. F. Havel and M. E. Snow, A new method for building protein conformations from sequence alignments with homologues of known structure, J. Mol. Biol., 217, 1991, 1-7.

27. S. Srinivasan, C. J. March, and S. Sudarsanam, An automated method for modeling proteins on known templates using distance geometry, Protein Science, 2, 1993, 277-289.

28. E. S. Huang, R. Samudrala, and J. W., Ponder, Distance geometry generates native-like folds for small helical proteins using the consensus distances of predicted protein structures, Protein Science, 7, 1998.

29. G. A. Williams, J. M. Dugan, and R. B. Altman, Constrained global optimization for estimating molecular structure from atomic distances, J. Com-

put. Biol., 8, 2001, 523-547.

30. B. Hendrickson, Conditions for unique graph realizations, SIAM J. Comput., 21, 1992, 65-84.

31. B. Hendrickson, The molecule problem: Exploiting structure in global optimization, SIAM J. Optim., 5, 1995, 835-857.

32. W. Glunt, T. L. Hayden, S. Hong, and J. Wells, An alternating projection algorithm for computing the nearest Euclidean distance matrix, SIAM J. Mat. Anal. Appl., 11, 1990, 589-600.

33. W. Glunt and T. L. Hayden and M. Raydan, Molecular conformations from distance matrices, J. Comput. Chem., 14, 1993, 114-120.

34. J. Mor and Z. Wu, Global continuation for distance geometry problems, SIAM J. Optim., 7, 1997, 814-836.

35. J. Mor and Z. Wu, Distance geometry optimization for protein structures, J. Global Optim. 15, 1999, 219-234.

36. Z. Zou, R. H. Byrd, and R. B. Schnabel, A stochastic/perturbation global optimization algorithm for distance geometry problems", J. Global Optim., 11, 1997, 91-105.

37. A. Kearsly, R. Tapia, and M. Trosset, Solution of the metric STRESS and SSTRESS problems in multidimensional scaling by Newton's method, Computational Statistics 13, 1998, 369-396.

38. M. Trosset, Applications of multidimensional scaling to molecular conformation, Computing Sciences and Statistics 29, 1998, 148-152.

39. A. Le Thi Hoai and T. Pham Dinh, Large scale molecular optimization from distance matrices by a d.c. optimization approach, SIAM J. Optim., 4, 2003, 77-116

40. P. Biswas, T. Liang, K. Toh, and Y. Ye, A SDP based approach to anchor-free 3D graph realization, Department of Management Science and Engineering, Electrical Engineering, Stanford University, Stanford, California, 2007.

41. A. Grosso, M. Locatelli, and F. Schoen, Solving molecular distance geometry problems by global optimization algorithms, J. Comput. Opt. and Appl., 2007, to appear.

42. Q. Dong and Z. Wu, A linear-time algorithm for solving the molecular distance geometry problem with exact inter-atomic distances, J. Global Optim., 22, 2002, 365-375.

43. Q. Dong and Z. Wu, A geometric buildup algorithm for solving the molecular distance geometry problem with sparse distance data, J. Global Optim., 26, 2003, 321-333.

44. D. Wu and Z. Wu, An updated geometric buildup algorithm for solving the molecular distance geometry problem with sparse distance data, J. Global Optim., 37, 2007, 661-673.

45. D. Wu, Z. Wu, and Y. Yuan, Rigid vs. unique determination of protein structures, Optimization Letters, (published online, DOI: 10.1007/s11590-007-0060-7), 2007.

46. A. Sit, Z. Wu, and Y. Yuan, A stable geometric buildup algorithm for the solution of the distance geometry problem using east-squares approximation, 2007, submitted.

47. G. H. Golub and C. F. van Loan, Matrix Computations, Johns Hopkins University Press, 1989.

48. M. Sippl and H. Scheraga, Solution of the embedding problem and decomposition of symmetric matrices, Proc. Natl. Acad. Sci. USA, 82, 1985, 2197-2201.

49. M. Sippl and H. Scheraga, Cayley-Menger coordinates, Proc. Natl. Acad. Sci. USA 83, 1986, 2283-2287.

50. M. R. Garey and D. S. Johnson, Computers and Intractability: A Guide to the Theory of NP-Completeness, W. H. Freeman & Co., 1979.

51. J. Mor and Z. Wu, e-Optimal solutions to distance geometry problems via global continuation, in Global Minimization of Non-Convex Energy Functions: Molecular Conformation and Protein Folding, P. M. Pardalos, D. Shalloway, and G. Xue, eds., American Mathematical Society, 1996, 151-168.

52. H. X. Huang and Z. A. Liang, and P. Pardalos, Some properties for the Euclidean distance matrix and positive semi-definite matrix completion problems, J. Global Optim., 25, 2003, 3-21.

THE DIFFERENTIAL GEOMETRY OF PROTEINS AND ITS APPLICATIONS TO STRUCTURE DETERMINATION

ALAIN GORIELY

Program in Applied Mathematics and Department of Mathematics,
University of Arizona, Tucson, AZ 85721, USA
E-mail: goriely@math.arizona.edu

ANDREW HAUSRATH

Department of Biochemistry and Molecular Biophysics,
University of Arizona, Tucson, AZ 85721, USA

SÉBASTIEN NEUKIRCH

Laboratoire de Modélisation en Mécanique, UMR 7607
CNRS & Université Pierre et Marie Curie, Paris, France

Understanding the three-dimensional structure of proteins is critical to understand their function. While great progress is being made in understanding the structures of soluble proteins, large classes of proteins such as membrane proteins, large macromolecular assemblies, and partially organized or heterogeneous structures are being comparatively neglected. Part of the difficulty is that the coordinate models we use to represent protein structure are discrete and static, whereas the molecules themselves are flexible and dynamic. In this article, we review methods to develop a continuous description of proteins more general than the traditional coordinate models and which can describe smooth changes in form. This description can be shown to be strictly equivalent to the traditional atomic coordinate description. We apply the method to three major areas. The first is structure determination. The second is structural modeling, where the capability of smoothly deforming structures allows the exploration of possible forms to create models for classes of proteins. The third is to use coarse-grained mechanical models of proteins to predict the structures, motions, and conformational rearrangements in large oligomeric protein complexes.

1. Introduction

The accelerating growth of structural biology has created an enormous amount of information which we are only beginning to interpret. Today structural biology is approaching a comprehensive taxonomy of soluble globular proteins, and to make further progress it is imperative to address frontier areas which include membrane proteins, intrinsically disordered proteins, large macromolecular assemblies, and partially organized or heterogeneous structures such as cytoskeletal assemblies and amyloid fibers. These types of problems lie at the experimental frontier, as they severely tax the capabilities of existing physical techiques. Less widely appreciated is that these problems also lie at an intellectual frontier, as we currently lack a language to rigorously describe continuously variable or inexactly defined structures. This review presents some recent efforts to use a *continuous* representation of protein structure to model and gain some insight into protein structure. The formalism employed is based on the differential geometry of continuous curves- the fold of a protein is considered as a curve which follows the path of the protein backbone. We have developed methods to interconvert continuous curves and discrete atomic coordinates as a way to create and manipulate protein models[1,2]. In this review we present new tools based on continuous methods to investigate and utilize experimental structures and for examining relationships between the known structures and fold families. The review addresses three main objectives:

(1) **Structure determination:** How to devise methods for obtaining the geometric parameters needed to specify a curve (and its associated atomic model) from experimental data for the purpose of structure determination by electron microscopy, NMR, and crystallography.

(2) **Study of Fold Continua for Structural Prediction and Comparison:** How to explore the possibilities simple protein architectures offer by variation of the underlying geometrical parameters to investigate the relations between known folds and as a novel means of predicting molecular structure.

(3) **Continuum Mechanics of Biological Structure:** How to model conformational changes associated with biological function in terms of deformations of elastic bodies.

The mathematical methods developed for creating smooth deformations of protein models can be used to investigate intrinsically disordered regions, evolutionary structural change, distributed conformational changes associated with binding, and accommodation of structures to quaternary structure rearrangements.

The starting point is to represent and model the protein backbone by continuous curves or surfaces. A basic mean of comparison between math-

ematical structures is to create transformations or mappings between them
and then to investigate the properties of the mappings. In particular, properties of continuity and differentiability are crucial for such investigations.
The natural representation of molecular structure with atomic coordinates
is discrete and so precludes the application of continuous methods. However, continuous supersets which contain the discrete points corresponding
to atomic positions can be operated on in this manner. Use of this alternative continuous representation allows the application of powerful analytical
and geometric methods to answer questions about protein folds and conformations and their relationships through study of their corresponding space
curves and surfaces.

It is critical not to lose sight of the atomic details, which are the foundation for understanding the properties of proteins and provide much of the
underpinning for our current mechanistic view of biological processes. The
challenge is to integrate both continuous and discrete representations to
best understand the mathematical, physical, chemical, and biological properties of proteins and other macromolecules, using the appropriate viewpoint and theoretical tools for a particular question.

In Section 2 the relevant mathematics for protein curve construction
and extraction of curve parameters are described, and various avenues for
further mathematical developments will be indicated. Section 3 describes
how the small number of geometric parameters needed to describe simple
protein architectures allows the systematic search of an entire fold space.
Section 5 applies the capabilities of continuous deformation of folds from
Section 3 and the optimization of curve parameters from Section 2 to the
problem of protein design.

2. Methods

The path of a protein backbone is often represented as a series of line segments connecting the alpha-carbon atoms. It could also be represented as
a smooth curve passing through the same points. In general, regular three-
dimensional curves can be completely specified by their *curvatures* which
describe the local bending and twisting of the curve along its length. The
local description in terms of curvatures and the global description in terms
of spatial coordinates are entirely equivalent. For the particular case of
proteins, we have developed specific methods to construct the curvatures of
curves that follow the path of protein backbones, and to construct coordinate models of proteins from such curves. The interplay between the local

and global descriptions and in particular the modulation of curvatures to control the three-dimensional shape of protein models plays a central role for the work described here. The rest of this section gives some technical details on how to construct curves from data sets and, conversely, how to build protein models with idealized geometry from curvatures. The next sections explore various applications of this formalism to structure determination, protein fold exploration, and protein design.

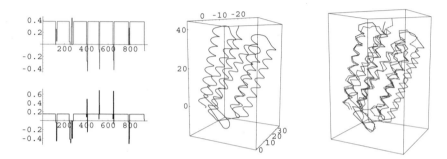

Figure 1.   Construction of curves from curvatures. A curve (C.) is constructed from its curvature-torsion profile (A. and B.). Note that the *signed* curvature (real) is used here rather than the curvature itself (assumed positive). D. The position of the C_α atoms obtained from the curve and the experimentally determined C_α coordinates of bacteriorhodopsin are superimposed. This example utilized 96 curve parameters to specify the curve, where as 3 coordinates are required for each of the 228 atoms in the C_α trace.

Let $\mathbf{r} = \mathbf{r}(s)$ be a curve in $\mathbb{R}^3$ parameterized by its arc-length s. At each point s on the curve, one can define (assuming sufficient regularity of the curve) a local general orthonormal basis $\{\mathbf{d}_1(s), \mathbf{d}_2(s), \mathbf{d}_3(s)\}$ by defining the orientation of the vector $\mathbf{d}_3$ with respect to the tangent vector $\mathbf{r}' = v_1\mathbf{d}_1 + v_2\mathbf{d}_2 + v_3\mathbf{d}_3$ to the curve. Since the vectors $\{\mathbf{d}_1(s), \mathbf{d}_2(s), \mathbf{d}_3(s)\}$ are orthonormal, their evolution is governed by

$$\frac{\partial \mathbf{d}_1}{\partial s} = k_3\mathbf{d}_2 - k_2\mathbf{d}_3, \quad \frac{\partial \mathbf{d}_2}{\partial s} = k_1\mathbf{d}_3 - k_3\mathbf{d}_1, \quad \frac{\partial \mathbf{d}_3}{\partial s} = k_2\mathbf{d}_1 - k_1\mathbf{d}_2. \quad (1)$$

That is, $D' = DK$, where D is the matrix whose columns are the basis vectors, ( )$'$ denotes the derivative with respect to s, and K is the skew-symmetric matrix

$$K = \begin{bmatrix} 0 & -k_3 & k_2 \\ k_3 & 0 & -k_1 \\ -k_2 & k_1 & 0 \end{bmatrix} \quad (2)$$

This general description of local bases for curves allows to define both the shape of the curve but also the evolution of a triad of orthonormal vectors attached to it through the specification of a vector of *curvatures* $\mathbf{k} = (k_1, k_2, k_3)$ and a vector of *basis orientations* $\mathbf{v} = (v_1, v_2, v_3)$. This description becomes more familiar if we specialize the general basis by defining $\mathbf{d_3}$ as the tangent vector (that is, $(v_1, v_2, v_3) = (0, 0, 1)$ and $\mathbf{d_1}$ as the normal vector). In which case, Eq. (1) becomes the Frenet equations and the curvatures are $(k_1, k_2, k_3) = (0, \kappa, \tau)$ where κ and τ are, respectively, the curvature and torsion of the curve at the point s. Curvature and torsion can be determined from a given C^3 curve $\mathbf{r}(s)$ by standard differential geometry identities. The *curvatures* of $\mathbf{r}(s)$ are the entries of the Darboux vector $\mathbf{k} = (k_1, k_2, k_3)$ and they can only be determined up to a phase factor φ that describes the phase difference between the normal vector and the first basis vector $\mathbf{d_1}(s)$. The general basis has many advantages due to its compact form, the possibility to define local bases for less regular curves or to assign to the phase factor additional information (such as the twist and shear of a ribbon, or the material property of a tube surrounding the central curve). For clarity sake, in this review, we mostly restrict our analysis to the Frenet frame and define $D = [\mathbf{n}(s), \mathbf{b}(s), \mathbf{t}(s)]$ as the normal, binormal and tangent vectors and $\mathbf{k} = (0, \kappa, \tau)$ where κ is the signed curvature (defined on the real rather than strictly positive) and τ is the usual torsion. The exact description of protein backbones with piecewise helical curves in the next section will require the use of the general basis.

2.1. *Curvatures to curve*

If the curvatures $\mathbf{k}$ and basis orientation $\mathbf{v}$ are given, the curve can be readily obtained by integrating Eq. (1) together with $\mathbf{r}' = \sum a_i \mathbf{d}_i$. These equations form a system of 12 linear non-autonomous equations for the basis vectors D and curve $\mathbf{r}$ that can be written in a compact form by introducing the vector $\mathbf{Z} = (d_{1,x}, d_{2,x}, d_{3,x}, d_{1,y}, d_{2,y}, d_{3,y}, d_{1,z}, d_{2,z}, d_{3,z}, x, y, z)^T$ in which case, the linear system reads

$$Z' = MZ, \quad \text{with} \quad M = \begin{bmatrix} K^T & 0 & 0 & 0 \\ 0 & K^T & 0 & 0 \\ 0 & 0 & K^T & 0 \\ V_1 & V_2 & V_3 & 0 \end{bmatrix}, \tag{3}$$

where K is defined above and V_i is the 3-matrix whose only non-zero entry is row i with value $\mathbf{v}$. The integration of Eq. (3) as a function of s provides both the curve position but also the evolution of the basis. In the particular case where Frenet frame is used $\mathbf{v} = (0,0,1)$ and for given curvature κ and torsion τ, the curve is reconstructed. A theoretical example of such a construction is given in Figure 1 and a more detailed fit of an actual protein is shown on Fig. 2.

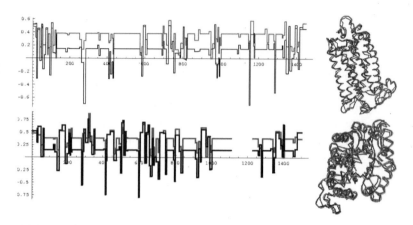

Figure 2. Left) curvature profiles, and Right) Polyhelix models of the primarily α-helical protein bovine rhodopsin (1U19.pdb above) and the mixed α/β protein bacterial luciferase (1BRL.pdb, below.) A disordered section in the latter is missing from the model, resulting in a gap in the curvature profile.

2.2. Curves from atomic models

The first problem is to obtain curves describing the protein backbone from a set of C_α coordinates. One of such constructions can be found in Richardson ribbon diagrams, where curves, ribbons and helices are used to build a three-dimensional picture of the protein backbone. However, this remarkable construction, obtained by spline fitting, is mostly used for visualization purposes and its mathematical formulation is not a faithful representation of the protein that can be used for other purposes[3,4]. Therefore, one first needs to obtain an exact curve representation of the protein backbone, that is a curve that passes exactly through each C_α atom. Obviously, this can be done in many different ways since there are infinitely many choices of curves passing through a given set of points. A possible choice is through the use of polynomial or rational splines by considering the set of N first atoms and find the spline of degree d through this set, then sequentially build another spline for the next N atoms (with possible overlap). Depending on the degree of smoothness required on the curve the parameters N

and d can be adjusted within the proper algorithm[5]. This representation offers an exact description of the protein backbone but does not carry much information on its global structure. In effect, it replaces one discrete set of points by a discrete set of curves through subsets of points.

A more useful way to represent the C_α trace is to find a piecewise helical curve (referred to hereafter as a *polyhelix*) through the points, that is a continuous curve built out of connected helices. Many authors have considered the problem of fitting helices though sets of points in space. This problem arises in protein structure[6], engineering design of cables and springs[7], and nuclear and particle physics for particle tracking[8]. To obtain an exact representation by a polyhelix, 4 consecutive C_α atoms are considered and a unique helix can be constructed (see for instance[9]), the helix is characterized by its first C_α, curvature and torsion, and axis. To pinpoint the position of the other C_α, three arclengths are required. Together, it amounts to 12 data points corresponding to the 12 atomic coordinates, providing a 1-1 map between atomic data and helical data. The construction proceeds by considering the fourth to seventh C_α's for the next helical piece and so on (See Figure 3). This construction does not provide a purely local representation of the curve since extrinsic data (position of the axis in space) is required. However, using the general basis described above, a complete local representation of the curve can be obtained by specifying for each local piece the following data for the ith helix starting at the atom number $3i + 1$: a constant curvature vector $\mathbf{k}^{(i)}$, the orientation of the basis given by a constant vector $\mathbf{v}^{(i)}$ and the arclength positions of the atoms on the helix $S^i = \{s_{3i+2}, s_{3i+3}, s_{3i+4}\}$. This represents 9 data for each successive 3 atoms. The change of orientation of the axis is characterized by the change in the vector $\mathbf{v}$ between the ith and $(i+1)$th helices. Computationally, the protein backbone and the position of the C_α is fully characterized by a list of triplets $H^{(i)} = \{\mathbf{v}^{(i)}, \mathbf{k}^{(i)}, S^{(i)}\}$ and the positions are recovered in extrinsic coordinates by integrating Eq. (3). While this operation seems to be a daunting task, it actually amounts to straightforward matrix algebra due to the exact analytical solution of these differential equations in the cases where curvatures and orientations are constant (see Section 2.4 on polyhelices). It is important to note that this representation is general and not restricted to the analysis of proteins with alpha-helices. A representation of atomic data in terms of polyhelices has many advantages: it is purely local in nature and so exploits the natural geometry of the protein, the curvatures carry global information on the curve through curvatures and torsion and therefore allows for direct identification of regions of interest (for instance alpha-helices or different types of turns), and modulation of these curvatures over long distances identifies long-range structure (*e.g.* bending and curving of helices, twist of beta-sheets, etc...). To a certain extent, different authors have explored the local geometry of existing proteins using similar approaches with alternative formulations[10]. This construction is the starting point of our analysis. The relevant aspect of local representation is that it helps connect experimental data to structure determination and modeling performed through the use of curve geometry as presented in the next Sections. There, idealized geometries based on polyhelices are used to explore possible folds in parameter space.

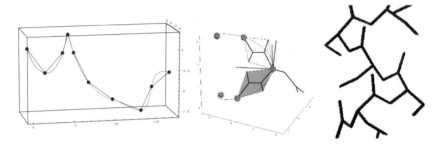

Figure 3. A. Construction of curves from coordinates: A continuous polyhelix curve of 4 segments constructed from points. B: Construction of coordinates from curves: The local basis on a curve centered on the 3rd C_α and selected atoms expressed in that local coordinate system. C: Section of an idealized polyserine helix constructed with EDPDB compared with the model constructed in the local coordinates as in section 2.3.

Polyhelices provide an exact local representation of the protein backbone. However, in many studies, one may be interested in describing non-local properties of these structures. For instance, one may be interested in representing an alpha-helix that may not be strictly helical (due to bending or super-twisting) by one single helix. The problem is then to fit a given structure (helix, plane, ...) through a set of points. This can be done through some averaging process on the exact local representation or directly by fitting through least-square computations. An example of helix fitting is shown in Figure 1 where the curvature-torsion profile was optimized by fitting helical segments to the C_α coordinates from the protein bacteriorhodopsin. There are many different outstanding mathematical and computational issues associated with the problem of fitting a helix[11], or a cylinder[12] through which will are addressed in separate papers.

2.3. *Atomic models from curves*

Curves can be used to describe proteins and construct models in many different ways. At the basic level, the curve can represent the backbone and C_α can be superimposed by imposing that they are located on the curves at determined position s. In particular, at suitable values of s, the local coordinate system has its origin at the C_α positions, which is a natural choice for the coordinate system in which to express the atomic coordinates for the remaining atoms in the residue. A set of local coordinate system $\mathbf{a} = \{a_1, a_2, a_3\}$ represents the point $\mathbf{p}_a = \mathbf{r}(s) + \sum_{i=1}^{3} a_i \mathbf{d}_i(s)$ in the external coordinates. Conversely, any point $\mathbf{p}_a(s)$ has local coordinates $a_i = (\mathbf{p}_a - \mathbf{r}(s)) \cdot \mathbf{d}_i(s)$, $i = 1, 2, 3$. We have converted and tabulated the local coordinates for all the rotamers from[13] which allows the construction of the alpha-helical regions of protein models from curves with idealized geometry, and which can also be used, albeit with some distortion of the backbone atomic arrangements, in the rest of the protein.

2.4. *Polyhelices*

A particularly simple choice of curves is obtained by choosing the curvatures to be piecewise constant so that the curves are piecewise helical. These polyhelices can be used to map precisely atomic coordinates to continuous curves. Conversely, they can also be used to study and classify large families of proteins with idealized geometries. The advantages of this representation are threefold; first, it is consistent with the representation from the atomic coordinates making the comparison with experimental data straightforward; second, large families of proteins can be represented by few parameters and the exploration of fold spaces can be achieved with minimal effort; third, the computation of polyhelices can be reduced to simple linear algebra, making it computationally exact and reliable. The computation of polyhelices is achieved by integrating Eq. (3). Since M is now a constant matrix, the solution of this system is given by

$$Z(s) = A(\kappa, \tau; s)Z(s_0) \tag{4}$$

where $A(\kappa, \tau; s) = e^{(s-s_0)M}$ is the matrix exponential. Matrix A can be computed exactly and its entries are linear combinations of trigonometric and polynomial function of s with coefficients depending on κ and τ. A polyhelix with N helices is completely characterized by a list of curvatures and length: $P = \{(\kappa^{(i)}, \tau^{(i)}, L^{(i)}), i = 1..N\}$ and an initial position and basis orientation $Z^{(0)}$. The jth helix on the curve is given by the last three components of the vector Z^j:

$$Z^{(j)} = A(\kappa^{(j)}, \tau^{(j)}; s)Z^{(j-1)} = \prod_{k=1}^{j} A(\kappa^{(j)}, \tau^{(j)}; s)Z^{(0)}, \quad L_{j-1} \le s \le L_j,$$

(5)

where $L_j = \sum_{k=1}^{j} L^{(k)}$. Examples of such computations are given in Figure 1. This analytic expression for the curves provides an efficient way to explore fold space as shown in the next Sections.

2.5. *Embedding methods*

While α-helices lend themselves to a linear description, beta-sheets are inherently a linear but 2-dimensional structure. It is therefore natural to use a 2-dimensional embedding of a curve to describe them. The construction proceeds in 2 stages as illustrated in Figure 4. The linear character of the chain is accommodated by a description of the backbone as a plane curve specified by its curvature. The second stage is to map the curve into a two-dimensional surface plane in a three-dimensional space. The particular embedding chosen defines the surface characteristics of the beta-sheet model and can be described by the classical Darboux frame field representation. Here again, the atoms can be represented in the local Darboux frame. The choice of mapping is restricted by the constraints on bond angles and distances and possible beta-architecture with such constraint can be systematically explored within this framework (this and related ideas on possible architectures are discussed in Section 4).

More explicitly, the construction proceeds in the following steps.

(1) **Polyarc plane curves:**

A plane curve $c_p(s) = \{u(s), v(s)\}$ is described up to rotations and translations in terms of its plane curvature profile $\kappa_p(s)$. Any plane curve with constant curvature κ_p is a circular arc with radius $1/\kappa_p$, and a plane curve with zero curvature is a straight line. The Frenet equations in the plane are described in terms of the tangent and normal vectors $\mathbf{t}_p(s)$ and $\mathbf{n}_p(s)$ as the ODE system $\mathbf{c}'_p = \mathbf{t}$, $\mathbf{t}'_p = \kappa_p \mathbf{n}_p$, and $\mathbf{n}'_p = -\kappa_p \mathbf{t}_p$. Denoting the initial basis as $Z_0 = \{t_u, n_u, t_v, n_v, c_u, c_v\}$, a plane curve and coordinate system may be cast as a matrix equation and its solution obtained in closed form as a matrix exponential.

$$Z' = NZ \quad Z(s) = B(\kappa_p, s)Z_0 \quad \text{where} \quad B(\kappa, s) = e^{Ns}. \quad (6)$$

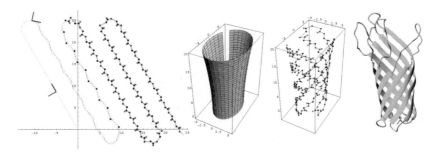

Figure 4. A. A plane curve and its local coordinate system: the side-to-side alternation of the beta-strands is accommodated by expression of the backbone plane atomic coordinates as a function of a sine wave expressed in that coordinate system B. A surface in three dimensions, schematically representing the form of a flared, asymmetric beta-barrel. C. Atomic model resulting from embedding the plane from (A) onto the surface from (B) D. Ribbon diagram of ompA (PDB code 1BXW). Figure and fit by Katie White.

Explicitly

$$
N = \begin{pmatrix}
0 & \kappa & 0 & 0 & 0 & 0 \\
-\kappa & 0 & 0 & 0 & 0 & 0 \\
0 & 0 & 0 & \kappa & 0 & 0 \\
0 & 0 & -\kappa & 0 & 0 & 0 \\
1 & 0 & 0 & 0 & 0 & 0 \\
0 & 0 & 1 & 0 & 0 & 0
\end{pmatrix} \tag{7}
$$

$$
B(\kappa, s) = \begin{pmatrix}
\cos \kappa s & \sin \kappa s & 0 & 0 & 0 & 0 \\
-\sin \kappa s & \cos \kappa s & 0 & 0 & 0 & 0 \\
0 & 0 & \cos \kappa s & \sin \kappa s & 0 & 0 \\
0 & 0 & -\sin \kappa s & \cos \kappa s & 0 & 0 \\
\frac{\sin \kappa s}{\kappa} & \frac{1-\cos \kappa s}{\kappa} & 0 & 0 & 1 & 0 \\
0 & 0 & \frac{\sin \kappa s}{\kappa} & \frac{1-\cos \kappa s}{\kappa} & 0 & 1
\end{pmatrix} \tag{8}
$$

If the curvature profile is specified with a list of pairs $Q = \{(\kappa^{(i)}, L^{(i)}), i = 1..N\}$, then the vector Z is obtained with

$$
Z^{(j)}(s) = B(\kappa^{(j)}, s - s_0^{(j)}) \cdot \prod_{k=1}^{j-1} B(\kappa^{(k)}, L^{(k)}) \cdot Z^{(0)}, \quad s_0^{(j)} \le s \le s_0^{(j-1)} \tag{9}
$$

where $s_0^{(j)} = \sum_{k=1}^{j-1} L^{(k)}$. This expression for polyarc curves has a similar structure to the polyhelix construction in that it provides a recursive parametric expression for the plane curve and its local coordinate system.

(2) **Plane Curves on Surfaces:**

Given a plane curve, it can be embedded into three dimensions by mapping the region of the plane which it occupies onto a surface. Any combination of a plane curve and a surface will generate a space curve. This construction is the strategy proposed here for modeling beta-sheet structure.

A general plane curve which can be represented as a map $c : \Re \to \Re^2$, whereas a surface is $S : \Re^2 \to \Re^3$. Therefore the space curve $C : \Re \to \Re^3$ can be written as the composition $C = S \circ c$. We use polyarcs as the explicit map from $\Re \to \Re^2$. Choice of the map $\Re^2 \to \Re^3$ depends on the particular type of structure being modeled. A general means to describe a surface is a 2-dimensional parametrization. $S_3(u, v) = \{X(u, v), X(u, v), Z(u, v)\}$. Therefore the space curve obtained by mapping a plane curve in 2 dimensions to its corresponding space curve in three dimensions can be written $C_3(s) = \{X(C_2(s)), Y(C_2(s)), Z(C_2(s))\}$. One well-understood class, surfaces of revolution ([14], Chapter 20) is especially applicable to beta barrels. More generally, the flexible and compact representation of curves as polyhelices can be utilized to specify a class of surfaces (termed polysheets) useful for representing β-sheets. Because this type of surface is constructed in terms of polyhelices, it is easily integrated into a polyhelix description to allow for models of mixed alpha and beta architecture. (Although beta strands can be represented within the polyhelix construction as seen in Figure 2, the geometry of the sheet is not explicitly used.)

3. Fold Space Exploration

The set of protein folds is a subset of the set of all possible space curves that can be constructed by standard differential geometry tools. By investigating the set of possible curves, we can find within it the possible protein folds. The key problem is to identify among all curves the ones that may constitute the path of a protein backbone. Mathematically, the problem amounts to finding functions to score the potential of a curve to take the shape of a protein. The main idea is then to explore continuous families of curves defined by a set of parameters and isolate good protein candidates that correspond to points within that parameter space. Small parameter spaces which describe simple protein architectures can be exhaustively sampled. The ability to explore the entire realm of possibilities inherent in a particular architecture makes it possible to see relations between folds that may not be apparent. The method has applications to protein structure prediction, genome interpretation, and homology modeling. Sequences can be threaded onto curves obtained by a systematic or guided parameter space search. Postulated folds may also serve as protein design targets. Finally, the ability to "interpolate" or "extrapolate" from existing folds may allow prediction of new folds (and explicit construction of their coordinate models) before they are experimentally observed.

3.1. *Protein Quality Functions*

Most curves in space could not be realized as paths of protein backbones, because they have impossibly tight bends, unrealistically straight segments, have regions that approach too close to other regions of the curve, or are too loosely packed to have sufficient interactions to remain folded. However, there are some curves that satisfy all those criteria. A fundamental question is to identify simple criteria (geometric and physical) to quantify whether a given curve might be realizable as protein backbone conformations if the right amino acid sequence could be found. To address this question, we introduce the idea of *protein quality functions*, to quantify the potential of a curve to be realized as a protein fold. The *curvature space* is the space of parameters defining a a family of curves (for instance, the family of helices is a three-dimensional space defined by curvature, torsion, and length) and a protein quality function is defined at each point of the curvature space and takes real values. A contour plot of a quality function over the curvature space would have islands in regions that correspond to protein-like curves (for instance, α-helices would score very high in the family of helices and a small island would be centered around the ideal value of curvature and torsion for α-helices). Once such a function has been identified, it is possible to investigate questions about the density of folds in fold space, or the connectedness of fold space (*i.e.* are regions of protein folds connected or widely separated?). What are the possible choices for quality functions? Clearly, to conduct a search over large regions of the fold space, the quality functions should be easy to evaluate.

A simple protein quality function can be expressed as a ratio of a term that expresses curve compactness and a term that penalizes a curve which approaches itself too closely. To quantify compactness, the notion of contact order[15] serves nicely. Given a set of points $\{P(s_k)\}$ on a curve, two points $P(s_i)$ and $P(s_j)$ form a contact when within a prescribed contact distance in space. The contact order is the sequence distance $|s_j - s_i|$ averaged over all contacts. Contact order is large for curves in which many pairs of points distant on the curve are close in space and so serves as a simple quantitative measure of compactness.

$$C_O = \frac{1}{LN} \sum_{}^{N} |s_j - s_i| \tag{10}$$

where N is the number of contacts and L the number of points. However, a curve that is too compact will approach too close to itself. Defining a clash as a pair of points which are closer than a prescribed clash distance in space, curves with self-intersections are severely penalized by using the quality function:

$$Q_1 = \frac{C_O}{2^M} \tag{11}$$

where C_O is the contact order of the curve and M the number of clashes. Examination of the distances between C_α coordinates of several repeat proteins suggested the use of 9 Angstroms as the contact distance and 4 Angstroms as the clash distance. An advantage of this quality function is that it can be used both on C_α coordinates from real proteins and on points obtained from curves.

There many different choices for quality functions emphasizing different features of a candidate curve. Such functions could be evaluated on smooth curves which may have advantages for theoretical investigations or on discrete sets of points obtained from the curve, which have the advantage of ease of comparison with experimentally obtained coordinate sets. Other interesting possibilities include the use of the global radius of curvature[16] providing both a local estimate of curvature and a global estimate of self-contact; simplified versions of energy functionals as used in homology modeling and structure prediction[17,18] ; family of Vasiliev knots invariants for protein which have already shown great promise for classification purposes[19,20]; or statistics of distributions of curve parameters in curves fitted to experimentally determined coordinates.

3.2. The fold spaces of polyhelices

As an example of fold space exploration we use the quality function Q_1 defined above to study the fold space of some polyhelical families. As discussed in Section 2, a curve consisting of N helical segments can be specified by a list of N {curvature,torsion,length} triples and the fold space is a $3N$-dimensional which can be systematically explored with the function Q_1. This space is very convenient to model proteins with α-helices since a few parameters are necessary to describe the main building blocks of the proteins.

First, a simple application of the use of the quality function to rank the turns connecting two α helices is shown in Figure 5. The curvature/torsion profile and its corresponding helical hairpin are shown. The turn is parameterized by the two triples $\{\{\kappa_1, \tau_1, l_1\}, \{\kappa_2, \tau_2, l_2\}\}$. This 6-dimensional fold space can be searched exhaustively. Contour plots (in which light colors indicate high (favorable) values of Q_1 function) as a function of κ_2 and τ_2, for different values of l_2 are shown for 2 different choices of κ_1, τ_1, and l_1 in Figure 5.

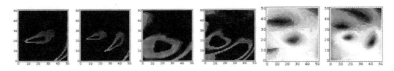

Figure 5. A and B. Contour plot of the PQ function (κ_2 vs. τ_2) for $l_2 = 3.0$ and $l_3 = 5.0$ C and D. A similar plot using different κ_1, τ_1, and l_1 values. E. and F. Contour plot of the best value of the quality function for any choice of κ_2, τ_2, and l_2 plotted as a function of κ_1, and τ_1. Here l_2 is 3.0 (E) and 5.0 in (F).

The presence of islands and plateaus indicates that only certain combinations of curvature and torsion gives rise to reasonable turns. By selecting points within the white regions, a list of turns which specify high-scoring protein-like helical hairpin curves can be collected in a library of candidates for connecting turns. Once this library is built, one can proceed hierarchically with the search for protein candidate by finding helical repeats where the connecting turns are given by helical hairpin from the library.

As a more complex example, consider the various types of helical repeat proteins which share a common architecture of (helix$_1$-turn$_1$-helix$_2$-turn$_2$)N. Curves corresponding to this architecture can be specified with 14 parameters. (The curvature and torsion of an alpha helix are fixed, so only 2 parameters for the lengths of the helices are needed. Each turn is described by 2 $\{\kappa, \tau, l\}$ segments.) A curve is determined by a point the 14 dimensional fold space. A systematic search of this space is still a daunting undertaking and some simplifying assumptions are necessary. We

Figure 6. On the left is a ribbon diagram of beta-catenin, and on the right is the ribbon diagram of coordinates constructed from the curve- in between is their superposition with beta-catenin and the curve-derived model.

assume that the sections helix$_1$-turn$_1$-helix$_2$ and helix$_2$-turn$_2$-helix$_1$ as helical hairpins. This reduces the search to a 4-dimensional space, over the two (continuous) helix lengths and two turns from the (discrete) helical hairpin list. For high-scoring curves, we construct polyalanine atomic models and overlay them on experimentally determined repeat protein coordinate sets (PDB entries 1i7x,1b89,and 1b3u. We will expand the comparison and use more sensitive methods of structure comparison[21]). An example of a "hit" from this search is shown in Figure 6, in which a curve close to the armadillo repeat protein β-catenin (PDB code 1i7x) was obtained. Quite remarkably, this result shows that a construction solely based on simple geometric principles can capture the form of existing proteins accurately, and that simple quality functions can be used to search rapidly through the curve specification parameter space and identify protein-like curves. Suggestively, some regions of the fold space have high quality values and yet describe curves that do not resemble any known proteins. Some appear to have plausible packing arrangements- examples are shown in Figure 6. The left three show examples of a family which can be described as a stack of antiparallel coiled-coils. The right three show examples of a family which most closely resembles leucine-rich repeats but which have a second helix in place of the beta-strand. Once interesting regions of a fold space have been identified with one quality functions, it can be further explored by using other quality functions providing, in effect, a series of filters for plausible proteins.

This methodology has been demonstrated with helical repeat proteins due to the small number of parameters needed to describe curves with such architecture. However, the same idea is applicable both to non-repeat proteins and also to beta- or mixed alpha-beta architectures by using different (or mixed-) representations of curves such as the one given in Section 2.5 for beta sheets.

3.3. *Modeling helical bundle and β-barrel membrane proteins*

Experimental structure determinations of membrane proteins are more difficult than those for soluble proteins, and there is a continuing need for improved modeling methods applicable to membrane proteins[22]. Most known membrane proteins fall into two structural categories: helical bundles and β-barrels. These classes of proteins are well suited to our continuous descriptions. The helical bundle category is conveniently represented with polyhelices (Fig. 2). The β-barrel models are constructed using the embedding procedure described in Section 1.3, and using the Darboux frame for atomic model construction (Fig. 7).

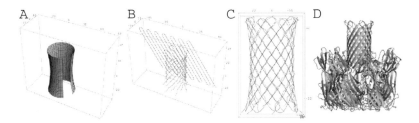

Figure 7.   Embedding of plane curve onto surface to model β-barrel proteins. A) Surface of revolution B) An antiparallel plane curve in the x-z plane shown with C_α trace of the regular β-barrel section from α-hemolysin (7AHL.pdb Resulting fit to these C_α coordinates of the plane curve from B) embedded into the surface from A). D) Ribbon diagram of α-hemolysin for comparison. (Note this molecule is trimeric, whereas only a single plane curve is used to model the barrel in this example. More complex curves and surfaces are used for modeling less regular β-barrels.)

For these protein architectures, we will use these efficient parametrizations of structure to explore the classes of likely folds. Backbone coordinate models of the possible β-barrel structures with different numbers of strands and different sheet registers will be constructed[23]. Similarly, helical bundles with different numbers of helices and different patterns of membrane insertion and helical arrangement will be constructed (cf[24]). These will be scored by geometric criteria to devise concrete examples of likely membrane protein architectures. Since membrane proteins tend to have a simpler form than globular proteins a coarse sampling of the possible structures can of provide useful starting models for other applications. For instance, plausible folds can be used for modeling purposes by threading protein sequences onto them. This may be useful in conjunction with different structural biology methods described in section 2, in that a small number of experimental constraints on the structure may be sufficient to distinguish between certain classes of models. If more detailed information becomes available, such generic structures will serve as structural templates for curvature-based optimization.

4. Protein Design

The goal of *de novo* design of proteins is to create sequences that fold into a desired three-dimensional structure[25,26]. The continuous representation of proteins is of great utility for design work as it allows the overall specification of protein architecture without requiring that all the atomic details be considered at the outset. It separates the geometric problem of finding a suitable backbone from the problem of consistent atomic interactions. Therefore, it provides tools to specify a protein scaffold that may not resemble natural protein folds.

The primary determinant of globular protein folds is encoded in the binary pattern of hydrophobic and polar residues which defines the hydrophobic core and hydrophilic surface of the protein[27,28,29]. Recognizing the fundamental character of the hydrophobic core for structure specification, the computational protein design field initially focused on means to design cores compatible with existing natural scaffolds[30]. In its simplest form, the problem of protein design is the problem of selecting the best sequence which fills the interior volume of a given scaffold. The need for criteria to distinguish among different sequences that satisfy the crude inside/outside constraint but which led to stable but non-unique structures[31] has led to the development of sophisticated energy functions that take into account the many physical interactions specifying a unique folded protein structure[32]. By incorporating these finely-tuned energy functions with the power and efficiency of new combinatorial algorithms[33], core-packing algorithms have matured to the point where realizable designs have become routine. However, it has become increasingly clear that protein backbones are not rigid scaffolds but rather can move in unexpected ways in response to changes in core composition[34,35]. The problem of how to incorporate such backbone freedom into protein design algorithms, and to control the computational cost of exploring these extra degrees of freedom, is the main challenge to the field protein design field[36].

The greatest appeal of a geometric approach to protein design is that it separates the specification of structure from the validation of structure. The geometric representation can be used to construct new plausible models, and different energy functionals can be used to rank them. By separating the two problems, the constraints of sequence and structure can be looked at as independent, and either can be varied to best satisfy the other. In essence, one can ask the question "What is the best sequence for this particular curve?" as well as "What is the best curve for this particular sequence?" By iterating between the two problems, a solution for which both sequence and structure are mutually optimal can be determined .

The continuous representation provides a theoretical foundation for backbone design as well as natural and inexpensive computational methods especially compared to the computations in dihedral angle space.

Recently hybrid methods which alternate between sequence rearrangement and energy minimization to allow backbone relaxation have shown great promise[37] We propose to generalize these ideas by adapting the continuous description to allow simultaneous minimization of an externally specified energy function with respect to curve parameters and side-chain rotamers as a way to find mutually optimal sequence and structure solutions.

4.1. *Creation of structure specification and optimization tools*

An important part of protein design is the development of algorithms for optimization of models optimizing user-supplied design criteria (such as a the minimization of an energy function) by adjusting curve parameters, and the identification of an interior volume needed to construct a hydrophobic core. Considered in combination, these two constraints will identify suitable scaffolds for design targets. Section 5.2 will discuss how to create sequences that best conform to these scaffolds.

The optimization of curves and curve-derived coordinate models by variation of curve parameters is based on the the same underlying methods as the ones described of Section 3 where agreement with experimental data rather than with design constraints was required. From an atomic perspective, the design task is to create energy functions balancing the relative contributions of the different types of interactions (*e.g.* electrostatic, Vanderwaal's, hydrophobic) involved in stabilizing folded proteins. From the coarse-grained perspective given by the continuous description, simple geometric criteria based on protein quality functions have considerable discriminating power. The main idea is therefore to use the optimization algorithms of Section 3 with a protein quality function of Section 4 to identify suitable design targets which can then be used to construct detailed atomic models as in section 5.2.

The starting point for sequence design is the determination of the binary hydrophobic/hydrophilic pattern in the primary sequence[38] A design target conformation can be characterized by a curve. To determine a binary pattern along the curve, it is necessary to identify the interior volume enclosed by the curve. To do so, a simple method (See Figure 8) consists in locating C_α atom on the curve, and to construct a grid covering the extent of the C_α. For every point on the grid, the number of neighboring C_α atoms is determined. Choosing only the points which have many neighboring C_α atoms as the center of spheres, an approximate "interior" volume is obtained by taking the union of these spheres. Constructing the C_β atoms from the curve, the atoms facing the "interior" volume can be assigned a hydrophobic character by standard core-packing algorithm like DEE[39,33]. Subsequently, the curve can be modified so as to bury more or fewer side-chains according to desired criterion such as buried hydrophobic surface or volume[40,41].

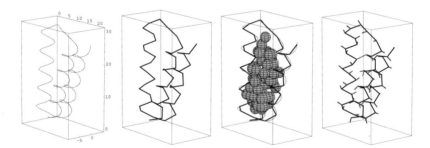

Figure 8. Determination of binary patterning. A: A three-helix bundle curve. B: C_α atoms on the curve. C: "Inside" volume as the union of spheres centered at points near (within 8 Angstroms) sufficiently many (more than 12) C_α atoms. D: C_β atoms constructed from the curve located within the "inside" volume.

5. Continuum Mechanics of Biological Structures

The methods described in the previous sections can describe protein models accurately but are of geometric nature and exist without reference to physical assumptions about molecules. In this section we build on this foundation, refining the models by incorporating physical considerations to relate structural and energetic properties of molecules and moreover make experimentally testable predictions. We will use classical elasticity theory for this purpose. Primarily we will seek analytical solutions by employing the semi-inverse method of Saint-Venant, although with modern computational resources, numerical solution of the elastic field equations can be applied when needed as an alternative.

For the purposes of structural modeling, elastic deformations can be used to describe large-scale, distributed conformational changes. But importantly, these responses are described in terms of changes in body coordinate systems, which can be made to coincide with the local coordinate frames we have used for construction of atomic models. Thus elasticity provides a natural formalism for devising physical theories which make structural predictions. For instance the elastic energy of filamentous structures can be described with the equation

$$E_{elastic} = \frac{1}{2} \int_0^L \left(B_1(s)\kappa_1^2(s) + B_2(s)\kappa_2^2(s) + B_3(s)\kappa_3^2(s) \right) ds \qquad (12)$$

where B_1, B_2, and B_3 are the elastic constants and the κ_i are the deviations from equilibrium values of the curvatures. Additional terms may be included, for instance to include nonlocal effects or to obtain force-extension curves for study of mechanical responses.

5.1. *Continuum elastic theory of coiled-coils*

Within this formalism, a coiled-coil can be modeled as conjoined elastic filaments with elastic constants $B_1 = B_2 = B$ which describe resistance to bending and $B_3 = C$ which describes resistance to twisting. In a coiled-coil configuration, the center line $\mathbf{r}(s)$ of each filament itself is a helix. The axis is along z, the radius is written R, the pitch is $2\pi R/\tan\theta$ and the super-helical angle θ is the complement of the pitch angle. We parametrise the (helical) center line as:

$$\mathbf{r}(s) = \begin{pmatrix} +R \, \sin\psi(s) \\ -R \, \cos\psi(s) \\ s \, \cos\theta + z_0 \end{pmatrix}, \qquad \psi(s) = \frac{\sin\theta}{R}s + \psi_0 \qquad (13)$$

where $\psi(s)$ is the equatorial angle in the (x, y) plane. The (constant) curvature and torsion of the super-helical axis are $\kappa = \sin^2\theta/R$ and

$\tau = \sin\theta\cos\theta/R$. The Frenet and Cosserat frame vectors are obtained from this parametrization by standard identities and can so be used to construct atomic models.

Coiled-coils are held together by interactions along one face of their constituent helices, and this interaction face can be parametrized in terms of the Cosserat directors and the parameter $\hat{\tau}$ which describes the twisting of the interaction face along the surface of the alpha helix. Constraining these interaction surfaces to be joined provides a structural constraint, and minimizing Equation 12 subject to this constraint yields the equilibrium conformation of the coiled-coil in terms of a relation between the elastic constants B and C, the equilibrium super-helical angle θ_0 and the interface parameter $\hat{\tau}$.

$$-\frac{2B\sin^3\theta_0\cos\theta_0}{C\cos 2\theta_0} = \sin\theta_0\cos\theta_0 - \hat{\tau}R \tag{14}$$

This elastic theory[42] agrees well with structures of leucine zipper coiled-coils (Table 1) and can be used to construct atomic models of large macromolecular complexes not easily accessible to experimental structural analysis.

GCN4	X-ray data				model	
	res./turn	rise/res.	R	2θ	$\hat{\tau}$ (rad/Å)	2θ
dimer	3.62	1.51 Å	4.9 Å	$-23.4°$	-0.039	$-22°$
trimer	3.60	1.53 Å	6.7 Å	$-26.8°$	-0.033	$-25°$
tetramer	3.59	1.52 Å	7.6 Å	$-26.0°$	-0.030	$-26°$

5.2. Modeling the open and closed states of the CusCFBA bacterial efflux complex

An application of the coiled-coil theory is to the bacterial metal efflux complex cusCFBA. This complex allows bacteria to survive high concentrations of toxic heavy metals such as copper by pumping them out of the cell[43]. Other members of this family of proteins are involved in bacterial drug resistance. The structure of this complex is shown schematically in Figure 9. CusC forms the channel, CusA is thought to function as a pump, and the peripheral subunits CusF and CusB are thought to effect the opening and closing of the channel in response to metal concentration[44]. A crystal structure of TolC[45] provides a structural model for the closed form of CusC in which the channel is blocked, and these authors postulated a distinct open conformation for this molecular complex. The particular bending and twisting which is needed to open and close an iris-like arrangement of helices is also well-described within our coiled-coil theory. This theory can also be used to devise atomic coordinate models of the states of the complex to investigate the mechanism of action of the cusCFBA complex, and in particular the role of the periplasmic subunits cusF and cusB[46,44].

Our theory of coiled-coils[42] relies on a geometric relation between the interface residues by which the individual helices associate. A simple modification allows the constraint to be modified to devise barrel-like structures, which are open in the middle. But it is considerably more complicated to devise a constraint equally compatible with *two distinct states* which must be accommodated in an iris-like structure. In the simple coiled-coil theory an interaction surface on the individual helices (such as the stripe of leucine residues in a leucine zipper) defines the coiled-coil geometry. Our hypothesis for the general problem of forming an iris is that two distinct interaction surfaces (with different values of $\hat{\tau}$) must necessarily exist, which specify distinct coiled-coil geometries corresponding to the open and closed states. In the context of the cusCFBA system, the periplasmic subunits cusB and cusF could bind to one or the other interaction surface so as to control the state of the channel subunit cusC.

5.3. *Modeling oligomerization states of Adiponectin*

A second application of the coiled-coil theory is to the anti-diabetic signalling hormone adiponectin, which stimulates insulin sensitivity. This molecule circulates in the bloodstream in three distinct oligomerization states (trimer, hexamer and 18mer) but only the largest appears to be active in signalling[47,48]. The Acrp30 gene coding for adiponectin is organized into segments which code for a globular headpiece, a collagen-like domain, and a short tail region. The structure of the globular headpiece was determined by crystallography but the detailed structure of the remaining portion of the molecule in unknown. The collagen-like domain is necessary for the assembly of adiponectin monomers into trimers through formation of a triple-helical collagen I-type coiled-coil[49,50].

We hypothesize that the organizing principle by which the coiled-coil coil trimers are assembled into the hexameric and high-molecular weight

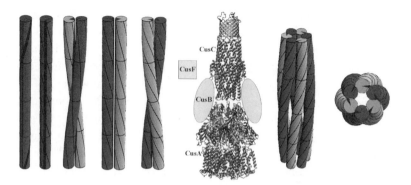

Figure 9. From left: Individual filaments of a coiled-coil in the A) unstressed state B) twisted state C-E) Different coiled-coil configurations constrained by their interaction surfaces. The equilibrium state requires a balance between bending and twisting of the filaments. F) Ribbon diagram representing the bacterial efflux complex. (Coordinates are from the closely related multidrug resistance system components TolC (1EK9.pdb G) Schematic models of an iris formed from a hexameric coiled-coil. (Note CusC is trimeric but its iris is formed by 12 helices.)

oligomerization states is a coiled-coiled-coil. Our elastic coiled-coil model is well suited to modeling not only the collagen triple helix, but also the higher-order forms by treating the collagen triple helix itself as a single elastic filament.

The first step in the mathematical approach consists in modeling adiponectin as different collagen strands twined together. This provides a family of possible models depending on the geometric parameters (radii, twist, etc.). Purely geometric requirements on the possible forms already strongly constrain the possible models. The second step is to identify within this large family of coiled structures, subfamilies for which collagen domains are complementary (see Figure 11). This is done by defining a suitable energy that can be minimized within the family of coils. The third and crucial step is to consider specific adiponectin molecules and use constraints from experimental data to refine the structure. The experimental data comes from various sources. The stoichiometry of the three states has been accurately determined by analytical ultracentrifugation and estimates of the radial and axial dimensions of the different species are available from electron micrographs. This information provides a basis for modeling the oligomers, but more subtle details such as the relative twists of the individual molecules in the oligomers or the juxtaposition of sidechains requires higher-resolution or site-specific information.

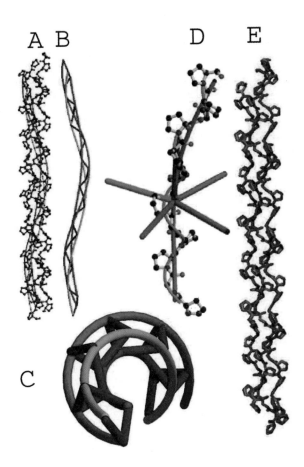

Figure 10. **Construction of coordinate models of collagen using Cosserat frames.** **A**) X-ray crystallographic model (blue) of $(Gly - Pro - Pro)_{10}$ (1K6f.pdb) with helical curves (red, green, and yellow) fitted to C_α coordinates of one chain. **B**) C_α trace of chain from **A**) with fitted helical curves. The green curve corresponds to the Gly positions, red to the X positions, and yellow to the Y positions in the Gly-X-Y pattern **C**) Top view of **B**) **D**) The cyan curve is a coiled-coil curve passing through the C_α positions, and the orange curve corresponds to the axis of the cyan curve. The Cosserat frame attached to the orange curve ($\mathbf{d}_1$, cyan; $\mathbf{d}_2$, magenta; $\mathbf{d}_3$, blue) is obtained via a rotation of the Frenet frame ($\mathbf{t}$, blue; $\mathbf{n}$, green; $\mathbf{b}$, red). The cyan vector twists about the orange curve so as to pass through each C_α position in turn. The particular Cosserat frames where the $\mathbf{d}_1$ vector passes through a C_α position can be used to construct coordinate models. **E** Complete coordinate model constructed using the Cosserat frames for each amino acid is shown in red, superimposed on the experimentally determined coordinates from 1K6F.pdb, in blue. The rmsd on all atoms is 0.3 Å.

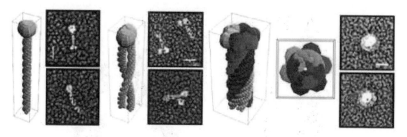

Figure 11. Models of the oligomerization states of adiponectin compared to electron micrographs Left: trimer (a single coiled-coil) Center:) Hexamer (two trimeric coiled-coils wound around each other to form a coiled-coiled-coil) Right:) Octadecamer (a distinct coiled-coiled-coil composed of 6 trimeric coiled-coils)

6. Conclusions

We have presented here various results regarding the use of simple differential geometry to model protein structures. The classical approach to model proteins is in term of discrete models based on atomic coordinates. These models have shown to be very successful for a variety of problems from identifying folds to protein functions. However, we believe that they are limited in many respects. First, computationally, they require considerable effort and ingenuity and will always be limited by computer speed and memory. In this regard, it appears clearly that there is a need for hybrid methods that combine both discrete and continuous models. We expect that the lines of research presented here will be relevant to solve these problems. Second, from a conceptual standpoint, much can be gained from understanding proteins as continuous flexible objects; the full power of continuum mechanics will allow us to gain insight into some basic phenomena such as energy transfer in the ATP-synthase or the mechanical response of fibrous proteins.

Acknowledgments

This material is based in part upon work supported by the National Science Foundation under grants No. DMS-0604704 and DMS-IGMS-0623989 to A.G. and a BIO5 Institute Grant to A.G. and A.H. We also acknowledge many discussions and ongoing collaborations with Tsushuen Tsao and Megan McEvoy. The beta barrel figures were produced by Katie White in collaboration with A.H.

References

1. A. C. Hausrath and A. Goriely. Repeat protein architectures predicted by a continuum representation of fold space. *Protein Science*, 15(4):753–760, 2006.

2. A. C. Hausrath and A. Goriely. Continuous representations of proteins: Construction of coordinate models from curvature profiles. *Journal of Structural Biology*, (In press), 2006.

3. R. Koradi, M. Billeter, and K. Wuthrich. Molmol: A program for display and analysis of macromolecular structures. *Journal of Molecular Graphics*, 14(1):51, 1996.

4. P. J. Kraulis. Molscript - a program to produce both detailed and schematic plots of protein structures. *Journal of Applied Crystallography*, 24:946–950, 1991.

5. S. Asaturyan, P. Costantini, and C. Manni. Local shape-preserving interpolation by space curves. *Ima Journal of Numerical Analysis*, 21(1):301–325, 2001.

6. J. A. Christopher, R. Swanson, and T. O. Baldwin. Algorithms for finding the axis of a helix: fast rotational and parametric least-squares methods. *Computers Chem.*, 20:339–345, 1996.

7. M. J. Keil and J. Rodriguez. Methods for generating compound spring element curves. *J. Geometry and Graphics*, 3:67–76, 1999.

8. R. Frühwirth, A. Strandlie, and W. Waltenberger. Helix fitting by an extended riemann fit. *Nuclear Instruments and Methods in Physics Research A*, 490:366–378, 2002.

9. H. Sugeta and T. Miyazawa. General methods for calculating helical parameters of polymer chains from bond lengths, bond angles, and internal rotation angles. *Biopolymers*, 5:673–679, 1967.

10. M. Bansal, S. Kumar, and R. Velavan. Helanal: A program to characterize helix geometry in proteins. *Journal of Biomolecular Structure & Dynamics*, 17:811–820, 2000.

11. Y. Nievergelt. Fitting helices to data by total least squares. *Computer Aided Geometric Design*, 14:707–718, 1997.

12. O. Devillers, F. P. Mourrain, and P. Trebuchet. Circular cylinders through four or five points in space. *Discrete Comput Geom*, 29:83–104, 2003.

13. S. C. Lovell, J. M. Word, J. S. Richardson, and D. C. Richardson. The penultimate rotamer library. *Proteins-structure Function Genetics*, 40(3):389–408, 2000.

14. Alfred Gray. *Modern differential geometry of curves and surfaces with Mathematica*. CRC Press, Boca Raton, 2nd edition, 1998.

15. K. W. Plaxco, K. T. Simons, and D. Baker. Contact order, transition state placement and the refolding rates of single domain proteins. *Journal of Molecular Biology*, 277(4):985–994, 1998.

16. O. Gonzalez and J. H. Maddocks. Global curvature, thickness, and the ideal shapes of knots. *Proc. National Acad. Sciences United States Am.*, 96(9):4769–4773, 1999.

17. T. Lazaridis and M. Karplus. Effective energy functions for protein structure

prediction. *Current Opinion in Structural Biology*, 10(2):139–145, 2000.

18. F. Melo, R. Sanchez, and A. Sali. Statistical potentials for fold assessment. *Protein Science*, 11(2):430–448, 2002.

19. P. Rogen and B. Fain. Automatic classification of protein structure by using Gauss integrals. *Proc. Natl. Acad. Sci. USA*, 100:119–124, 2003.

20. P. Rogen and H. Bohr. A new family of global protein shape descriptors. *Math. Biosc.*, 182:167–181, 2003.

21. L. Holm and C. Sander. Protein-structure comparison by alignment of distance matrices. *Journal of Molecular Biology*, 233(1):123–138, 1993.

22. A. Oberai, Y. Ihm, S. Kim, and J. U. Bowie. A limited universe of membrane protein families and folds. *Protein Science*, 15(7):1723–1734, 2006.

23. A. G. Murzin, A. M. Lesk, and C. Chothia. Principles determining the structure of beta-sheet barrels in proteins .2. the observed structures. *J. Mol. Biol.*, 236(5):1382–1400, 1994.

24. J. U. Bowie. Helix-bundle membrane protein fold templates. *Protein Science*, 8(12):2711–2719, 1999.

25. R. B. Hill, D. P. Raleigh, A. Lombardi, and N. F. Degrado. De novo design of helical bundles as models for understanding protein folding and function. *Accounts of Chemical Research*, 33(11):745–754, 2000.

26. N. Pokala and T. M. Handel. Review: Protein design - where we were, where we are, where we're going. *Journal of Structural Biology*, 134(2-3):269–281, 2001.

27. W. A. Lim and R. T. Sauer. Alternative packing arrangements in the hydrophobic core of lambda-repressor. *Nature*, 339(6219):31–36, 1989.

28. N. C. Gassner, W. A. Baase, and B. W. Matthews. A test of the "jigsaw puzzle" model for protein folding by multiple methionine substitutions within the core of t4 lysozyme. *Proceedings of the National Academy of Sciences of the United States of America*, 93(22):12155–12158, 1996.

29. S. Dalal, S. Balasubramanian, and L. Regan. Protein alchemy: Changing beta-sheet into alpha-helix. *Nature Struct. Biol.*, 4(7):548–552, 1997.

30. J. W. Ponder and F. M. Richards. Tertiary templates for proteins - use of packing criteria in the enumeration of allowed sequences for different structural classes. *Journal of Molecular Biology*, 193(4):775–791, 1987.

31. S. F. Betz, D. P. Raleigh, and W. F. Degrado. De-novo protein design - from molten globules to native-like states. *Current Opinion in Structural Biology*, 3(4):601–610, 1993.

32. D. B. Gordon, S. A. Marshall, and S. L. Mayo. Energy functions for protein design. *Current Opinion in Structural Biology*, 9(4):509–513, 1999.

33. J. Desmet, M. Demaeyer, B. Hazes, and I. Lasters. The dead-end elimination theorem and its use in protein side-chain positioning. *Nature*, 356(6369):539–542, 1992.

34. E. P. Baldwin, O. Hajiseyedjavadi, W. A. Baase, and B. W. Matthews. The role of backbone flexibility in the accommodation of variants that repack the core of t4-lysozyme. *Science*, 262(5140):1715–1718, 1993.

35. B. H. M. Mooers, D. Datta, W. A. Baase, E. S. Zollars, S. L. Mayo, and B. W. Matthews. Repacking the core of t4 lysozyme by automated design.

Journal of Molecular Biology, 332(3):741–756, 2003.

36. J. R. Desjarlais and T. M. Handel. Side-chain and backbone flexibility in protein core design. *Journal of Molecular Biology*, 290(1):305–318, 1999.

37. B. Kuhlman, G. Dantas, G. C. Ireton, G. Varani, B. L. Stoddard, and D. Baker. Design of a novel globular protein fold with atomic-level accuracy. *Science*, 302(5649):1364–1368, 2003.

38. G. A. Lazar and T. M. Handel. Hydrophobic core packing and protein design. *Current Opinion in Chemical Biology*, 2(6):675–679, 1998.

39. L. L. Looger and H. W. Hellinga. Generalized dead-end elimination algorithms make large-scale protein side-chain structure prediction tractable: Implications for protein design and structural genomics. *Journal of Molecular Biology*, 307(1):429–445, 2001.

40. T. J. Richmond. Solvent accessible surface-area and excluded volume in proteins - analytical equations for overlapping spheres and implications for the hydrophobic effect. *Journal of Molecular Biology*, 178(1):63–89, 1984.

41. C. E. Kundrot, J. W. Ponder, and F. M. Richards. Algorithms for calculating excluded volume and its derivatives as a function of molecular-conformation and their use in energy minimization. *Journal of Computational Chemistry*, 12(3):402–409, 1991.

42. S. Neukirch, A. Goriely, and A. C. Hausrath. A continuum elastic theory of coiled-coils with applications to the mechanical properties of fibrous proteins and energy transduction by the atp synthase. *Biophysical Journal*, (Submitted), 2006.

43. G. Grass and C. Rensing. Genes involved in copper homeostasis in Escherichia coli. *J. Bacteriology*, 183(6):2145–2147, 2001.

44. I. R. Loftin, S. Franke, S. A. Roberts, A. Weichsel, A. Heroux, W. R. Montfort, C. Rensing, and M. M. McEvoy. A novel copper-binding fold for the periplasmic copper resistance protein cusf. *Biochem.*, 44(31):10533–10540, 2005.

45. V. Koronakis, A. Sharff, E. Koronakis, B. Luisi, and C. Hughes. Crystal structure of the bacterial membrane protein ToOC central to multidrug efflux and protein export. *Nature*, 405(6789):914–919, 2000.

46. J. T. Kittleson, I. R. Loftin, A. C. Hausrath, K. P. Engelhardt, C. Rensing, and M. M. McEvoy. Periplasmic metal-resistance protein CusF exhibits high affinity and specificity for both Cu-I and ag-i. *Biochem.*, 45(37):11096–11102, 2006.

47. T. S. Tsao, H. E. Murrey, C. Hug, D. H. Lee, and H. F. Lodish. Oligomerization state-dependent activation of NF-kappa B signaling pathway by adipocyte complement-related protein of 30 kDa (acrp30). *J. Biological Chem.*, 277(33):29359–29362, 2002.

48. T. S. Tsao, E. Tomas, H. E. Murrey, C. Hug, D. H. Lee, N. B. Ruderman, J. E. Heuser, and H. F. Lodish. Role of disulfide bonds in Acrp30/adiponectin structure and signaling specificity - Different oligomers activate different signal transduction pathways. *J. Biological Chem.*, 278(50):50810–50817, 2003.

49. K. Kobayashi and T. Inoguchi. Adipokines: Therapeutic targets for metabolic syndrome. *Current Drug Targets*, 6(4):525–529, 2005.

50. G. W. Wong, J. Wang, C. Hug, T. S. Tsao, and H. F. Lodish. A family of Acrp30/adiponectin structural and functional paralogs. *Proc. National Acad. Sciences United States Am.*, 101(28):10302–10307, 2004.

A NEW NEUROSURGICAL TOOL INCORPORATING DIFFERENTIAL GEOMETRY AND CELLULAR AUTOMATA TECHNIQUES*

ALEXANDER R. OSHMYANSKY

Centre for Mathematical Biology
University of Oxford
24-29 St. Giles Oxford, OX1 3LZ, UK
E-mail: alexander.oshmyansky@stx.ox.ac.uk

PHILIP K. MAINI

Centre for Mathematical Biology
Oxford Centre for Integrative Systems Biology
University of Oxford
24-29 St. Giles Oxford, OX1 3LZ, UK
E-mail: maini@maths.ox.ac.uk

Using optical coherence imaging, it is possible to visualize seizure progression intraoperatively. However, it is difficult to pinpoint an exact epileptic focus. This is crucial in attempts to minimize the amount of resection necessary during surgical therapeutic interventions for epilepsy and is typically done approximately from visual inspection of optical coherence imaging stills. In this paper, we create an algorithm with the potential to pinpoint the source of a seizure from an optical coherence imaging still. To accomplish this, a grid is overlaid on optical coherence imaging stills. This then serves as a grid for a two-dimensional cellular automation. Each cell is associated with a Riemannian curvature tensor representing the curvature of the brain's surface in all directions for a cell. Cells which overlay portions of the image which show neurons that are firing are considered "depolarized". The cellular automation is then run with the following rules:

---

*This work is supported by the Marshall Aid Commemoration Commission.

(1) At each step all squares in contact with a depolarized square become depolarized if $|\nabla_u v| * t \leq c$. ∇_u is the covariant derivative in the direction of vector u. The vector u is a unit vector in the direction the depolarization from the original depolarized square. v is a tangent vector to the brain manifold at a touching square. t is the total time in terms of time steps that has been spent at a square. c is a constant given by the speed of the propagation of the wavefront.

(2) If a square depolarizes its neighboring squares, it becomes repolarized.

(3) A repolarized square cannot be depolarized again for t_r time steps, given by a the neuronal refractory period.

While the simulation is running, the depolarizing "wavefront" of cells converges on to a few specific cells which we hypothesis correspond to the epileptic focus. Simulations on several parts of the brain are run, comparisons are made to actual optical coherence imaging visualizations, and a tool is proposed for use intraoperatively during therapeutic epilepsy surgery.

1. Seizure Modelling

There has been significant effort put into modelling seizure progression. A seizure is generally defined as a transient disturbance of cerebral functioning due to an abnormal paroxysmal neuronal discharge in the brain[1]. The majority of effort has been focused on modelling results obtained from electroencephalograms (EEGs). Models have been created utilizing partial differential equation (PDE) techniques[2], stochastic differential equations[3], nonlinear dynamics models[4], and a variety of other methods. As a potential application of this work, an ideal model would prove an effective tool to supplement qualitative clinician assessment of EEG data for diagnostic purposes. Although this has not yet been achieved, significant progress has been made.

In this paper we focus a modelling data on seizure progression obtained from optical coherence imaging. The end purpose of this research is for therapeutic rather than diagnostic use. During the process of optical coherence imaging, changes in the optical spectroscopic properties of neocortical tissue are directly visualized[5]. The changes in spectroscopic properties are directly correlated to changes in cerebral hemodynamic properties such as blood oxygenation and blood volume. As cerebral hemodynamics can serve as a proxy for neuronal activity, as it does in functional magnetic resonance imaging, optical coherence imaging effectively results in visualization of neuronal activity on the brain surface.

A recent application of this technique has been to visualize seizures occurring intraoperatively[5]. Epilepsy is a condition of recurrent seizures which often originate from a focus of pathologic neurons in the cerebral cortex[1]. Epilepsy which is refractory to medical therapy is often treated by surgical resection of epileptic foci. However, it is often difficult to locate an epileptic focus for resection. This often results in much more tissue than is necessary being removed during therapeutic surgery (a common epilepsy surgery is temporal lobectomy in which the entire temporal lobe of a patient is removed when it is believed an epileptic focus is present on the lobe) or too little being removed which eliminates the therapeutic benefit of surgical intervention. Optical coherence imaging allows direct qualitative inspection of seizure initiation and progression from an epileptic focus, thus allowing better identification of the boundaries of foci.

The aim of this paper is to model seizure progression on the surface of the brain with a simple wave model on top of a curved manifold. It is hypothesized that this model can be directly applied to intra-operative optical coherence imaging data to effectively predict in a more quantitative fashion the origin of a seizure and the boundaries of an epileptic focus. This would then allow more effective resection within epilepsy surgery.

2. Combining Geometry with Cellular Automata

In creating a model that would be most effective intra-operatively, one had to be created that would allow for easy selection of initial conditions. Essentially, we want to eventually allow surgeons to take an image obtained through imaging of intrinsic optical signals (IIOS, another term for optical coherence imaging) and select easily yet still accurately which groups of neurons are activated on the image. Ideally, an algorithm based on our model would then determine the boundaries of an epileptic focus.

To select groups of activated neurons on an image, we propose overlaying a grid on an image obtained by IIOS. Cells which are partially filled are then labelled as "depolarized." It should then be easy to run a cellular automata[6] simulation on the grid after creating rules for how seizure progression should behave. In fact, there is already an extensive literature on simulating neuronal activity using cellular automata[7] though usually not in so direct a fashion.

However, the surface of the brain is of course not flat and therefore a flat grid would obviously be ineffective for modelling the cerebral or cerebellar surface without some modification. The modification we propose is to designate each cell in the simulation with a "curvature" given by a Riemannian curvature tensor.

The concept of curvature on a Riemannian manifold is abstract but remarkably versatile. As a brief review, a manifold is a space such that an open neighborhood around each point in the space is locally Euclidean (further information on Riemannian geometry can be found in Petersen[8]). A Riemannian manifold more specifically is a manifold with a defined inner product at each point. The Riemannian curvature tensor can be used to define the curvature of a Riemannian manifold. If u and v are coordinate vector fields and w is a manifold, the Riemannian curvature tensor can be defined by:

$$R(u,v)w = \nabla_u \nabla_v w - \nabla_v \nabla_u w \qquad (1)$$

where ∇ is the covariant derivative, a generalization of the derivative which gives a derivative along the tangent vectors of a manifold. Effectively, the tensor thus gives all information about the curvature of a point in a single quantity which does not rely on local coordinates.

Being able to refer to curvature with a single quantity is quite useful for our purposes. Background independence (lack of dependence on a local coordinate system) also proves to have utility. By defining a curvature at each cell in a cellular automata simulation, one can approximate the curvature of a given manifold. Rules of a cellular automata simulation can then be easily modified to determine geodesic motion on a curved "cellular manifold" rather than on a flat cellular automata grid.

3. Using Cellular Manifolds to Model Seizures

From observation of images obtained by IIOS and basic knowledge of neural physiology[9], we will create rules for how a depolarized cell should interact with other cells on a flat grid. In order to mimic neuronal dynamics grossly, all cells surrounding a depolarized cell should become depolarized at the next time step after a wave of depolarization has had time to fully propagate through a cell. After a given time period, the depolarized cell should begin to "repolarize" and enter a refractory period during which it cannot be depolarized.

It reality, some portions of a propagating depolarization wavefront on the cortex would of course move faster than others given individual axon and dendrite lengths and connectivity of individual neurons. However, at the level of coarsity that is being dealt with in any given grid cell, thousands of neurons are being represented simultaneously. We feel that it is safe to ignore this effect as it should prove minimal at the coarsity described.

The following rules are thus proposed for the cellular automata that shall be run:

(1) At each step all squares in contact with a depolarized square become depolarized if $|\nabla_u v| * t \leq c$. ∇_u is the covariant derivative in the direction of vector u. The vector u is a unit vector in the direction the depolarization from the original depolarized square. v is a tangent vector to the brain manifold at a touching square. t is the total time in terms of time steps that has been spent at a square. c is a constant given by the speed of the propagation of the wavefront.

(2) If a square depolarizes its neighboring squares, it becomes repolarized.

(3) A repolarized square cannot be depolarized again for t_r time steps, given by a the neuronal refractory period.

Note that all vectors v are given by the Riemann curvature tensor at a given point. Using the definition of the Riemann curvature tensor in local coordinates x^μ (written in Einstein summation notation):

$$R^\rho{}_{\sigma\mu\nu} = dx^\rho (R(\partial_\mu, \partial_\nu)\partial_\sigma) \tag{2}$$

the tangent vectors v are merely the coordinate vector fields $\partial_\mu = \partial/\partial x^\mu$ while vectors u correspond with the coordinates x^ρ that we choose based on the direction of depolarization wavefront propagation.

4. Example and Qualitative Comparison to Actual Data

Data for direct comparison to simulations is not yet available to the authors at this time. In lieu of this, we show a simplified example of the proposed technique and show images obtained by IIOS during a seizure near the same area from available literature[10] for comparison. The similarities in depolarization progression seen on visual inspection are clearly observed.

To illustrate the proposed technique, we overlay a grid on a plain intra-operative photograph and run a cellular automata with the proposed rules starting from an arbitrary cell deemed to be the epileptic focus. Image 1 is taken at the very beginning of the Sylvian fissure which is underneath the middle cerebral artery which extends from the bottom left to the center of the image. For the sake of simplicity in illustration, curvature will be calculated only at the Sylvian fissure and only unit curvature values will be used. (That is, $\nabla_u v$ will always equal either 1 or -1).

For the simulation, cells filled in yellow on image 2 have $\nabla_u v = 1$ and cells filled in blue have $\nabla_u v = -1$. On image 3, the actual simulation is shown. Green cells are depolarized in at each time-step.

An actual sequence from optical coherence imaging of a seizure is shown in image 4. The similarity to the pattern of spread of the depolarization wavefront can be qualitatively observed. The spread pattern is very similar to that predicted from the simplified simulation. Plans for more direct quantitative analysis of simulations are discussed in the next section.

5. Future Work

Given the methodology proposed in the paper, a practical neurosurgical tool will ultimately be developed. Ideally, this tool will calculate brain curvature from pre-operative scans (MRI or CT) for each individual patient. Surgeons will be able to select areas seen to be depolarized on images taken by IIOS and the tool will calculate the borders of epileptic foci.

Present work is focusing on creating a generic phantom which may be used for the majority of patients without necessitating access to imaging. Once a phantom is complete, further work will compare directly modelled seizures with those actually visualized. This will directly assess the quality of the proposed algorithm.

There is also a significant amount of interesting mathematics that can likely be extracted from the concept of a cellular manifold. Work without direct intent of an application will be done to further expand on this idea.

Acknowledgments

The authors would like to thank the Marshall Aid Commemoration Commission for their support of this research. We would also like to thank the students and staff of the Oxford Centre for Mathematical Biology for their input and insight. Additionally, we would like to the thank Dr. Michael Haglund of the Duke University Neurosurgical Department for demonstrating the utilization of IIOS during epilepsy surgery and for his support.

References

1. L. M. Tierney, Jr., S. J. McPhee, M. A. Papdakis, Current Medical Diagnosis and Treatment, Lange (2004).
2. M. Bertalmio, L. T. Cheng, S. Osher and G. Sapiro, *Journal of Coputational Physics* **174(2)**, 759 (2001).
3. T. D. Frank and P. J. Beek, *Phys. Rev. E* **264**, 021917 (2001).
4. C. J. Stam, *Clin. Neurophysiol.* **116(10)**, 2266 (2005).
5. M. M. Haglund and D. W. Hochman, *Epilepsia* **45**, 43 (2004).
6. S. Wolfram, Cellular Automata and Complexity. Westview Press (2002).

7. L. O. Chua, L. Yang, *IEEE Transactions on Circuits and Systems* **37(10)** 1257 (1988).
8. P. Petersen, Riemannian Geometry, Springer (2006).
9. D. Purves, G. J. Augustine, D. Fitzpatrick, and W. C. Hall, Neuroscience, Sinauer Associates (2004).
10. N. Pouratian, A. F. Cannestra, N. A. Martin, A. W. Toga, *Neurosurg Focus*, **13(4)** (2002). *Available at* http://www.medscape.com/viewarticle/443946_2.

Figures

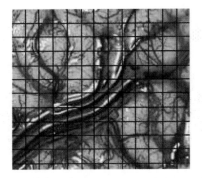

Figure 1. Intraoperative picture of the beginning of the Sylvian Fissure. Image by Robert F. Spetzler, M.D., Barrow Neurological Institute, Phoenix, AZ, USA.

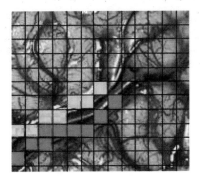

Figure 2. Highlighted Squares represent curvature of brain. At yellow squares $\nabla_u v = 1$ and at blue squares $\nabla_u v = -1$.

Time Step

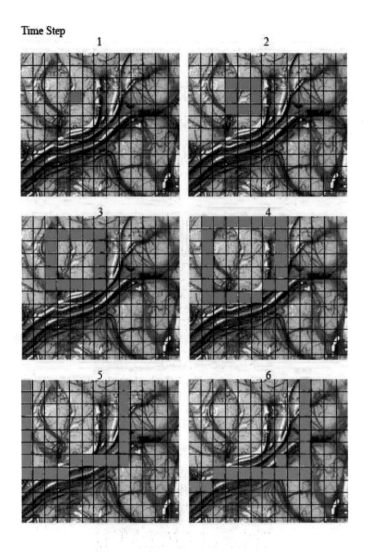

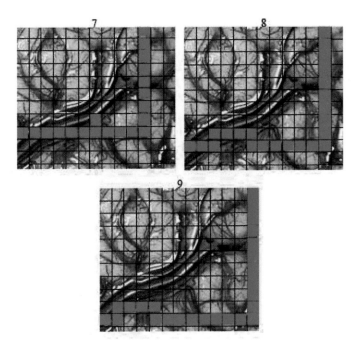

Figure 3. Cellular automata simulation of seizure progression. Cells filled green are depolarized.

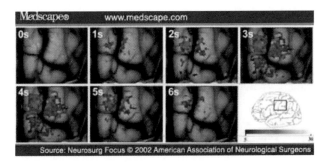

Figure 4. Optical imaging stills from a seizure triggered near the same area as the simulation was run in. Not the similar spread pattern to the simulation up to 4 sec. at which point the seizure begins to resolve. Image from[10].

MODELING THE GROWTH AND INVASION OF GLIOMAS, FROM SIMPLE TO COMPLEX: THE GOLDIE LOCKS PARADIGM

RUSSELL ROCKNE, ELLSWORTH C. ALVORD JR., P. J. REED, KRISTIN R. SWANSON

Department of Pathology, University of Washington
1959 N.E. Pacific St., Seattle WA 98195, USA

As with all mathematical modeling, the scope of the question to be explored determines the scope of the most appropriate model. The case is no different for the modeling of primary brain tumors (gliomas), ranging from too simple, not accounting for the major feature of gliomas (extensive invasion), to too complicated, with too many variables and no easy way to translate from culture media *in vitro* to brain tissue *in vivo*. We settle on a "just right" approach which utilizes currently available magnetic resonance imaging (MRI) to estimate two defining characteristics, net rates of proliferation (ρ) and diffusion (D). Most importantly, these parameters are predictive of clinical behavior, and can be tailored to individual patients *in vivo* and in real time. These two rates combine to generate a linear radial growth pattern of the MRI visible portion of each glioma. Further, we introduce a novel method for the calculation of glioma invasion through grey and white matter.

1. Introduction

The history of cancer modeling is long and complicated. The modeling of gliomas as a uniquely different variety of cancer is much shorter, with models ranging from simple to highly complex. Gliomas are primary brain tumors which are known for their ability to aggressively and diffusely invade surrounding tissue, while staying within the confines of the central nervous system. Gliomas account for nearly 50% of all primary brain tumors, and are almost always fatal, with survival times ranging from 6 months to many years from the time of diagnosis, roughly correlated with histological characteristics which can vary widely, leading to a I-IV grade scale from the World Health Organization (WHO)[1,2].

Not unlike the 3 bears of the Goldie Locks fairytale, the spectrum of models for gliomas range from overly simple to unwieldy complex: neither extreme leads to broad applicability or accurately reflect the clinically observable reality. Simple models of exponential growth neglect the invasive nature and spatial distribution of the disease, and systems of several, even dozens of differential equations produce huge parameter spaces that are all but impossible to resolve within the context of an individual patient. When limiting our focus to the problem of interpreting information available to doctors caring for these patients via clinical imaging, we illustrate a model that may be "just right:" simple enough for parameter values to be calculated from as few as two magnetic resonance images (MRI) routinely obtained during the course of glioma diagnosis and treatment, yet complicated enough to model the differential, heterogeneous migration of glioma cells throughout the brain.

We present the extremes of glioma modeling from the earliest and simplest studies by Collins et al.[3] using information obtained from chest x-rays to measure tumor volume-doubling time, to the more recent work by Zhang et al.[4] with spheroid-based models of molecular interaction complexity. In between is our reaction-diffusion modeling approach originally developed by Murray[5] and colleagues to measure rates of tumor proliferation (ρ) and diffusion (D), initially by Tracqui et al.[6] at the level of resolution provided by computerized tomography (CT), including the effects of surgical resection of various extents. The model has been expanded by Swanson et al. to the level of resolution provided by MRI, including the effects of radiation therapy[7,8]. Further, we present a new addition to the model: a novel technique for calculating glioma migration velocities through the complex architecture of the grey and white matter of the brain.

2. Telescoping Levels of Magnification: From Simple to Complex

Glioblastoma multiforme (GBM) or WHO grade IV tumors are the most invasive of the gliomas and also have the shortest life expectancy: 6—12 months on average. Due in part to the location in the brain at diagnosis, invasive procedures and assessment of response to treatment rely heavily on medical imaging such as magnetic resonance (MRI), computerized tomography (CT) and positron emission tomography (PET). However, these imaging modalities leave much of the invading, leading-edge of the tumor invisible and therefore often omitted in targeted therapy, and provide a distinct opportunity for a mathematical model to provide insight on disease progression. Nevertheless, these imaging modalities along with general clinical assessment and tissue obtained from biopsy are the predominant data

130

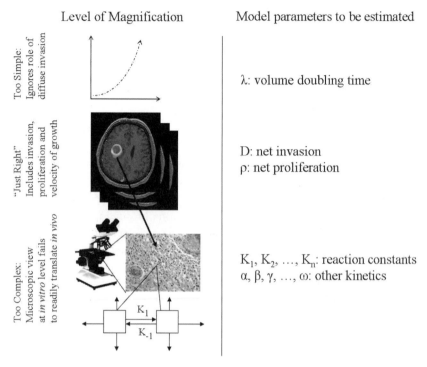

Level of Magnification

Model parameters to be estimated

Too Simple: Ignores role of diffuse invasion

"Just Right" Includes invasion, proliferation and velocity of growth

Too Complex: Microscopic view at *in vitro* level fails to readily translate *in vivo*

λ: volume doubling time

D: net invasion
ρ: net proliferation

$K_1, K_2, ..., K_n$: reaction constants
$\alpha, \beta, \gamma, ..., \omega$: other kinetics

Figure 1. Telescoping levels of magnification: the modeling of observable phenomena on a range of scales and the parameters involved.

available from which doctors are able to make decisions on patient treatment and response to therapy. Therefore, any model which is to simulate glioma growth and invasion at the clinical level must account for and reconcile with tumor presentation and progression on medical imaging.

2.1. Porridge is too cold: The simple approach

It is exactly this macroscopic, clinically-based modeling approach which led Collins *et al.*[3] to postulate a constant volume-doubling growth rate based on serial x-ray visualizations of metastatic neoplasms of the lungs. It is a logical step then — from a mathematical point of view — to consider the tumor cell population (c) to have a growth rate proportional to the cell population:

$$\frac{\partial c}{\partial t} = \lambda c, \tag{1}$$

where λ represents the per-time growth rate and its well known cellular or volume-doubling time: $t_{dbl} = \lambda/\ln(2)$. Although the preliminary work by Collins et al.[3] was not specifically aimed at modeling glioma growth, a tremendous amount of research has considered the pros and cons of this formulation as applied to practically all types of cancers, including gliomas. The most extensive study was by Blackenberg et al.[9], who correlated volume-doubling times with histologic grade and survival. This simple line of thought continues to pervade current quantification of glioma growth. As recently as 2000, Haney et al.[10] reported glioma growth rates in terms of volume-doubling and halving times using full 3-dimensional reconstructions of gross tumor volumes (GTV) obtained from serial MRIs. Variations on this theme, including logistic and gompertzian growth models, break down at once for invasive tumors when considering the simple question: what about the invading cells that are not visible on clinical imaging? How can we measure the growth rate of the total cell population N when only a percentage of the cells is visible? Even if we could assume a constant proportion between visible bulk and invisible invading cells, where we could postulate some fraction of λ as applicable to the visible bulk, we would have no assurances that this proportion would be constant within any one patient or from one patient to another.

2.2. *Porridge is too hot: The complicated approach*

On the other extreme are models similar to those presented by Zhang et al.[4] and Stein et al.[11] seeking insight into *in vivo* dynamics by modeling processes on the molecular level with tumor spheroids grown *in vitro*. Briefly, the model that Zhang et al.[4] investigate relies on more than 20 equations with greater than 65 parameters to explore the role of EGFR in a cancer cell's decision whether to migrate or proliferate. For answering questions at the molecular level, (e.g. tumor spheroids on gels in dishes) experimental models like these may be appropriate and lead to insight regarding molecular mechanisms. One can argue that such unwieldy models may be able to generate nearly any desired result, especially when bounds on key parameters are not known biologically. Further, for answering questions at the clinical level, it is difficult to imagine a means of parameterizing these large complex models for use in individual patients. On that note, Stamatakos et al.[12] have proposed a more clinically focused model for treatment response which includes the cell cycle as well as molecular abnormalities. This model summarily ignores the extensive invasion of gliomas and can only be used to hypothetically simulate virtual tumor behavior, suggesting that the predictive power in individual patients is minimal, again, because of the large number of parameters that need to be estimated for each patient.

2.3. Porridge is just right: The middle approach

The most useful models are those that can be predictive, specifically, those that keep in mind the type of question that is intended to be answered by the model. In other words, the most effective model is one that can be tested while suggesting new hypotheses. In between the simplicity of exponential growth and the complexity of inter-cellular dynamics is the only patient-specific model we know of that is able to provide clinically valuable metrics for invasion and treatment response: a reaction-diffusion equation which consists of two dominant parameters, both of which can be calculated for an individual patient from only two pre-treatment MRIs.

Motivated by the observation that gliomas are more locally invasive than other tumors, Murray began investigating a PDE model famously formalized by Fisher[13] which characterizes glioma growth by net rates of diffusion (D) and proliferation (ρ) in word form as:

$$\begin{array}{ll} \text{Rate of change} & = \quad \text{net diffusion} + \text{net proliferation} \\ \text{in glioma cell density} & \qquad \text{of glioma cells} \end{array}$$

By assuming Fickian diffusion of glioma cells Murray[5] formalized the equation as follows:

$$\frac{\partial c}{\partial t} = \nabla \cdot (D \nabla c) + \rho c \left(1 - \frac{c}{k} \right). \tag{2}$$

This is of course the classic Fisher-Kolmogoroff equation, the analytical properties of which are well known, including Fisher's approximation[13]: more on that later. Here $c = c(x, t)$ denotes the tumor concentration at time t and spatial location x, D is the diffusion coefficient representing the net motility of tumor cells in units cm^2/day, and ρ is the net proliferation rate in units day^{-1}. A zero-flux boundary condition is imposed to prevent cells from leaving the finite, bounded brain, with an initial condition of concentration $c_0 = c(x, 0)$. It is the distinct ability of this model to capture the glioma invading diffusely into the normal appearing surrounding tissue well beyond what is visible on clinical imaging, and is at the same time the distinct failure of other models which ignore the distribution of glioma cells throughout the brain. Other forms of the diffusion term, e.g. with compact support, have been considered but considering the extensive invasion of gliomas throughout the brain evidenced in experimental and clinical studies the infinite propagation implicit in a Fickian diffusion term is a natural choice.

The first estimates for model parameters D and ρ came from Tracqui et al.[6], who calculated the diffusion and proliferation rates of glioma cells of a recurrent malignant astrocytoma (WHO grade III) visualized by serial CTs during the patient's last year yielding $D = 0.0013\text{cm}^2/\text{day}$ and $\rho = 0.012/\text{day}$. Woodward et al.[14] estimated these values to be a representative average for all high grade (III, IV) gliomas, and about 10 times that of low grade (I, II) gliomas.

The extent of glioma invasion visible on clinical imaging modalities such as MRI, positron emission topography (PET) and CT is limited by thresholds of detection, rendering much of the invasive leading edge of the disease invisible to the clinical eye. Currently, MRI is used as the primary tool in assessing extent of invasion and response to therapy in gliomas and is frequently performed in the following three modalities: T1 or T2-weighted and fluid-attenuated inversion-recovery (FLAIR).

T1-weighted MRI is "enhanced" with the contrast agent gadolinium, with abnormalities visualized as areas of increased contrast by intravenously injected gadolinium (T1Gd) passing through the blood-brain barrier (BBB). In normal brain, there is no (minimal) penetration of gadolinium through the BBB, but in malignant tumors, where the BBB is broken due to the immature and malformed blood vessels in the tumor mass, gadolinium leaks into the surrounding tissue. Regions of contrast-enhancement on T1Gd MRI reveal the location of vascular structures of the tumor, which is primarily considered to be its primary mass, but not the peripherally invading isolated tumor cells, which are at such low concentrations that the BBB, though porous enough to allow water to pass into the tissue as edema, is not sufficiently porous to allow penetration of gadolinium.

T2-weighted (T2) and FLAIR MRI modalities reveal increased water concentration or edema (extra-vascular interstitial fluid), which has been shown to also contain isolated tumor cells at low but significant concentrations[15,16]. These regions do not enhance on T1Gd, as edema and vascularization are not synonymous in tumor biology. Generally, the T2 abnormality completely encompasses the T1Gd abnormality.

Using parameter ranges obtained from Tracqui et al.[6], along with the evidence that suggests glioma cells migrate with greater velocity through white matter of the brain than through grey[17], Swanson reformulated the model to allow the diffusion coefficient to depend on the tissue environment[18]:

$$\frac{\partial c}{\partial t} = \nabla \cdot (D(x)\nabla c) + \rho c \left(1 - \frac{c}{k}\right), \qquad (3)$$

where $D(x) = D_g$ for x in grey matter and $D(x) = D_w$ for x in white matter and k is taken to be approximately 10^8 cells/cm^3, based on an average cell diameter of 10μm. A virtual brain atlas obtained through work by the BrainWeb[19] group allowed for precise segmentation of the 3-dimensional environment into regions of cerebral spinal fluid (CSF) grey and white matter. Initial studies assumed a ratio of $D_w/D_g = 5$, but as we will show, a wide range can be expected and in fact measured within a given patient. Such differences produce quite asymmetric patterns of growth *in vivo* and *in silico*, cf. Figure 2b.

It is well known that Fisher's equation permits traveling wave solutions in one and two (planar) spatial dimensions[5,13]. However, for 3 spatial dimensions (e.g. spherical symmetry) a traveling wave does not exist but solutions approach a wave-like solution (asymptotically) such that a given cell density contour will grow nearly linearly in time (cf. Figure 2a). Swanson *et al.*[20,21] has illustrated that even in the complex 3 dimensional architecture of the human brain, a wave-like solution exists with a speed approximately that of Fisher's approximation: $v = 2\sqrt{D\rho}$[13], figure 2.

Confirmation of the hypothesis that the velocity of radial expansion is constant during much of the clinically relevant time range first came from Swanson, who observed a patient with a high grade glioma (HGG) and serial MRIs without intervening treatment[20] and then from a series of 27 patients with low grade gliomas (LGG) followed with serial MRIs for up to 15 years[22]. The HGG radial velocity was 12mm/year and the average LGG velocity was 2mm/year, a ratio of approximately 6:1, not far from the 10:1 radio predicted by Woodward *et al.*[14].

2.3.1. *Patient-specific glioma growth dynamics: invasion (D) and proliferation (ρ)*

With Fisher's approximation, we need only one additional equation involving D and ρ in order to parameterize an individual glioma. Clearly a minimum of two MRI observations is required to compute a velocity, and for each observation we can expect at least one pair of T1Gd and T2 MRI measurements (which are almost always performed in the same imaging ses-

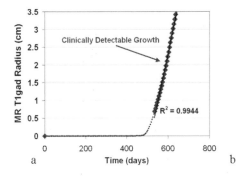
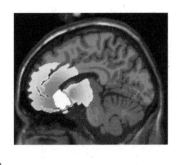

a   Time (days)   b

Figure 2. Virtual glioblastoma. a) Radius of the simulated T1Gd MRI (bright white region in b) over time which highlights the applicability of our assumption that the 3D approximation to a constant radial velocity is reasonable during the clinically detectable part of the tumor's growth. b) A virtual glioblastoma for which the white region approximates the T1Gd abnormality and the bright grey scale area illustrates glioma invasion invisible on MRI. The outer edge is not a zero glioma cell density contour thus there exists invasive glioma cells peripheral to even the outermost edge of the abnormality.

sion). Because a tumor concentration gradient exists between T1Gd and T2 abnormalities, T1Gd and T2 MRI thresholds of detection are taken to be fixed fractions of the total carrying capacity and are estimated in order to create reasonable cell gradients relative to the number of cells invisible on both T1Gd and T2. The result is an approximation for the ratio D/ρ, which yields the necessary second equation to obtain D and ρ, supposing that the diffusion is held constant $D(x) = D$ and roughly represents the average rate of diffusion through grey and white matter, consistent with the application of homogenization theory to the heterogeneous brain tissue[23].

At this point, we have succeeded in providing a model which is both simple enough to be parameterized in real time using only data obtained from routine clinical imaging, yet complicated enough to encapsulate differential, heterogeneous migration. Additionally, the model has been extended to include the effects of surgical, chemo- and radiation therapies, although those model extensions will not be presented here[7,24−29].

2.3.2. Differential migration: a new way to eat porridge of two distinct densities

Further refining our notion of velocity and differential migration, let us think about individual cells or undefined "lumps" of glioma tissue, and how they move with time. Since we know that the rate of migration is greater in white matter than the rate in grey matter ($D_w > D_g$), we would expect to observe a cell traveling faster in white matter than a comparable cell in grey matter (as is observed *in vivo* and *in vitro*). The observable velocity of a moving cell over a specific period of time V will then be dependent upon the amount of white matter and grey matter the cell is traveling through. We define V_w to be the velocity in white matter and V_g as the velocity in grey matter. It is our hypothesis that the net observable velocity V is a linear combination of V_w and V_g such that $V = \kappa V w + \gamma V g$ where κ and γ correspond to the relative amounts of white matter and grey matter, respectively, encountered along the path traveled by a tumor cell from the first to second time point.

Now consider the necessary abstraction of this concept to clinically observable tumor progression on MRI. Since MRI pixel intensity does not correlate with cell density, we cannot assume that the progression of pixels on the perimeter of the enhancing region from one time point to the next represents the migration of any specific subpopulation of cells. We therefore consider an attempt to measure the rate at which the clinically observable tumor edge advances through time.

Statistical parametric mapping (SPM) is an image analysis package in MATLAB developed by members and collaborators of the Wellcome Department of Imaging Sciences at the University College London, and is capable of importing MRI images and decomposing them into probability maps of grey matter and white matter and able to give estimates as to which area of a brain is most likely grey matter or white matter. Using this information, one can define the relative amounts of grey matter and white matter along a specific path, defining the values of κ and γ. We define the MRI-observable tumor velocity V as

$$V = \frac{\text{pathlength}}{t_2 - t_1}$$

where $t_2 > t_1$. Considering the heterogeneity of the brain tissue it is clearly unreasonable for us to assume a path normal to the original perimeter, let us consider many such paths from location x on the inner perimeter to location y on the outer perimeter. We can now construct a linear system of equations representing n paths,

$$
\begin{bmatrix}
\kappa_1 & \gamma_1 \\
\vdots & \vdots \\
\kappa_{n-1} & \gamma_{n-1} \\
\kappa_n & \gamma_n
\end{bmatrix}
\begin{bmatrix}
V_w \\
V_g
\end{bmatrix}
=
\begin{bmatrix}
V_1 \\
\vdots \\
V_{n-1} \\
V_n
\end{bmatrix}
\tag{4}
$$

and solve for V_w and V_g using linear least squares, since clearly the system is over-determined. These values of V_w and V_g can then be used to calculate D_w and D_g by assuming a constant value for ρ and using Fisher's approximation for velocity.

To define a path, one must answer two questions: first, given any location on the inner surface, how do we know which location to connect it to on the outer surface in order to define the path? Second, how do we define the path which should connect a given point on the inner surface to its most probable ending point on the outer surface in order to assure that the tumor front as a whole advances in an outward direction?

We begin by treating the pixels of our MRI image data as unique states in a Markov process[30]. Let each pixel define a state capable of being occupied by tumor. Further, we define the pixels on the surface corresponding to the later time point as absorbing states. We define all pixels between the later time point surface and the earlier time point surface, including the earlier time point surface pixels, as transient states. We next construct a matrix of transition probabilities for all defined states. Transition probabilities from one state to neighboring states are defined by distance and gradient functions which act to approximate outward diffusion with a variable tendency a towards moving in directions of higher white matter concentration. Specifically let us define the probability of moving from state i to state j as:

$$
p_{ij} = \frac{(P_{ij}(d) + aP_{ij}(w))}{a + 1},
\tag{5}
$$

where

$$
P_{ij}(d) = \frac{d_j d_i^{-1}}{\sum_j d_j d_i^{-1}} \quad \text{and} \quad P_{ij}(w) = \frac{w_j w_i^{-1}}{\sum_j w_j w_i^{-1}}.
$$

Here d represents the distance of a state to the inner surface and w the white matter (WM) concentration of a given state, and a is a weighting factor for preferential motility towards a state consisting of more white matter. $P_{ij}(d)$ therefore gives a greater probability of movement to a neighboring

138

state that is further from the inner surface. $P_{ij}(w)$ gives a greater probability of movement towards a state of higher WM concentration relative to the current state's WM concentration. These two individual probabilities are then combined using a weighted average. We can state our transition matrix in the following form[30]:

$$T = \begin{bmatrix} Q & R \\ 0 & I \end{bmatrix},$$ (6)

where Q is the sub-set of transition probabilities for moving from a transient state to another transient state, and R is the set of probabilities for moving from a transition state to an absorbing state. In this form, a first step analysis can be preformed, thereby computing M:

$$M = \left((I' - Q)^{-1} \right) R.$$ (7)

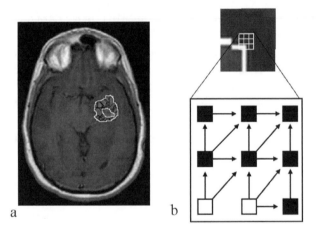

Figure 3. a) Axial T1Gd MRI illustrating the inner and outer perimeters of the imaging abnormality at first and second observation, respectively, and various paths of travel computed in between. b) Magnification and illustration of pixels as transition states along with available paths of travel in a Markov process.

M defines the probability of being absorbed by a specific absorbing state when starting in a specific transient state. This information is then used to associate points on the outer surface with points on the inner surface by choosing the state on the outer surface with the greatest probability of absorption. Next, we need some method of defining the path traveled between each point on the inner surface and its corresponding ending point of greatest probability on the outer surface. For this purpose we employ Dijkstra's Algorithm[31] which uses the data inherent to the transition matrix T and returns the minimum spanning path between a starting and ending point. By inversion of non-zero probabilities, the minimum path corresponds to the path of greatest probability of travel. In two dimensions, the paths predicted by this approach look like those shown in figure 3a, where the white perimeters represent the tumor's location at two time points, and dark lines between represent a sub-set of computed paths.

The results of applying this method to a tumor in three dimensions, and solving random sub-systems of paths for values of V_w and V_g produce histograms as shown in figure 4. Random sub-systems are solved to show the variation inherent to velocity in grey and white matter. This histogram supports the assumption that movement is greater in white matter than in grey matter.

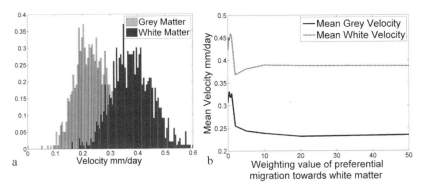

Figure 4. a) Histogram of velocity (mm/day) distribution in grey and white matter for a large number of sample paths with $a = 5$. b) Mean velocity in grey and white matter as a function of the preferential WM migration factor a.

As the percent of WM along a path increases, variance in observable velocity also increases. The underlying reason for these results is likely the influence of anisotropic diffusion in the white matter of the brain. Tumor cells are known to preferentially migrate in the direction of the white matter tracts in the brain. This observed variance acts as a problem hindering the accuracy of our results.

The sensitivity of the resulting mean grey and white matter velocities to variable values of a were analyzed for values of a ranging over $\{0, 0.1, 0.2, 0.3, 0.4, 0.5, 0.75, 1, 2, 5, 10, 20, 50\}$. Though a general trend seems to be observable, the range of variation should be kept in mind. The standard deviation in the computed mean grey velocities is 0.4 cm/day and the standard deviation in the computed mean white velocity is 0.3 cm/day.

3. Conclusion/Summary

We have presented a small, selected sampling of a wide range of approaches to modeling glioma growth and invasion from the very simple to the very complex. On the simple side, a constant exponential volume-doubling time ignores the spatial distribution of cells within the brain, specifically the diffuse invasion, a critical characteristic of gliomas which sets them apart from other forms of cancer. On the complicated side, models which attempt to encompass so many processes that their accurate parameterization is years away if even possible at all are often tested *in vitro* with no logical or foreseeable application to *in vivo* dynamics. In between is our model which is simple enough for the two necessary parameters to be calculated with a single patient using data routinely collected during treatment, yet complicated and flexible enough to model anisotropic progression of the MRI-detectable tumor front through time, and to incorporate the effects of surgical, chemo- and radiation therapies. As Murray has emphasized, "the principle use of any theory is in its predictions"[5], and iterative comparisons of real and virtual situations are essential. Our approach yields a modeling technique which immediately provides metrics for tumor proliferation and invasion, as well as response to therapy for individual patients *in vivo*, and in real-time and on the level of clinical observation. Although we have chosen examples on the far extreme ends of simplicity and complexity, we know of no other model in between which both accurately and satisfactorily is able to quantify the growth or invasion rates of gliomas from data routinely obtained *in vivo*, and despite the limited nature of medical imaging to capture complex biological processes and micro-environment, it is the only view available to clinicians while the patient is still alive. For these reasons, we believe our model to be "just right" in the balance between simplicity and complexity.

Acknowledgments

KRS gratefully acknowledges the support of the McDonnell Foundation, Dana Foundation, NIH/NCI, Shaw Professorship in Investigative Neuropathology, and the Academic Pathology Fund.

References

1. Alvord, E. C., Jr. Shaw, C. M.: The pathology of the aging human nervous system, Lea & Febiger, Philadelphia (1991)

2. Kleihues, P., Louis, D. N., Scheithauer, B. W., Rorke, L. B., Reifenberger, G., Burger, P. C. and Cavenee, W. K.: The who classification of tumors of the nervous system. J Neuropathol Exp Neurol, **61**, 215-25; discussion 226-9 (2002)

3. Collins, V. P., Loeffler, R. K. and Tivey, H.: Observations on growth rates of human tumors. Am J Roentgenol Radium Ther Nucl Med, **76**, 988-1000 (1956)

4. Zhang, L., Athale, C. A. and Deisboeck, T. S.: Development of a three-dimensional multiscale agent-based tumor model: Simulating gene-protein interaction profiles, cell phenotypes and multicellular patterns in brain cancer. J Theor Biol, **244**, 96-107 (2007)

5. Murray, J. D.: Mathematical biology ii. Spatial models and biological applications, Springer-Verlag, New York (2003).

6. Tracqui, P., Cruywagen, G. C., Woodward, D. E., Bartoo, G. T., Murray, J. D. and Alvord, E. C., Jr.: A mathematical model of glioma growth: The effect of chemotherapy on spatio-temporal growth. Cell Prolif, **28**, 17-31 (1995)

7. Swanson, K. R., Rockne, R., Rockhill, J. K. and Alvord, E. C., Jr.: Mathematical modeling of radiotherapy in individual glioma patients: Quantifying and predicting response to radiation therapy. AACR Annual Meeting. Los Angeles, CA. (2007)

8. Harpold, H., Alvord, E. C., Jr. and Swanson, K. R.: The evolution of mathematical modeling of glioma proliferation and invasion. J Neuropathol Exp Neurol, **66**, 1-9 (2007)

9. Blankenberg, F. G., Teplitz, R. L., Ellis, W., Salamat, M. S., Min, B. H., Hall, L., Boothroyd, D. B., Johnstone, I. M. and Enzmann, D. R.: The influence of volumetric tumor doubling time, DNA ploidy, and histologic grade on the survival of patients with intracranial astrocytomas. AJNR Am J Neuroradiol, **16**, 1001-12 (1995)

10. Haney, S. M., Thompson, P. M., Cloughesy, T. F., Alger, J. R. and Toga, A. W.: Tracking tumor growth rates in patients with malignant gliomas: A test of two algorithms. AJNR Am J Neuroradiol, **22**, 73-82 (2001)

11. Stein, A. M., Demuth, T., Mobley, D., Berens, M. and Sander, L. M.: A mathematical model of glioblastoma tumor spheroid invasion in a three-dimensional in vitro experiment. Biophys J, **92**, 356-65 (2007)

12. Stamatakos, G., Antipas, V. P. and Ozunoglu, N. K.: A patient-specific in vivo tumor and normal tissue model for prediction of the response to radiotherapy. Methods Inf Med, **46**, 367-75 (2007)

13. Fisher, R. A.: The wave of advance of advantageous genes, Ann Eugenics, (1937).

14. Woodward, D. E., Cook, J., Tracqui, P., Cruywagen, G. C., Murray, J. D. and Alvord, E. C., Jr.: A mathematical model of glioma growth: The effect of extent of surgical resection. Cell Prolif, **29**, 269-88 (1996)

15. Dalrymple, S. J., Parisi, J. E., Roche, P. C., Ziesmer, S. C., Scheithauer, B. W. and Kelly, P. J.: Changes in proliferating cell nuclear antigen expression in glioblastoma multiforme cells along a stereotactic biopsy trajectory. Neurosurgery, **35**, 1036-44; discussion 1044-5 (1994)

16. Kelly, P. J.: Computed tomography and histologic limits in glial neoplasms: Tumor types and selection for volumetric resection. Surg Neurol, **39**, 458-65 (1993)

17. Giese, A.Westphal, M.: Glioma invasion in the central nervous system. Neurosurgery, **39**, 235-50; discussion 250-2 (1996)

18. Swanson, K. R., Alvord, E. C., Jr. and Murray, J. D.: A quantitative model for differential motility of gliomas in grey and white matter. Cell Proliferation, **33**, 317-329 (2000)

19. Cocosco, C. A., Kollokian, V., K.-S., K. R. and Evans, A. C.: Brainweb: Online interface to a 3d simulated brain database. Neuroimage, **5**, S425 (1997)

20. Swanson, K. R.Alvord, E. C., Jr.: The concept of gliomas as a "Traveling wave": The application of a mathematical model to high- and low-grade gliomas. Can J Neurol Sci, **29**, 395 (2002)

21. Swanson, K. R.Alvord, E. C., Jr.: Serial imaging observations and postmortem examination of an untreated glioblastoma: A traveling wave of glioma growth and invasion. Neuro-Oncol, **4**, 340 (2002)

22. Mandonnet, E., Delattre, J. Y., Tanguy, M. L., Swanson, K. R., Carpentier, A. F., Duffau, H., Cornu, P., Van Effenterre, R., Alvord, E. C., Jr. and Capelle, L.: Continuous growth of mean tumor diameter in a subset of grade ii gliomas. Ann Neurol, **53**, 524-8 (2003)

23. Papanicolaou, G. C.: Diffusion in random media. In: Keller, J. B., McLaughlin, D. W. and Papanicolaou, G. C. (ed) Surveys in applied mathematics. Plenum Press, New York (1995)

24. Swanson, K. R.Alvord, E. C., Jr.: A 3d quantitative model for brain tumor growth and invasion: Correlation between the model and clinical behavior. Neuro-Oncol, **3**, 323 (2001)

25. Swanson, K. R., Alvord, E. C., Jr. and Murray, J. D.: Quantifying efficacy of chemotherapy of brain tumors with homogeneous and heterogeneous drug delivery. Acta Biotheoretica, **50**, 223-237 (2002)

26. Swanson, K. R., Alvord, E. C., Jr. and Murray, J. D.: Virtual brain tumours (gliomas) enhance the reality of medical imaging and highlight inadequacies of current therapy. British Journal of Cancer, **86**, 14-18 (2002)

27. Swanson, K. R., Alvord, E. C., Jr. and Murray, J. D.: Virtual resection of gliomas: Effects of location and extent of resection on recurrence. Mathematical and Computer Modeling, **37**, 1177-1190 (2003)

28. Swanson, K. R., Chakraborty, G., Rockne, R., Wang, C., Peacock, D. L., Muzi, M., E.C. Alvord, J., Krohn, K. and Spence, A. M.: A mathematical model for glioma growth and invasion links biological aggressiveness assessed by MRI with hypoxia assessed by FMISO-PET. 53rd Annual Meeting of the Society for Nuclear Medicine. (2007)

29. Swanson, K. R., Murray, J. D. and E.C. Alvord, J.: Combining radiological observations with a three-dimensional model to predict behavior of brain

tumors in real patients. SIAM Life Sciences and Imaging Sciences Conference. Boston, MA. (2002)
30. Taylor, H.Karlin, S.: An introduction to stochastic modeling, Academic Press, Chestnut Hill (1984).
31. Cormen, T., Leiserson, C., Rivest, R. and Stein, C.: Introduction to algorithms, MIT Press and McGraw-Hill, (2001).

ADVANCED-DELAY DIFFERENTIAL EQUATION FOR AEROELASTIC OSCILLATIONS IN PHYSIOLOGY

JORGE C. LUCERO

Department of Mathematics, University of Brasilia.
Brasilia DF 70910-900, Brazil
E-mail: lucero@unb.br

This article analyzes a mathematical model for some aeroelastic oscillators in physiology, based on a previous representation for the vocal folds at phonation. The model characterizes the oscillation as superficial wave propagating through the tissues in the direction of the flow, and consists of a functional differential equation with advanced and delay arguments. The analysis shows that the oscillation occurs at a Hopf bifurcation, at which the energy absorbed from the flow overcomes the energy dissipated in the tissues. The bifurcation value of the flow pressure increases linearly with the tissue damping and the oscillation frequency. Also, it is minimum when the phase delay of the superficial wave to travel along the tissues is π, and increases indefinitely when the delay tends to 0 and to 2π.

1. Introduction

Aeroelastic oscillatory phenomena appear commonly in physiology by the interaction of a flowing fluid, such as blood or air, with the surrounding elastic structure of tissues. Several of those phenomena result in the production of physiological sound. A good example is the oscillation of the vocal folds during phonation[1]. Under appropriate conditions, the airflow blowing through the glottis induces their oscillation. The oscillation, in turn, modulates the airflow, which, after interacting with the oral and nasal cavities, results in the sound that we perceive as voice. This aeroelastic mechanism of sound production is common to most mammals[2], and also to songbirds by action of their syrinx's membranes[3]. The same phenomenon is also responsible for the production of sound in other physiological systems, such as in blood arteries during sphygmomanometry[4], in the lips when playing a brass musical instrument[5], in the nostrils when blowing the nose[6], and in the soft palate when snoring[7].

Almost two decades ago, Titze[8] set forth the dynamical principles of the vocal fold oscillation. He proposed a mucosal wave model in which motion of the vocal fold tissues is represented as a surface wave propagating in the direction of the airflow. Since then, the original model and its several variations have been used in further studies of phonation dynamics, e.g., see Refs. 9–12, and have also been applied to the avian syrinx[13]. The model has been particularly useful to identify the threshold conditions that its various parameters must meet in order to start the oscillation. Considering the air pressure as control parameter, the oscillation is generated through at a Hopf bifurcation of the subcritical type. This bifurcation, in combination with a cyclic fold bifurcation, produces an oscillation hysteresis phenomenon clearly visible in voicing onset-offset patterns during running speech[11,14]. The bifurcation value of the air pressure, called the phonation threshold pressure, has been interpreted as a measure of ease of phonation, and proposed as a diagnostic tool for vocal health[15].

A drawback of the mucosal wave model is that it assumes a small time delay for the mucosal wave to travel along the tissues. In the case of the vocal fold oscillation, the time delay is in the same order of magnitude than the period of the oscillation[8], therefore the assumption provides a rather crude approximation. In a recent work[16], an extended version of the model to the general case of arbitrary time delays for the mucosal wave has been presented. From the extended model, a theoretical equation for the phonation threshold pressure was derived, which has better qualitative agreement with experimental data than previous expression. In the present article, the extended model is considered as a general representation of aeroelastic oscillatory systems in physiology, and its dynamics is further explored.

2. Mucosal Wave Model

Let us briefly review the mucosal wave model for the vocal folds[8], shown in Figure 1. Complete right-left symmetry is assumed, and motion of tissues is allowed only in the horizontal direction. A wave propagates through the superficial tissues, in the direction of the airflow (upward).

Letting ξ be displacement of the tissues from their rest position, and y the vertical distance from the midpoint of the glottis in the direction of the airflow, then the tissue wave has the general expression

$$\xi(y,t) = x(t - y/c), \tag{1}$$

where t is time, $x(t) = \xi(0,t)$ is the displacement of the tissues at the midpoint of the glottis and c is the wave velocity.

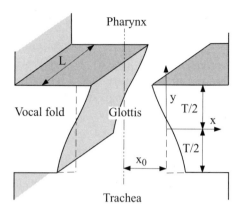

Figure 1. Mucosal wave model of the vocal folds.

We consider the simple case in which the vocal fold separation along the glottal height is constant, when they are at their rest position. In that case, the glottal cross sectional area a at the height y is $a = 2L(x_0 + \xi)$, where x_0 is the half-width at the rest position, and L is the vocal fold length. The glottal areas a_1 and a_2 at the lower $(y = -T/2)$ and upper $(y = T/2)$ edges of the vocal folds, respectively, are then

$$a_1 = 2L[x_0 + x(t + \tau)], \tag{2}$$

$$a_2 = 2L[x_0 + x(t - \tau)]. \tag{3}$$

where $\tau = T/(2c)$ is the time delay for the surface wave to travel half the glottal height.

The glottal aerodynamics is modeled by assuming that the subglottal pressure is constant and equal to the lung pressure P_L, and that the pressure at the exit of the glottis is the atmospheric pressure. From the lungs up to the exit of the glottis, the air flow is approximately frictionless, stationary, and incompressible. At the glottal exit, the flow detaches from the glottal wall and forms a jet stream, loosing almost all its energy by turbulence. Under such conditions, the mean glottal pressure P_g may be expressed by

$$P_g = \frac{P_L}{k_t} \left(1 - \frac{a_2}{a_1} \right). \tag{4}$$

where k_t is a transglottal pressure coefficient, and $a_1 > 0$ (open glottis).

The mechanical properties of the vocal fold tissues are lumped at the midpoint of the glottis, which yields the equation of motion

$$M\ddot{x} + B\dot{x} + Kx = P_g,$$
(5)

where M, B, and K are the mass, damping, and stiffness, respectively, per unit area of the vocal fold medial surface.

Introducing Eqs. (2), (3) and (4) into Eq. (5), we obtain finally the complete equation

$$M\ddot{x} + B\dot{x} + Kx = \frac{P_L}{k_t}\frac{x(t+\tau) - x(t-\tau)}{x_0 + x(t+\tau)}$$
(6)

with $x_0 + x(t+\tau) > 0$.

The same model has been applied to the avian syrinx[13], and may be equally extended to other physiological systems by using appropriate values of its parameters. Note that all the systems mentioned in the introduction consists of a flow (blood or air) passing through a constricted channel (formed by an artery, the lips, the nostrils, or the soft palate). To generalize the above model, and at the same time reduce the number of parameters, we introduce the new adimensional variable $u = x/x_0$, and the parameters $\alpha = B/M$, $\omega = \sqrt{K/M}$, $p = P_L/(k_t x_0 M)$, which yields the differential equation

$$\ddot{u} + \alpha\dot{u} + \omega^2 u = p\frac{u(t+\tau) - u(t-\tau)}{1 + u(t+\tau)},$$
(7)

with $1 + u(t+\tau) > 0$. Letting $v = \dot{u}$, the following equivalent bidimensional form is obtained

$$\begin{cases} u' = v, \\ v' = -\alpha v - \omega^2 u + p\dfrac{u(t+\tau) - u(t-\tau)}{1 + u(t+\tau)} \end{cases}$$
(8)

with $1 + u(t+\tau) > 0$.

3. Small τ Approximation

Equation (8) is a functional differential equation with advance and delay arguments ($t + \tau$ and $t - \tau$, respectively). It has one fixed point (rest position) at $(u, v) = (0, 0)$.

First, let us assume that the delay τ is small enough (this has been the standard assumption in previous studies, e.g., see Refs. 8–12), so that the advanced-delay terms may be approximated by the linearization

$$u(t \pm \tau) \approx u(t) \pm \tau v(t)$$
(9)

which reduces Eq. (8) to an ordinary differential equation

$$\begin{cases} u' = v, \\ v' = -\alpha v - \omega^2 u + \dfrac{2p\tau v}{1 + u + \tau v}, \end{cases} \tag{10}$$

with $1 + u + \tau v > 0$.

The eigenvalues λ of the Jacobian matrix at the rest position are given by the characteristic equation

$$\lambda^2 + (\alpha - 2p\tau)\lambda + \omega^2 = 0. \tag{11}$$

Let us consider p as the control parameter, because of its direct relation with the air pressure source. The characteristic equation has a pair of conjugate complex roots, which crosses the imaginary axis from left to right as p increases and crosses the bifurcation value $p_t = \alpha/(2\tau)$. At this value, a Hopf bifurcation occurs, in which the rest position changes its stability: it is a stable focus for $p \leq \alpha/(2\tau)$, and an unstable one at $p > \alpha/(2\tau)$. At the same time, a limit cycle is generated [17].

The type of Hopf bifurcation may be determined as follows[17]: Let us s be the signed distance along a line through the origin, and $P(s)$ be the Poincaré map for the focus. Further, let us $\sigma \equiv d'''(0)$ be the Lyapunov number for the focus, where $d(s) = P(s) - s$ is the displacement function. Then, if $\sigma \neq 0$, the origin is a weak focus and a Hopf bifurcation occurs at the bifurcation value of the control parameter, and the sign of σ indicates its type: it is supercritical for $\sigma < 0$, and subcritical for $\sigma > 0$. For a general planar analytic system

$$\begin{aligned} \dot{x} &= ax + by + p(x,y), \\ \dot{y} &= cx + dy + q(x,y), \end{aligned} \tag{12}$$

where $\Delta = ad - bc > 0$, $a + d = 0$, and the analytic functions $p(x,y) = \sum_{i+j \geq 2} a_{ij} x^i y^j$, $q(x,y) = \sum_{i+j \geq 2} b_{ij} x^i y^j$, the Lyapunov number is given by

$$\begin{aligned} \sigma = \frac{-3\pi}{2b\Delta^{3/2}} \Big\{ &\left[ac(a_{11}^2 + a_{11}b_{02} + a_{02}b_{11}) + ab(b_{11}^2 + a_{20}b_{11} + a_{11}b_{02}) \right. \\ &+ c^2(a_{11}a_{02} + 2a_{02}b_{02}) - 2ac(b_{02}^2 - a_{20}a_{02}) - 2ab(a_{20}^2 + b_{20}b_{02}) \\ &\left. - b^2(2a_{20}b_{20} + b_{11}b_{20}) + (bc - 2a^2)(b_{11}b_{02} - a_{11}a_{20}) \right] \\ &- (a^2 + bc)\left[3(cb_{03} - ba_{30}) + 2a(a_{21} + b_{12}) + (ca_{12} - bb_{21}) \right] \Big\}. \end{aligned} \tag{13}$$

In our case, for $p = \alpha/(2\tau)$, we have

$$\begin{cases} u' = v, \\ v' = -\omega^2 u - \alpha v u - \alpha \tau v^2 + \alpha u^2 v + 2\alpha \tau u v^2 + \alpha \tau^2 v^3 + \ldots \end{cases} \tag{14}$$

which produces

$$\sigma = \frac{3\pi\alpha}{2\omega}(1 + \alpha\tau + 3\omega^2\tau^2) > 0. \tag{15}$$

and therefore the bifurcation is subcritical.

4. General Case for Arbitrary τ

Let us consider now the general case, given by Eq. (8). Linearization around the equilibrium position yields

$$\begin{cases} u' = v, \\ v' = -\alpha v - \omega^2 u + p[u(t + \tau) - u(t - \tau)], \end{cases} \tag{16}$$

whose characteristic equation is

$$\lambda^2 + \alpha\lambda + \omega^2 - 2p\sinh(\lambda\tau) = 0 \tag{17}$$

For $p = 0$, Eq. (17) has the roots

$$\lambda = -(\alpha/2) \pm \sqrt{(\alpha/2)^2 - \omega^2}, \tag{18}$$

which have negative real parts.

For $p > 0$, Eq. (17) may have an indefinite number of roots. Let us assume a pair of imaginary roots $\lambda = \pm i\mu$. Substituting into Eq. (17), using the identity $\sinh(ix) = i\sin(x)$, and separating real and imaginary parts, we obtain

$$-\mu^2 + \omega^2 = 0 \tag{19}$$

$$\mu\alpha - 2p\sin(\mu\tau) = 0 \tag{20}$$

The first equation produces the oscillation angular frequency $\mu = \omega$. The value of p given by the second equation is the bifurcation pressure

$$p_t = \frac{\alpha\omega}{2\sin(\omega\tau)} \tag{21}$$

with $0 < \omega\tau < \pi$.

According to Rouché's Theorem[18], the roots of the characteristic equation depend continuously on the parameter p. Hence, for $0 \leq p < p_t$, all roots have negative real parts, and at $p = p_t$, a pair of roots become imaginary. We verify next that those roots cross the imaginary axis from left to right. Implicit differentiation of Eq. (17) produces

$$[2\lambda + \alpha - 2p\tau\cosh(\lambda\tau)]\frac{d\lambda}{dp} = 2\sinh(\lambda\tau) \tag{22}$$

Substituting $\lambda = \omega i$, $p = p_t$, given by Eq. (21), and separating the real part we obtain finally

$$\frac{d\,\mathrm{Re}\,(\lambda)}{dp}\bigg|_{p=p_t} = \frac{4\omega\sin(\omega\tau)}{\left\{\alpha^2\left[1 - \omega\tau\cot(\omega\tau)\right]^2 + 4\omega^2\right\}} > 0 \qquad (23)$$

for $0 < \omega\tau < \pi$. This is the transversatility condition, which proves that the roots cross the imaginary axis and therefore their real parts become positive.

The above results imply that the equilibrium position at $x = 0$ is stable for $p < p_t$, and unstable for $p > p_t$. Further, by the Hopf Bifurcation Theorem for functional differential equations [19], a limit cycle is generated at $p = p_t$.

5. Conditions for the Oscillation Onset

The oscillation threshold pressure is then given by Eq. (21). Let us $\delta = 2\omega\tau$ denote the phase delay for the surface wave to travel the whole constricted channel (recall that τ is the time delay to travel half that distance). We may rewrite the above equation as

$$p_t = \frac{\alpha\omega}{2\sin(\delta/2)}, \qquad 0 < \delta < 2\pi \qquad (24)$$

This equation tells us that the oscillation threshold pressure increases with the tissue damping α and the oscillation frequency ω, and depends on the phase delay δ following a cosecant characteristic, as shown in Fig. 2. Its minimum value occurs at $\delta = \pi$. Note also that, for $\tau \to 0$, $\sin(\omega\tau) \to \omega\tau$, and $p_t \to \alpha/(2\tau)$, as found in the previous section.

It is instructive to consider also the exchange of energy between the airflow and the tissues at the oscillation onset, as in Ref. 11.

The energy dissipated in the tissues is the work done by the damping force

$$W_\alpha = \oint_{\text{cycle}} \alpha v\,dx = \int_0^{2\pi/\omega} \alpha v^2\,dt \qquad (25)$$

In the vicinity of the bifurcation point, the oscillation may be described by the sinusoid $x = A\sin(\omega t)$, which produces

$$W_\alpha = \pi A^2 \alpha \omega \qquad (26)$$

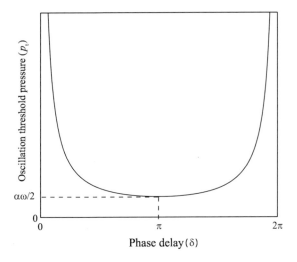

Figure 2. Oscillation threshold pressure vs. phase delay of the surface wave to travel along the tissues.

The energy absorbed from the airflow is the work done by the air pressure

$$W_p = \oint_{\text{cycle}} p_g dx = \int_0^{2\pi/\omega} p_g(t) v dt \tag{27}$$

where p_g is the right side of Eq. (7). In the vicinity of the bifurcation point, p_g may be approximated by its linear part, as done in Eq. (16), which, together with a sinusoidal approximation for $x(t)$, produces

$$W_p = 2\pi A^2 p \sin(\delta/2) \tag{28}$$

for $0 < \delta < 2\pi$. Letting $W_\alpha = W_p$, we obtain the oscillation threshold condition given by Eq. (21). For $p > p_t$, we have $W_p > W_\alpha$ (see Figure 3).

We may therefore say that the oscillation is fueled by a transfer of energy from the airflow to the tissues. When the air pressure increases and passes through the oscillation threshold value, the energy absorbed from the airflow is large enough to overcome the energy dissipated in the tissues, and so an oscillation of growing amplitude may start. The transfer of energy depends on the phase delay δ, following a sinusoidal characteristics. Its has a maximum at $\delta = \pi$, which is the same condition for a minimum at the threshold pressure p_t.

152

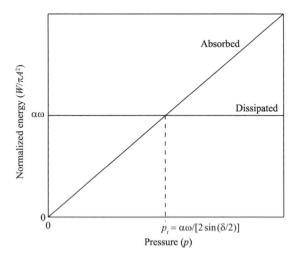

Figure 3.   Normalized absorbed and dissipated energy vs. flow pressure.

6. Conclusion

The above analysis has explored the dynamics of aeroelastic oscillations of some physiological systems. The results show that the oscillation appears at a Hopf bifurcation, where the rest position of the tissues becomes unstable and a limit cycle appears. However, identification of the type of the bifurcation has been left. In the case of small time delay, the bifurcation is of the subcritical type and causes an oscillation hysteresis phenomenon. Is it still subcritical at large time delays? This should be an interesting question for further research.

Acknowledgments

This work was supported by MCT/CNPq (Brazil).

References

1. I. R. Titze, *Principles of Voice Production* (Prentice-Hall, Englewood Cliffs, 1994).
2. I. Wilden, H. Herzel, G. Peters and G. Tembrock, *Bioacoustics* **9**, 171 (1998).
3. F. Goller and O. N. Larsen, *P. Natl. Acad. Sci. USA* **94**, 1487 (1997).
4. J. B. Grotberg and O. E. Jensen, *Annu. Rev. Fluid Mech.* **36**, 121 (2004).
5. S. Adachi and M. A. Sato, *J. Acoust. Soc. Am.* **99**, 1200 (1996).
6. D. H. Hodges, G. A. Pierce, O. A. Bauchau and M. J. Smith, *AE 6200: Aeroelasticity – Class Notes* (Georgia Institute of Technology, Atlanta, 2006).

7. Y. Aurégan and C. Depoiller, *J. Sound Vib.* **188**, 39 (1995).

8. I. R. Titze, *J. Acoust. Soc. Am.* **83**, 1536 (1988).

9. R. W. Chan and I. R.Titze, *J. Acoust. Soc. Am.* **119**, 2351 (2006).

10. C. Drioli, *J. Acoust. Soc. Am.* **117**, 3184 (2005).

11. J. C. Lucero, *J. Acoust. Soc. Am.* **105**, 423 (1999).

12. J. C. Lucero, *Comm. Math. Sci.* **3**, 517 (2005).

13. R. Laje and G. B. Mindlin, *Phys. Rev. E* **72**, 036218 (2005).

14. L. L. Koenig, W. E. Mencl, and J. C. Lucero, *J. Acoust. Soc. Am.* **118**, 2535 (2005).

15. I. R. Titze, S. S. Schmidt, and M. R. Titze, *J. Acoust. Soc. Am.* **97**, 3080 (1995).

16. J. C. Lucero and L. L. Koenig, *J. Acoust. Soc. Am.* **121**, 3280 (2007).

17. L. Perko, *Differential Equations and Dynamical Systema*, (Springer-Verlag, New York, 1991).

18. J. Dieudonné, *Foundations of Modern Analysis* (Academic Press, New York, 1960).

19. J. Hale, *Theory of Functional Differential Equations* (Springer-Verlag, New York, 1977).

GEOMETRY, ACTIVITY-DEPENDENT MECHANISMS, MEMBRANE KINETICS AND CHANNEL DENSITY DISTRIBUTION INTERPLAY IN SINGLE NEURON PLASTICITY. A COMPUTATIONAL STUDY

ENRICO CATALDO, MARCELLO BRUNELLI

Department of Biology – General Physiology Unit, University of Pisa,
Via San Zeno 31 Pisa, 56100, Italy

EVYATAR AV-RON, YIDAO CAI, DOUGLAS A. BAXTER

Department of Neurobiology and Anatomy,
The University of Texas-Houston Medical School
P.O. Box 20708 Houston, TX 77225, USA

Conduction of action potentials throughout the complex morphology of neurons may be modulated in an activity-dependent manner. Among modulatory mechanisms, afterhyperpolarization (AHP) plays an important role. To investigate how the AHP modulatory capabilities on transmission were dependent on the axonal geometry as well as on membrane properties such as channel kinetics, channel density distribution and membrane noise, multi-compartment computational neural models were built, using the neurosimulator SNNAP. Two kinetic schema for the sodium and potassium channels were compared. The simulations suggest that channel kinetics profoundly influence the AHP-dependent modulation of action potential conduction through points of impedance mismatch in the highly branched neurites of neurons.

1. Introduction

1.1. *Background*

The William James (1890) conjecture on the existence of a correlation between changes in behaviour and plastic changes in the nervous system has been widely confirmed by experimental studies realized primarily during the second half of the XX century. In addition, pursuing the suggestion that the sites of plasticity are the synapses and that learning corresponds to changes in the synaptic strength, most of the efforts have been directed toward elucidating mechanisms of synaptic transmission and plasticity[11]. Many of the experimental studies undertaken in this field

154

were motivated by Donald Hebb's conjecture regarding neural activity and plasticity: if two neurons fire often together, then the synaptic connection between them should be strengthened (i.e., *neurons that fire together wire together*). The generalised Hebb's rule[32] is that changes in synaptic strength depend on the correlation of firing activities between pre and post-synaptic neurons[11,13,22,30,40,51,52]. Activity-dependent synaptic plasticity is considered the most important neural correlate of learning and memory processes. However, there is a growing awareness that activity in a neural network can shape both the synaptic strengths and the intrinsic electrical properties of neurons within the network. Theoretical studies indicate that Hebb's rule alone (believed to be synapse-specific forms of plasticity) would destabilize the system on which it acts and empirical studies indicate that Hebb's rule is not a sufficient explanation for all of the observed examples of neural plasticity (e.g.,[36]). A problem resides in the feed-back nature of the mechanism, by which a neural network tends to saturate and loose its sensitiveness. Early theoretical studies proposed several mechanisms that may act in conjunction with the Hebb's rule, and thereby prevent saturation[22,24,30,42,43]. In addition, 'homeostatic' mechanisms recently have been discovered, which help ensure the stability of a system during Hebbian synaptic changes. Among them are: synaptic scaling, spike timing dependent plasticity, synaptic redistribution and neural excitability variation[1,24,42,43,47,66,67,68,69,71]. In addition to changes in synaptic efficacy, changes in neuronal excitability have been observed following learning (e.g.[3,9,10,19,29,45]; for reviews see[6,18]). The activity-dependent regulation of neuronal excitability is brought about by different processes, such as a spatial redistribution of the ionic channels or a recruitment of new membrane currents[8,20,21,23,28,44,46,49,53,57,72]. The afterhyperpolarization (AHP), generally due to the activation of K^+ conductances and to a Na^+/K^+ pump, is one of the regulatory mechanisms[2,8,12,17,27,33,35,55,57,61,62,63,64,65] and it can play different functional roles in various neurons[12,17,35,55]. For example, the AHP can influence the excitability and firing rates of cells, it can modulate synaptic transmission and plasticity, and it can contribute to conduction failure within branching neuritic structures[5,14,31,39,54]. In addition to the AHP, the geometry of a neuron can contribute to the spatial modulation of spike conduction throughout a cell[41,48,50,59,60]. The combination of an extensive neuritic arborization and activity dependent modulation of the membrane conductances can produce complex spatio-temporal filtering of the flow of information throughout a cell.

1.2. *Motivation and Goals*

Along the line of these investigations, in previous works[16,38,58] we showed, by means of multicompartement model of the mechanosensory T neuron of the leech, how AHP modulated transmission of action potentials through points of impedance mismatch, which are at sites of low-safety factor. Discharge of action potentials in leech T cells induce an afterhyperpolarization (AHP), through the activation of a Na^+/K^+ pump and a Ca^{2+}-dependent K^+ current. Those results suggested that AHP could reduce the number of action potentials transmitted at branching points, intrinsically modulating the neuronal activity.

To investigate how the AHP modulatory capabilities on transmission were dependent not only on the axonal geometry but also on other membrane properties such as channel kinetics, channel density distribution and membrane noise, we have built a reduced multi-compartment computational model of the cell, by using the neurosimulator SNNAP[4,6,73]. Two kinetics for the Na^+ and K^+ channels were compared, one reproducing the T cell electrical properties and the other a reference membrane with different action potential duration and pattern of discharge. The simulations suggested that channel kinetics profoundly determined the AHP modulatory capabilities. Depending on the channel kinetics, the transmission of action potentials was very sensitive or almost insensitive to the AHP value, the channel density distribution and membrane noise, respectively. The processing or conductive features of neurons seems to be determined in first instance by the channel kinetic of the membrane and secondarily by the axonal geometry and activity-dependent processes and noise.

2. Model

We developed a multicompartment model of a leech T neuron, representing the reduced version of a previously developed model[15,16], which was added to the ModelDB website (senselab.med.yale.edu), where it is possible to download and run the simulations and view parameters and equations. Reduced biological models are utilized primarily in computational studies to increase computational efficiency (e.g., the time step for integration was 10^{-5} sec in the reduced model as compared to 3×10^{-7} sec in the full model), and thereby decreasing the amount of time necessary to complete a simulation[70,73]. Our reduced model represented the anterior part of the T cell and consisted of 22 neuritic compartments and a somatic compartment (see Fig. 1(a)). The biophysical properties of the membrane were unchanged and the geometric sizes (see Table 1) were adapted to reproduce the characteristic pattern of discharge (Fig. 2(a)). The reduced model retained all the essential features of the complete one, such as the pattern of discharge, the shape and duration of the action potential and the value of the impedance mismatch along its length.

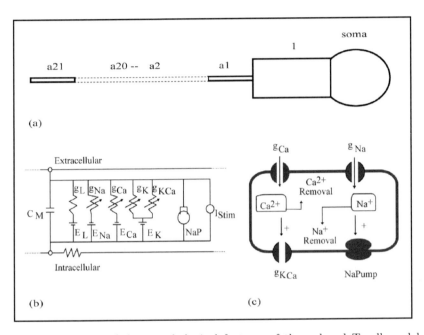

Figure 1. Diagram of the morphological features of the reduced T cell model. (a) Each compartment represents a Hodgkin-Huxley type electrical circuit, which contained the same capacitance and passive and active conductances in parallel (see (b)) of the complete model. The compartments are connected by axial resistances. The designation a indicates compartments anterior to the soma. The pattern of discharge and spike duration were very similar to those of the complete model. The load ratio at the 1a-1 site was 5.4. See Table 1 for geometrical details and parameters. (b) Equivalent electrical circuit of a single compartment. The membrane of each compartment was modeled as a membrane capacitance (C_M) in parallel with five ionic conductances and a pump. The specific capacitance was assumed to be 1 μF/cm^2. The intracellular resistivity (R_i) and the specific membrane resistance (R_m) were 200 Ωcm and 10000 Ωcm^2, respectively. (c) The modulatory pathways of the model. Intracellular pools of Ca^{2+} and Na$^+$ were included in the model. Contribution to the Ca^{2+} and Na$^+$ pools came from g_{Ca} and g_{Na}, respectively. Ca^{2+} and Na$^+$ were removed through a first-order kinetics process. The membrane conductances g_{Ca} and g_{Na} were positively (+) regulated by the intracellular pools of Ca^{2+} and Na$^+$, respectively.

Table 1. Geometric dimensions and capacitances of the compartments (increased by a factor 2 to account for membrane invagination). The surface areas specific capacitance of 1 $\mu F/cm^2$ was utilized.

Comp.	$D\,(\mu m)$	$L\,(\mu m)$	$A_m\,(\mu m^2)$	$C_m\,(pF)$
Soma	32	—	6434	64.34
1	25	50	7850	78.5
1a → 21a	2	36	450	4.5

Table 2. Serial conductances (μS) between compartments, obtained using a value of $200\Omega cm$ for the axial resistivity and the geometric dimensions of the compartments.

Comp.	μS
Soma-1	4.9
1-1a, 1a-2a, 2a-3a, 3a-4a, 4a-5a, 5a-6a, 6a-7a, 7a-8a, 8a-9a, 9a-10a, 10a-11a, 11a-12a, 12a-13a, 13a-14a, 14a-15a, 15a-16a, 17a-18a, 18a-19a, 20a-21a	0.0436

The value of the load ratio (5.4) at $a1 - 1$ site was within the range of the values utilized in the previous simulations. The values for the inter-compartment conductances are reported in Table 2.

The parameters describing the various ionic currents are reported in Table 3 and 4. Each compartment, endowed with the same voltage-dependent currents of the complete model[16], was modeled as an equivalent electrical circuit (Fig. 1(b)), in which some currents were regulated by intracellular Ca^{2+} and Na^+ (Fig. 1(c)). The membrane model contained a membrane capacitance (C_M), in parallel with two inward currents (Na^+ and Ca^{2+}), two K^+ currents, a leak and a pump current. The membrane currents were described by Hodgkin-Huxley type equations[22,70]. There was no inter-compartment ionic diffusion and the channel densities of all currents were taken uniform, initially.

The model was built by using the version 8 of the neurosimulator SNNAP (http://snnap.uth.tmc.edu/) (Simulator for Neural Networks and Action Potentials)[7,73], the Windows XP operating system and PC mi-

Table 3. Parameters of the membrane currents of the reduced T cell model membrane. E is the equilibrium potential. $\bar{g}$ is the maximum conductance in Eq. 2 (see Methods). h_A and s_A are respectively the half-activation voltage and the slope parameter of the activation function in Eq. 4. p is the power of the activation function in Eq. 2. $\tau_{A\max}$ and $\tau_{A\min}$ are respectively the maximum and minimum values of the activation time constant, $h_{\tau A}$ is the voltage at which the activation time constant is half maximal and $s_{\tau A}$ is the slope parameter of the activation time constant function in Eq. 5. h_B and s_B are respectively the half inactivation voltage and the slope parameter of the inactivation function and p is the power in Eq. 4. $\tau_{B\max}$ and $\tau_{B\min}$ are the maximum and minimum values of the inactivation time constant, $h_{\tau B}$ is the voltage at which the inactivation time constant is half maximal and $s_{\tau B}$ is the slope parameter of the inactivation time constant function in Eq. 5. $I_{K,Ca}$ and $I_{Na,P}$ were modulated by the intracellular concentrations of Ca^{2+} and Na^+, respectively.

Currents	I_{Na}	I_K	I_{Ca}	$I_{K,Ca}$	I_{leak}	$I_{Pump,Na}$
E (mV)	45	−62	60	−62	−48	−300
$\bar{g}$ (mS/cm^2)	350	90	2	23.6	0.1	2
h_A (mV)	−35	−22	−10			
s_A	−5	−9	−2.8			
p	3	2	1			
$\tau_{A\max}$ (msec)	0.1	6	0.6			
$\tau_{A\min}$ (msec)		1				
$h_{\tau A}$ (mV)		−10				
$s_{\tau A}$		10				
h_B (mV)	−50					
s_B ()	9					
p	2					
$\tau_{B\max}$ (msec)	14					

crocomputers.

The membrane potential of compartment $i(V_i)$ was calculated by the numerical solution (forward Euler method with fixed time step of 10^{-5} sec.) of the following system of coupled differential equations:

$$-C_i\frac{dV_i}{dt} = \sum I_{ion} + \sum g_{ij}(V_i - V_j) + I_{K,Ca} + I_{Pump,Na} - I_{stim} \qquad (1)$$

where C_i is the membrane capacitance (see Table 1), $\sum I_{ion}$ is the sum of voltage- and time-dependent ionic currents, $\sum g_{ij}(V_i - V_j)$ is the sum of currents due to voltage differences between the compartment i and the adjacent compartments $j(V_i - V_j)$ and conductance between the compartments

160

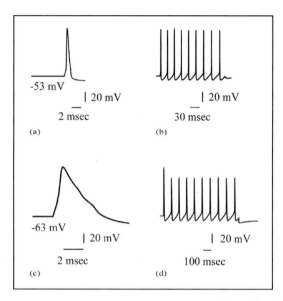

Figure 2. Two neural models. (a)-(b) reduced T cell and (c)-(d) generic cell. The two neural models differ in the duration of the action potential and in the pattern of discharge. They differ in the excitability as well. (b) the train of action potentials was obtained by injecting a depolarizing current of 0.4 nA intensity and 200 msec duration. (d) the train of action potentials was obtained by a current of 4 nA intensity and 1 sec duration.

(g_{ij}) (see Table 2), $I_{K,Ca}$ is the Ca^{2+}-dependent K^+ current, $I_{Pump,Na}$ is the current generated by the Na^+ pump (see below), and I_{stim} is an external stimulus. Each voltage- and time-dependent ionic current is given by the equation: $I_{ion} = g_{ion}(V_m, t)(V_m - E_{ion})$, where $g_{ion}(V_m, t)$ represents the voltage- and time-dependent conductance and E_{ion} is the equilibrium potential associated with each conductance. Each voltage- and time-dependent conductance was represented by a Hodgkin-Huxley-type formulation:

$$g_{ion} = \bar{g}_{ion} A_{ion}^p(V_m, t) B_{ion}(V_m, t) \tag{2}$$

where $\bar{g}_{ion}$ is the maximum value of g_{ion}, $A_{ion}^p(V_m, t)$ and $B_{ion}(V_m, t)$ represent the voltage- and time-dependent activation and inactivation functions, respectively, associated with g_{ion}, and p is an integer power to which A_{ion} is raised. The values of $A_{ion}(V_m, t)$ and $B_{ion}(V_m, t)$ were obtained by solving the differential equations:

$$\frac{dA_{\text{ion}}}{dt} = \frac{A_{\infty(\text{ion})}(V_m) - A_{\text{ion}}}{\tau_{A(\text{ion})}(V_m)} \quad ; \quad \frac{dB_{\text{ion}}}{dt} = \frac{B_{\infty(\text{ion})}(V_m) - B_{\text{ion}}}{\tau_{B(\text{ion})}(V_m)} \tag{3}$$

where $A_{\infty(\text{ion})}(V_m)$ and $B_{\infty(\text{ion})}(V_m)$ are the voltage-dependent steady-state values of the activation and inactivation, respectively, and $\tau_{A(\text{ion})}(V_m)$ and $\tau_{B(\text{ion})}(V_m)$ are the voltage-dependent time constants of the activation and inactivation, respectively.

The values of $A_{\infty(\text{ion})}(V_m)$, $B_{\infty(\text{ion})}(V_m)$, $\tau_{A(\text{ion})}(V_m)$ and $\tau_{B(\text{ion})}(V_m)$ were determined from the general algebraic equations:

$$A_{\infty(\text{ion})}(V_m) = \frac{1}{1 + \exp\left\{\frac{V_m - h_{A(\text{ion})}}{s_{A(\text{ion})}}\right\}}$$

$$\tag{4}$$

$$B_{\infty(\text{ion})}(V_m) = \left(\frac{1}{1 + \exp\left\{\frac{V_m - h_{B(\text{ion})}}{s_{B(\text{ion})}}\right\}}\right)^p$$

$$\tau_{A(\text{ion})}(V_m) = \frac{\tau_{A(\max)(\text{ion})} - \tau_{A(\min)(\text{ion})}}{1 + \exp\left\{\frac{V_m - h_{\tau A(\text{ion})}}{s_{\tau A(\text{ion})}}\right\}} + \tau_{A(\min)(\text{ion})} \tag{5}$$

$$\tau_{B(\text{ion})}(V_m) = \frac{\tau_{B(\max)(\text{ion})} - \tau_{B(\min)(\text{ion})}}{1 + \exp\left\{\frac{V_m - h_{\tau B(\text{ion})}}{s_{\tau B(\text{ion})}}\right\}} + \tau_{B(\min)(\text{ion})} \tag{6}$$

where $h_{A(\text{ion})}$ is the half-activation voltage, $s_{A(\text{ion})}$ is the slope parameter of the activation function, $h_{B(\text{ion})}$ is the half inactivation voltage, $s_{B(\text{ion})}$ is the slope parameter of the inactivation function, p is an integer power, $\tau_{A(\max)(\text{ion})}$ and $\tau_{A(\min)(\text{ion})}$ are respectively the maximum and minimum values of the activation time constant, $h_{\tau A(\text{ion})}$ is the voltage at which the activation time constant is half maximal; $s_{\tau A(\text{ion})}$ is the slope parameter of the activation time constant function; $\tau_{B(\max)(\text{ion})}$ and $\tau_{B(\min)(\text{ion})}$ are the maximum and minimum values of the inactivation time constant, respectively; $h_{\tau B(\text{ion})}$ is the voltage at which the inactivation time constant is half maximal; $s_{\tau B(\text{ion})}$ is the slope parameter of the inactivation time constant function. The parameters used in the present study are presented in Table 3.

The model incorporated two intracellular ion pools: one for Ca^{2+} and another for Na^+, which in turn modulated two membrane currents (see below). The dynamics for these pools were described by first order processes[26]. For each given compartment i, the concentration of an ion (i.e., $[\text{ion}]_i$) in a pool was obtained by the solution of the following differential equation:

$$\frac{d[\text{ion}]_i}{dt} = \frac{K_{(\text{ion})(i)}\left(-I_{(\text{ion})(i)}\right) - [\text{ion}]_i}{\tau_{(\text{ion})(i)}} \tag{7}$$

where $[\text{ion}]_i$ is the intracellular concentration of a given ion (i.e., either Ca^{2+} or Na^+) in compartment i (in arbitrary units), $K_{(\text{ion})(i)}$ is a scaling factor that determines what proportion of an ionic current (i.e., $I_{(\text{ion})(i)}$) contributes to the intracellular pool, and $\tau_{(\text{ion})(i)}$ is the time constant for the removal of ions from the pool. Although Eq. (7) represents the concentration of a ion in arbitrary units, it is possible to estimate the absolute concentration given the volume of each compartment and Faraday's constant. For example, one arbitrary unit of concentration of a monovalent ion is the equivalent of $\sim 60\mu M$ in the somatic compartment (assuming a steady ionic current of 1 nA and $K_{(\text{ion})} = 1$) see[70]. In the present model, diffusion of ions between compartments was not described. In addition, it was assumed that the removal of Na^+ via the electrogenic Na^+ pump (i.e., $I_{\text{Pump,Na}}$) was small relative to other factors. Thus, the material balance for both the Ca^{2+} and Na^+ were described by Eq. (7).

Empirical estimates of the dynamics of ion pools have focused on the soma of T cells[25,56]. To adjust the scaling factor $K_{(\text{ion})(i)}$ and time constant $(\tau_{(\text{ion})(i)})$ to reflect the morphological features of compartment i, it was necessary to relate the volume (V) and surface area (S) of each compartment to those of the soma by the relationships:

$$K_{(\text{ion})(i)} = \left(\frac{S_{(i)}/S_{(\text{soma})}}{V_{(i)}/V_{(\text{soma})}}\right) K_{(\text{ion})(\text{soma})}$$

$$\tag{8}$$

$$\tau_{(\text{ion})(i)} = \frac{S_{(\text{soma})}}{S_{(i)}} \tau_{(\text{ion})(\text{soma})}$$

For the soma compartment, the values for these parameters were: $\tau_{(\text{Ca})(\text{soma})} = 0.416$ sec, $K_{(\text{Ca})(\text{soma})} = 0.033$, $\tau_{(\text{Na})(\text{soma})} = 16$ sec, and $K_{(\text{Na})(\text{soma})} = 0.016$. The calculated values for each compartment, using equations (8) are presented in Table 4.

The final value for the somatic compartment was eventually rescaled to $\tau_{(\text{Ca})(\text{soma})} = 1.25$ sec and $K_{(\text{Ca})(\text{soma})} = 0.1$, to take into account additional factors influencing the Ca^{2+} dynamics, such as the release from the intracellular Ca^{2+} stores.

Table 4. Parameters used in Eq. 7 to describe the amount and the dynamics of the intracellular concentrations of Ca^{2+} and Na^+, in each compartment. To calculate these parameters Eqs. (8) were used.

Compartments	$K_{(Ca)}$	$\tau_{(Ca)}$ (msec)	$K_{(Na)}$	$\tau_{(Na)}$ (sec)
Soma	0.033	416	0.016	16
1	0.082	1466	0.013	18.76
a1 → a21	1.422	117	0.227	1.5

The intracellular pools of ions, in turn, regulated two membrane currents: one the Ca^{2+}-dependent K^+ current ($I_{K,Ca}$) and the other attributed to the Na^+ pump ($I_{Pump,Na}$).

For each given compartment i, $I_{K,Ca(i)}$ and $I_{Pump,Na(i)}$ were described by:

$$I_{ion(i)} = \bar{g}_{ion(i)} C_{ion} \left([ion]_i, t \right) \left(V_{m(i)} - E_{ion} \right) \tag{9}$$

where $\bar{g}_{ion}$ is the maximum conductance for a given ionic current, $C_{ion}([ion]_i, t)$ represents the concentration- and time-dependent activation function, and E_{ion} is the equilibrium potential associated with the current. The values of $C_{ion}([ion]_i, t)$ were obtained by solving the differential equation:

$$\frac{dC_{ion}}{dt} = \frac{[ion]_i - C_{ion}}{\tau_{C(ion)}} \tag{10}$$

where the intracellular concentration a given ion (either $[Ca^{2+}]i$ or $[Na^+]i$) was obtained from Eq. (7) and $\tau_{C(ion)}$ was 10 msec for $I_{K,Ca}$ and 100 msec for $I_{Pump,Na}$. Thus, the variable C_{ion} described the relationship between the concentration of a given ion and the magnitude of a given membrane current.

3. Results

3.1. Simulation of the excitability and action potentials of the reduced T cell model

The reduced model had all the electrical features nearly indistinguishable from those of the complete T cell model[16]. The resting membrane potential of the model neuron was $\cong -53$ mV and the input resistance was $\cong 40$ MΩ. The spiking properties of the model are reported in Fig. 2(a)-(b). By injecting a brief depolarizing current (1 msec, 5 nA) into the somatic compartment, a single action potential was fired (Fig. 2(a)). The action potential voltage peak was ~ 41 mV, the total amplitude of the spike was ~ 95 mV and the duration (as the time between the peak of the spike and the point corresponding to a voltage of 10% of the voltage peak), was ~ 0.7 msec. The pattern of discharge of the model neuron was tested by injecting a prolonged (200 msec) depolarizing current (0.4 nA) into the somatic compartment (Fig. 2(b)). The simulated burst showed adaptation of the spiking frequency, due to the presence of activity-dependent mechanisms ($I_{\text{K,Ca}}$ and $I_{\text{Pump,Na}}$). The reduced T cell model behaved similarly to the full model of the T cell[16], which was developed for studying the role of AHP in conduction failure. We utilized the more manageable reduced model to further investigate the issue of conduction failure (see below).

3.2. Activity-dependent spike transmission failure from peripheral neuritic compartment to somatic compartment: influence of membrane kinetics

After having built the reduced model of the T cell, we implemented another neural model, starting from the reduced one and changing only the kinetics of the K^+ and Na^+ currents with the values reported in Table 5, leaving unchanged all the other features. We refer to the new model as generic model. The generic model differed in the pattern of discharge and the shape and duration of the action potential. In particular, the generic model responded to the injection of depolarizing currents in a very different manner (Fig. 2(c)-(d)): by injecting a brief depolarizing current (1 msec, 5 nA) into the somatic compartment, a single action potential was fired (Fig. 2(d)). The action potential voltage peak was ~ 35 mV, the total amplitude of the spike was ~ 98 mV and the duration (as the time between the peak of the spike and the point corresponding to a voltage of 10% of the voltage peak), was ~ 2 msec. The pattern of discharge of the generic model neuron was tested by injecting a prolonged (1 sec) depolarizing current (4 nA) into the somatic compartment (Fig. 2(c)). The excitability of the generic model neuron was lower than the excitability T cell model and the duration of its action potential was ~ 3 times longer.

The generic cell model was built at the aim of investigating how neural excitability and the spike duration altered the conduction failure at site of low safety factor. In our model, a site of low safety factor is represented by the point 1a-1, at which a smaller neuritic branch (1a) joins a bigger one (1).

Table 5. Parameters of the membrane sodium and potassium currents of the Generic cell model membrane. See legend Table 3.

Currents	I_{Na}	I_{K}
E (mV)	50	−80
$\bar{g}$ (mS/cm^2)	16	64
h_A (mV)	−35	−35
s_A	−5	−9
p	3	2
$\tau_{A\text{max}}$ (msec)	6	80
$\tau_{A\text{min}}$ (msec)	1	8
$h_{\tau A}$ (mV)	−55	−20
$s_{\tau A}$	4	10
h_B (mV)	−55	
s_B ()	5	
p	1	
$\tau_{B\text{max}}$ (msec)	20	
$\tau_{B\text{min}}$ (msec)	5	
$h_{\tau B}$ (mV)	−50	

To compare the behaviors of the reduced T cell and of the generic cell model with respect to spikes conduction beyond 1a-1 when an AHP has been induced, we ran a series of simulations, consisting in eliciting trains of spikes in the distal compartment (a21) (Fig. 1(a)) and monitoring arrival of these spikes in the soma, as we systematically varied the stimulus frequencies and amplitude of the AHP. Previous works showed that spike activity in a T cell induces an AHP, which in turn, affects the excitability of the cell[13]. The activity-dependent enhancement of the AHP may induce conduction failure at points of low safety factors (i.e.: branch points and varicosities). In a previous paper[16], a systematic computational analysis of the relationship between the activity-dependent enhancement of the AHP and the spike frequencies that reach and are transmitted through branching points was undertaken. Along the same line of investigation, in the present paper we focused on other sides of the same issue. Action potentials were elicited in the distal neuritic branch of both T cell and generic cell model and conduction of these spikes throughout the cell was monitored. Spikes were elicited in compartment 21a at different frequencies: between 2 and

100 Hz, for the T cell and between 2 and 20 Hz for the generic cell (the choice of different intervals was dictated by the different excitability of the two model neurons, which determined the maximun spike frequency supported). For each frequency of stimulation in 21a, the frequency of spikes that was transmitted to the soma was measured in the absence of any AHP. Thereafter, an AHP was induced by eliciting bursts of spikes in the soma. By varying the intensity of somatic stimulation, the magnitude of the AHP was varied from 3 to 18 mV. An example of this stimulation protocol is illustrated in Fig. 3. In Fig. 3(a), the frequency of activity in 21a of the T cell model was 10 Hz. Prior to the induction of the AHP, there is no reduction in the frequency of spike activity in the soma. Spike conduction to the soma was affected by the induction of the AHP, however. After the induction of the AHP, only six of the ten spikes were conducted to the soma of the T cell model. In contrast to the T cell model, the generic cell model exhibited conduction failure prior to the induction of the AHP (Fig. 3(b)). After the induction of the AHP, with the same frequency (10 Hz) of activity stimulated in the compartment 21a, the mean frequency of spikes that were transmitted to the soma was no further reduced (Fig. 3(b)).

Thus, for a given frequency, activity-dependent enhancement of the AHP acted to reduce the fidelity of spike conduction in one case (Fig. 3(a)), and impedance mismatch acted as low-pass filter, independently by AHP, in another case (Fig. 3(b)).

Figure 4 summarizes results from all of the simulations. The frequencies of spikes that are transmitted to the soma are plotted as functions of the input frequencies (i.e., frequency of activity in the 21a) and the amplitude of the AHP. In addition, for each model, two values for the load ratio at the 1a-1 branch point were examined: 5.4 (Fig. 4(a)-(c)) and 2.2 (Fig. 4(b)-(d)). An action potential can fail to propagate from one branch to another of the same axon when the current supplied by the excited branch is insufficient to charge the membrane capacitance of the follower branch(es). For homogeneous membrane, the failure could happen when a small branch is connected to larger branch, and the current passing out of the small branch cannot drive the greater electrical load of the large branch above threshold level. The load ratio provides a measure of the probability to propagate an action potential beyond a branch point. The electrical load of a cable is generally represented by its input resistance (R_{Input}). If, for example, we have a main axon p (parent), which bifurcates in two post-branch axons $d1$ and $d2$ (daughters), we can calculate a number, i.e., LR, which is defined as the ratio between the input resistance of the parent (R_p)

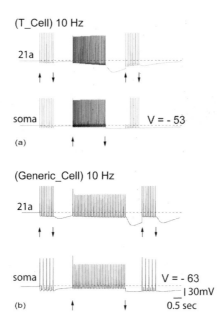

Figure 3. Activity- and frequency-dependent conduction failure of action potentials transmitted from the distal compartment 21a to the soma. Conduction of spike activity from the distal compartment 21a was examined before and after the induction of an AHP in the two different cell models, at the same frequency of activity. The AHP was induced by a brief, high-frequency burst of activity in the soma. (a) spikes were elicited in the compartment 21a of the reduced T cell model at 10 Hz. At this relatively low frequency of spike activity, the conduction of spikes from 21a to the soma was 1 : 1 (first burst of spikes). In presence of an AHP (final burst of spikes), the mean frequency of spikes that were transmitted to the soma was reduced. (b) spikes were elicited in the compartment 21a of the generic cell model, at 10 Hz. Due to impedance mismatch at 1a-1 branching point, the mean frequency of spikes that were transmitted to the soma was ∼ 5 Hz. The transmission of spikes from compartment 21a to the soma was not further degraded by the induction of the AHP. Arrows indicate the onset and offset of stimuli and the compartment into which the stimulus was injected.

to the (equivalent) input resistance of the daughters (R_{eq}):

$$\text{load ratio} = \frac{R_p}{R_{eq}}, \quad \text{with}: \ R_{eq} = \frac{R_{d1} R_{d2}}{R_{d1} + R_{d2}}.$$

The expression for the load ratio is general and is valid for every membrane (homogeneous or not) and for every structure. In the case of load ratio = 5.4, in T cell model (Fig. 4(a)), in the absence of an AHP, spike

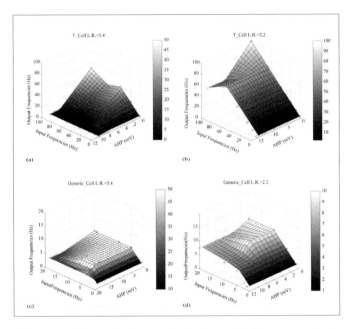

Figure 4. Summary of how the interaction between the frequency of spike activity and the AHP regulates conduction failure. The simulations illustrated in Fig. 3 were repeated using several frequencies for spike activity in 21a and inducing an AHP up to 18 mV, as measured in the soma compartment. The results were plotted with the two horizontal axes representing the frequency of spike activity in 21a and the amplitude of the AHP and the vertical axis representing the mean frequency of spike activity that was recorded in the soma. L.R. (load ratio) = 5.4: (a) when the amplitude of the AHP was zero, spikes were transmitted 1 : 1 to soma at frequencies < 40 Hz. At frequencies > 40 Hz, the signal from 21a to the soma was degraded. As the amplitude of the AHP increased, the signal from 21a to the soma degraded further, until eventually conduction was completely blocked. (c) full conduction, for low frequency spike trains, not dependent by the AHP value. At higher frequencies, there is a block of some action potentials, never completely blocked. LR = 2.2: same protocol. (b) in this case, the transmission of spike activity was much less influenced by the AHP. In addition, transmission of action potentials was never completely blocked. (d) spike conduction improved, in a manner less affected by the AHP.

activity in the periphery was transmitted to the soma with 1 : 1 fidelity at frequencies up to 40 Hz. Beyond 40 Hz, the number of spikes that reach the soma was reduced. Spiking activity at frequencies > 40 Hz was reduced by conduction failure at the 1a-1 branching point. Thus, this branching point acted as a low pass-filter. As the amplitude of the AHP was increased, the fidelity of spike transmission from the periphery decreased, and once again the higher frequencies of activity were affected to a greater degree. Ultimately, spike transmission failed completely when the amplitude of the AHP was > 10 mV. For the same load ratio = 5.4, in generic cell model (Fig. 4(c)), in the absence of an AHP, spike activity in the periphery was transmitted to the soma with 1 : 1 fidelity at frequencies up to 5 Hz. Beyond 5 Hz, the number of spikes that reach the soma was reduced. Spiking activity at frequencies > 5 Hz was reduced by conduction failure at the 1a-1 branching point. Also in this case, this branching point acted as a low pass-filter. However, as the amplitude of the AHP was increased (up to 18 mV), the fidelity of spike transmission from the periphery decrease but never below 2 Hz. When the load ratio was taken equal to 2.2 (Fig. 4(b)-(d)), the fidelity of transmission was higher. For both models, low frequencies of spike activity were unaffected and higher frequencies (of different sizes for the two models), never completely failed even with increased AHP.

3.3. *Activity-dependent spike transmission failure from peripheral neuritic compartment to somatic compartment: influence of ionic channel density distribution*

A basic assumption in both T cell and generic cell models it that channel density distribution was homogeneous. We investigated whether relaxing this condition, some changes could be found in our simulations.

We changed the density of Na^+ and K^+ channels in the compartment 1, in the direction of improving conduction. In one case, the constant factor of the Na^+ conductance was 260% of the initial value and in the other case the constant factor of the K^+ conductance was 33% of the initial value. These changes did not affected conduction failure of the generic cell model (data not shown). Whereas, the same changes applied to the T cell model (load ratio = 5.4), produced better conduction. The simulation protocol was the same as in Fig. 4. The results of the simulations are reported in Fig. 5. Fig. 5(a) and (b) summarizes the simulations results obtained after having increased the Na^+ conductance and decreased the K^+ conductance of the compartment 1, respectively. Comparing Fig. 5 to Fig. 4(a), it can be observed an improved spike conduction, for frequencies up to 40 Hz and AHP up to 10 mV.

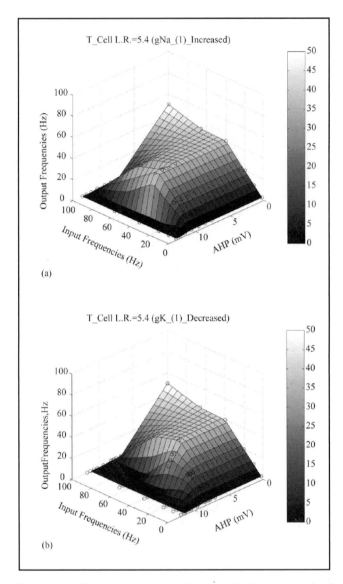

Figure 5. Summary of how the interaction between the frequency of spike activity and the AHP regulates conduction failure, with non homogeneous channel density distribution. Same protocol as in Fig. 4. (a) the maximal Na$^+$ conductance of the compartment 1 was increased to 260% of its initial value. The increased Na$^+$ conductance enhanced conduction (compare to Fig. 4(a)). (b) the maximal K$^+$ conductance of the compartment 1 was decreased to 33% of its initial value. In this case the conduction was also enhanced (compare to Fig. 4(a)).

3.4. Activity-dependent spike transmission failure from peripheral neuritic compartment to somatic compartment: influence of synaptic noise

The last part of our investigation focused on an analysis of the effects of channel and synaptic *noises* on spike transmission. Generally, neurons are subjected to background synaptic inputs, which can have a significant impact on the processing of the neural information. The T cell membrane potential, as recorded in the soma, is characterized by a certain amount of fluctuations mainly due to channel stochasticity, on the order of fraction of mV in intensity and of few msec in duration and to synaptic inputs, of the order of few mV and one — two hundred msec duration. To replicate the observed fluctuations in the resting membrane potential of the T cell, we introduced in compartment $a2$ of the reduced T cell model two kinds of noise.

One was implemented by introducing some randomness in the leak conductance and the other by adding a synaptic input to the same compartment. The parameters of the noises were chosen solely to match the known measured fluctuations of the membrane voltage. To take into account the channel noise, we added to the leak conductance of the compartment $a2$ a random fluctuation in the conductance following a Gaussian distribution and whose value was renewed every 200 integration time steps.

In addition, to take into account the synaptic noise, we added to the same compartment the following synaptic current $I_{syn} = G(V - E_{syn} = gA(V + 10)$, where $G = 0.08 \; \mu$ the maximal synaptic conductance and A, the time-dependent activation function of the conductance whose value is between 0 and 1, ($A = aY$, where a is an appropriate constant) was described by the equation

$$\frac{d^2Y}{dt^2} + \frac{2}{\tau}\frac{dY}{dt} + \frac{1}{\tau^2}Y = \frac{1}{\tau^2}X \,,$$

with $\tau = 82$ msec and the forcing function X equal to 1, when the presynaptic neuron was spiking and equal to 0, when the presynaptic neuron was not spiking. The synapse was driven by a fictitious presynaptic neuron that was endowed with a randomly varying conductance, which in a stochastic manner, took the membrane potential above the threshold for transmitter release. The presynaptic neuron had a capacitance of 1 nF, a *gleak* conductance of 10 ns and reversal potential of 10 mV and the maximum value of the random conductance $\pm 2000\%$ *gleak*, renewed every 500 integration time steps.

172

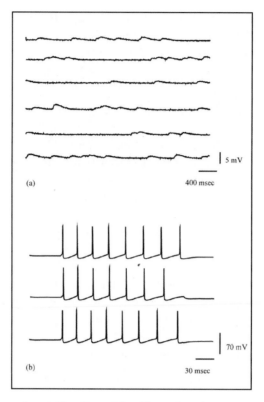

5 mV

(a)   400 msec

70 mV

(b)   30 msec

Figure 6. Noisy reduced T cell model. Channel and synaptic noises (see text for details) were added to the reduced T cell model and the effects of these new features on the neuron behavior were examined. (a) recordings of the fluctuating resting somatic membrane potential during several seconds. (b) three trains of action potentials were elicited by injecting identical depolarizing currents (0.4 nA). Note the variability of the discharge patterns, both in the interspike intervals and in the number of spikes.

After having added these new features to the model, we first ran a series of simulations to examine their effects on the resting membrane potential and on the pattern of discharge. We recorded the somatic potential for several seconds and the results are illustrated in Fig. 6(a).

The simulations illustrate the membrane voltage fluctuations. Moreover, it is important to note that the fluctuations did not elicit action potentials, which is in agreement with experimentally observations.

In a second series of simulations, we injected a depolarizing current (0.4 nA) in the somatic compartment and recorded the responses. The

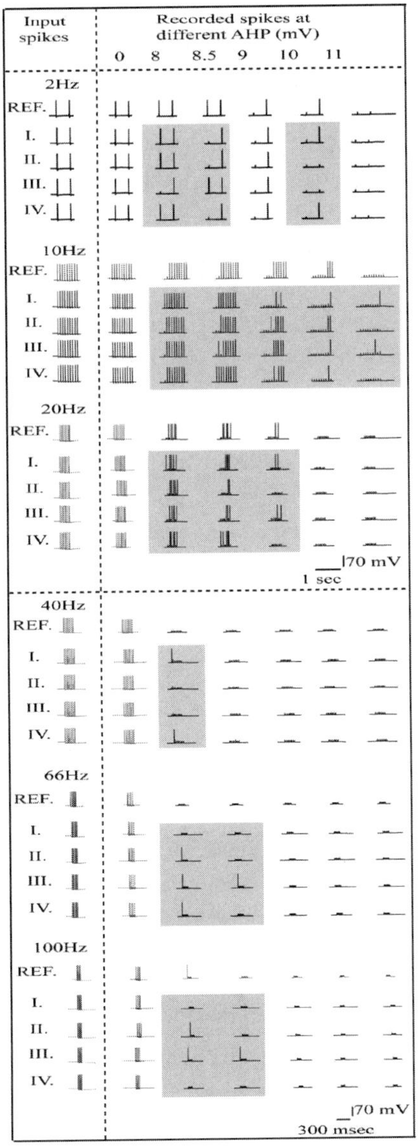

Figure 7. How spike transmission is affected by noise. The protocol of stimulation was the same as that used for producing results in Figs. 4 and 5. For each frequency of the spike train leaving the distal compartment 21a, the simulation was repeated four times and the results were shown in the I, II, III and IV rows. The shadings indicate the frequencies and the AHP at which the noise interfered with the spike transmission to the soma.

stimulus was repeated three times and the results are illustrated in Fig. 6(b). The noise introduced a variability in the patterns of discharge, both in the inter-spike times and in the number of spikes, which is in agreement with experimentally observations in our laboratory. Finally, with the new neural model available, we started to investigate how transmission was affected by the presence of noise. We repeated the simulation protocol utilized for producing Figs. 4 and 5, but the results were presented differently in Fig. 7. Moreover, each stimulus was repeated four times. Hence, for each input frequency in compartment a21 and for each value of the AHP, the transmitted trains in somatic compartment were measured four times. The results for the frequencies 2, 10, 20, 40, 66 and 100 Hz are illustrated in Fig. 7.

The major part of the variability in the number of spikes recorded in the somatic compartment, at a given frequency, was observed at intermediate frequencies and at intermediate values of the AHP. But, for higher frequencies and for higher AHP, the variability declined. The same noise addition was tested in the generic cell model, but in this case no differences were observed (data not shown).

4. Discussion

4.1. *Activity-dependent modulation of spike transmission*

The results of the present study illustrate the interdependence of the neural geometry, membrane biophysical properties and electrical activity in regulating the conduction of action potentials throughout a neuron, and they are in agreement with previous works[15,16]. Our results illustrated that, for a given action potential (with some characteristic shape, peak amplitude and duration), the fidelity of conduction was greatly influenced by both the neuritic geometry and by activity-dependent modulation of membrane conductances (Figs. 3, 4, 5, 7). Moreover, as the effects of the activity-dependent mechanisms developed, the range of frequencies that passed (full conduction) was reduced and the conduction block occurred at lower spiking frequencies. In Fig. 4, we reported the relationship between the arrival and transmitted frequencies at the low safety factor point 1a-1 and the AHP measured in the soma. By varying the geometric safety factor, (i.e., by varying inter-compartment load ratios at point 1a-1), the conduction failure was consistently affected.

4.2. *Activity-dependent modulation of spike transmission: effects of membrane channel kinetic, channel density and noise*

In this study, which represents an extension of a previous work[16], we addressed another important issue, that is: at a given geometry and AHP, what is the role of the channel kinetics in the modulation of transmission at sites of low safety factor? The results of our simulations showed the central role played by the kind of membrane channels in these processes (Fig. 4). We compared two neural models, a reduced T cell (Fig. 3(a)) versus a generic cell (Fig. 3(b)), which had different kinetics for the sodium and potassium channels. The different kinetics yielded a cell with a broader spike and higher spike threshold (Fig. 3(b)). Simulations showed that the generic cell (Fig. 4(c)-(d)) was much less sensitive to the AHP value than the reduced T cell (Fig. 4(a)-(b)), for the transmission of action potentials. Therefore, the geometry and the value of the AHP *per se* are not enough to decide whether a neuritic cable will conduct or not some trains of action potentials.

We wanted to evaluate the impact of channel density non homogeneity on the conduction failure. The same series of simulations were run, after having increased the Na^+ conductance in the compartment 1 (Fig. 5(a)) and decreased the K^+ conductance in the same compartment 1 (Fig. 5(b)). The conduction improved in both cases, showing that particular distribution of channel densities might be a relevant parameter for spike conduction throughout non homogeneous neuritic cables. The same changes did not affect conduction for the generic model cell (data not shown).

Finally, we wanted to evaluate the impact of synaptic noise on the capability of transmission and thus we introduced noise into the membrane of the reduced T cell model. We ran a series of simulations which showed some, albeit not too dramatic, effects on the spike transmission for certain frequency intervals and AHP values.

Overall, the simulations suggested that channel kinetics profoundly determined the AHP modulatory capabilities. Depending on the channel kinetics, the transmission of action potentials was very sensitive or almost insensitive to the AHP value, the channel density distribution and membrane noise, respectively. The processing or conductive features of neurons seems to be determined in first instance by the channel kinetic of the membrane and secondarily by the axonal geometry and activity-dependent processes and noise.

Acknowledgments

This work was supported by National Institutes of Health grants R01 RR11626 and P01 NS38310.

References

1. L. A. Abbott and S. B. Nelson, *Nat. Neurosci. suppl.* **3**, 1178 (2000).
2. J. D. Angstadt and W. O. Friesen, *J. Neurophysiol.* **66**, 1858 (1991).
3. I. Antonov, I. Antonova, E. Kandel and R. Hawkins, *J. Neurosci.* **21**, 6413 (2001).
4. E. Av-Ron, J. H. Byrne and D. A. Baxter, J. Undergrad. Neurosci. Edu. 4: A40-A52 (2006).
5. D. Baldissera and B. Gustafsson, *Brain Res.* **30**, 431 (1971).
6. D. A. Baxter and J. H. Byrne, *Learn. Mem.* **13**, 669 (2006).
7. D. A. Baxter and J. H. Byrne, *In: Methods in Molecular Biology: Neuroinformatics* (Crasto CJ, ed) 127 (2007).
8. H. Bostock and P. Grafe, *J. Physiol.* **365**, 239 (1985).
9. B. Brembs, F. Lorenzetti, F. Reyes, D. A. Baxter and J. H. Byrne, *Science* **296**, 1624 (2002).
10. J. Brons and C. Woody, *J. Neurophysiol.* **44**, 605 (1980).
11. T. H. Brown, J. H. Byrne, K. S. LaBar, J. E. LeDoux, D. H. Lindquist, R. F. Thompson and T. J. Teyler, *In: An Introduction to Cellular and Molecular Neuroscience (Byrne JH, Roberts JL, eds)*, 499 (2004).
12. J. C. Brumberg, L. G. Nowak and D. A. McCormick, *J. Neurosci.* **20**, 4829 (2000).
13. B. D. Burrell and C. L. Sahley, *J. Neurosci.* **24**, 4011 (2004).
14. L. Cangiano, P. Wallen and S. Grillner, *J. Neurophysiol.* **88**, 289 (2002).
15. E. Cataldo, E. *Ph.D. Thesis University of Pisa*, (2005).
16. E. Cataldo, M. Brunelli, J. H. Byrne, E. Av-Ron, Y. Cai and D. A. Baxter, *J. Comput. Neurosci.* **18**, 5 (2005).
17. D. Contreras, I. Timofeev and M. Steriade, *J. Physiol.* **494**, 251 (1996).
18. T. Crow, *Learn. and Mem.* **11**, 229 (2004).
19. T. Crow and D. Alkon, *Science* **201**, 1239 (1978).
20. M. F. Cuttle, Z. Rusznàk, A. Y. Wong, S. Owens and I. D. Forsythe, *J. Physiol.* **534**, 733 (2001).
21. G. D. Daoudal and D. Debanne, *Learn. and Mem.* **10**, 456 (2003).
22. P. Dayan and L. Abbott, *Theoretical Neuroscience - Computational and Mathematical Modeling of Neural Systems* (2001)
23. D. Debanne and M. Russier, *J. Physiol.* **548**, 663 (2003).
24. N. S. Desai, *J. Physiol. Paris* **97**, 391 (2003)
25. P.W. Dierkes, P. Hochstrate and W. R. Schlue, *Brain Res.* **746**, 285 (1997).
26. I. Epstein and E. Marder, *Biol. Cybern.* **63**, 25 (1990).
27. B. Frankenhauser and A. L. Hodgkin, *J. Physiol.* **131**, 341 (1956).
28. A. Frick, J. Magee and D. Johnston, *Nat. Neurosci.* **7**, 126 (2004).
29. K. Gainutdinov, L. Chekmarev and T. Gainutdinova, *Neuroreport* 9, 517

(1998).

30. W. Gerstner and W. M. Kistler, *In: Spiking Neuron Models*, 361 (2002).

31. B. Gustafsson, *Acta Physiol. Scand. Suppl.* **416**, 1 (1974).

32. D. Hebb, *The Organization of Behavior: a Neuropsychological Theory* (1949).

33. J. K. S. Jansen and J. G. Nicholls, *J. Physiol.* **229**, 635 (1973).

34. W. James, *Principles of Psychology* (1890).

35. U. Kim and D. A. McCormick, *J. Neurophysiol.* **80**, 1222 (1998).

36. H. Lechner and J. H. Byrne, *Neuron* **20**, 355 (1998).

37. G. LeMasson, E. Marder and L. F. Abbott, *Science* 259, 1915 (1993).

38. P. Lombardo, R. Scuri, E. Cataldo, M. Calvani, R. Nicolai, L. Mosconi and M. Brunelli, *Neurosci.* **128**, 293 (2004).

39. C. Lüscher, J. Streit, P. Lipp and H. R. Lüscher, *J. Neurophysiol.* 72, 634 (1994).

40. R. Malenka and R. Nicoll, *Science* **285**, 1870 (1999).

41. Y. Manor, C. Koch and I. Segev, *Biophys. J.* **60**, 1424 (1991).

42. K. Miller, *Neuron* **17**, 371 (1996).

43. K. Miller and J. MacKay, *Neural Comp.* **6**, 100 (1994).

44. H. Misonou, D. P. Mohapatra, E. W. Park, V. Leung, D. Zhen, K. Misonou, A. E. Anderson and J. S. Trimmer, *Nat. Neurosci.* **7**, 711 (2004).

45. J. J. Moyer, L. Thompson and J. Disterhoft, *J. Neurosci.* **16**, 5536 (1996).

46. J. Munoz-Cuevas, H. Vara and A. Colino, *J. Physiol.* **558**, 527 (2004).

47. A. Nelson, C. Krispel, C. Sekirnjak and S. du Lac, *Neuron* **40**, 609 (2003).

48. I. Parnas and I. Segev, *J. Physiol.* **295**, 323 (1979).

49. S. A. Raymond, *J. Physiol.* **290**, 273 (1979).

50. J. Rinzel, *Ann. N. Y. Acad. Sci.* **591**, 51 (1990).

51. M. Rioult-Pedotti, D. Friedman and J. Donoghue, *Science* **290**, 533 (2000).

52. M. Rogan, U. Staubli and J. Le-Doux, *Nat.* **390**, 604 (1997).

53. P. Roper, J. Callaway, T. Shevchenko, R. Teruyama and W. Armstrong, *J. Comput. Neurosci.* **15**, 367 (2003).

54. P. Sah and J. Bekkers, *J. Neurosci.* **16**, 4537 (1996).

55. M. V. Sanchez-Vives, L. G. Nowak and D. A. McCormick, *J. Neurosci.* **20**, 4286 (2000).

56. W. R. Schlue, *J. Neurophysiol.* **65**, 736 (1991).

57. P. C. Schwindt, W. J. Spain and W. E. Crill, *J. Neurophysiol.* **61**, 233 (1989).

58. R. Scuri, P. Lombardo, E. Cataldo, C. Ristori and M. Brunelli, *Neurosci.* (2007).

59. I. Segev and E. Schneidman, *J. Physiol. Paris* **93**, 263 (1999).

60. I. Segev and M. London, *Science* **290**, 744 (2000).

61. P. G. Sokolove, *Biophys. J.* **12**, 1429 (1972).

62. P. G. Sokolove and I. M. Cooke, *J. Gen. Physiol.* **57**, 125 (1971).

63. R. C. Thomas, *Physiol. Rev.* **52**, 563 (1972).

64. S. M. Thompson and D. A. Prince, *J. Neurophysiol.* **56**, 507 (1986).

65. V. Torre, *J. Physiol.* **333**, 315 (1982).

66. G. G. Turrigiano, *TINS* **22**, 221 (1999).

67. G. G. Turrigiano and S. B. Nelson, *Curr. Op. Neurobiol* **10**, 358 (2000).

68. G. G. Turrigiano, K. R. Leslie, N. S. Desai, L. C. Rutherford and S. B.

Nelson, *Nat.* **391**, 892 (1998).

69. I. van Welie, J. A. van Hooft and W. J. Wadman, *PNAS* **101**, 5123 (2004).
70. W. Yamada, C. Koch and P. Adams, *Multiple channels and calcium dynamics. In: Methods in Neuronal Modeling (Koch C, Segev I, eds)*, 137 (1989).
71. D. Yang and P. Mu-ming, *Neuron* **44**, 23 (2004).
72. W. Zhang and D. J. Linden, *Nat. Rev. Neurosci.* **4**, 885 (2003).
73. I. Ziv, D. A. Baxter and J. H. Byrne, *J. Neurophysiol.* **71**, 294 (1994).

MODELS OF ALLEE EFFECTS AND THEIR IMPLICATIONS FOR POPULATION AND COMMUNITY DYNAMICS

LUDĚK BEREC

Department of Theoretical Ecology,
Institute of Entomology, Biology Centre ASCR,
Branišovská 31, 37005 České Budějovice, Czech Republic
E-mail: `berec@entu.cas.cz`

Allee effects are broadly defined as a decline in individual fitness at low population sizes or densities. Although the roots of the concept go back at least to 1920's, until recently, Allee effects eked out on the periphery of ecological theory, in the shade of the prominently discussed negative density dependence. The situation has changed dramatically in the last ten years or so, and we can find an ever increasing number of studies considering Allee effects from an ever increasing range of disciplines. Mathematical models have always been an important tool by which to assess impacts of Allee effects for population and community dynamics. Actually, much of what we know about Allee effects comes from mathematical models. Up to now, Allee effects have been examined in the context of most existing model structures, and significantly altered our picture of population and community dynamics based on assuming negative density dependence only. This essay concerns modelling Allee effects and presenting their major implications for population and community dynamics. It shows what types of model are available, how they can be used, and, most importantly, how they have contributed to our understanding of the dynamical consequences of Allee effects. This essay is a modified extract from the book "Allee effects in ecology and evolution", co-authored by Franck Courchamp, Luděk Berec and Joanna Gascoigne, which is going to be published by the Oxford University Press early in 2008.

1. Introduction

1.1. *What are Allee effects?*

Allee effects are broadly defined as a decline in individual fitness at low population sizes or densities[1,2]. Although Allee effects long lingered on in the shade of negative density dependence, we now face an ever increasing number of studies considering Allee effects from an ever increasing range of disciplines. These include population and community ecology, population genetics, evolutionary biology, conservation biology and population management, parasitology, biodiversity theory, and even social sciences. Plenty of mechanisms have already been shown to create Allee effects[3,4], including the need to find mates, cooperative anti-predator behaviour, environmental conditioning, and inbreeding.

There are two pairs of concepts that are extremely useful in classifying Allee effects[5,4]. Researchers distinguish component and demographic Allee effects, depending on whether the decline in individual fitness at low population sizes or densities concerns a fitness component or the overall fitness, respectively. Whereas component Allee effects need not always give rise to demographic Allee effects, presence of the latter always indicates, by definition, an underlying component Allee effect. Demographic Allee effects are further categorized as strong or weak (Fig. 1). A strong Allee effect results in a critical population threshold (or Allee threshold) below which the population crashes to extinction. On the other hand, the (overall) fitness decline due to a weak Allee effect still ensures a positive yet reduced per capita growth rate of small or sparse populations.

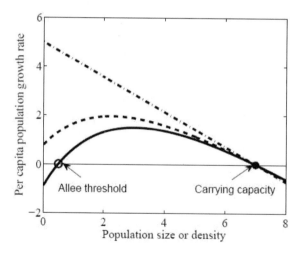

Figure 1. The per capita population growth rate as a function of population size or density for negative density dependence (dash-dot), weak Allee effect (dashed) and strong Allee effect (solid). Note a reduction in the per capita population growth rate in small or sparse populations if a demographic Allee effect is present. Strong Allee effects give rise to negative per capita population growth rates once the population size or density falls below the Allee threshold.

1.2. Role of models in understanding Allee effects

Mathematical models have always been an important tool by which to assess impacts of Allee effects for population and community dynamics. Actually, much of what we know about Allee effects comes from mathematical models. They provide formulas which help both formalize component and demographic Allee effects in the form of an equation and search for both types of Allee effect in empirical data. Also, models are inevitable at predicting wider population and community level implications of both component and demographic Allee effects, and thus help develop general ecological theory. On a more applied side, since Allee effects (predominantly) act in small or sparse populations and can give rise to critical population thresholds, they are a vital consideration in preventing rare, declining, endangered or fragmented populations from extinction. By providing predictions of the outcome of management actions in conservation (e.g. African wild dogs[6]), exploitation (e.g. red sea urchins[7]) or pest control (e.g. gypsy moths[8]), models save money where direct experimentation would be costly and allow decision-making where it would be unethical (such as any manipulation of very rare species). Allee effects can also be considered from an evolutionary perspective. One may ask whether populations have evolved to avoid or at least mitigate Allee effects. In addition, Allee effects themselves may act as a selection pressure, and members of populations subject to Allee effects may evolve different characteristics than those subject to negative density dependence. As it is extremely difficult to demonstrate any of these aspects empirically, mathematical models have played an essential role also here.

2. Component Allee Effects and Population Dynamics

This section considers the ways how component Allee effects can be modelled, how they can be used to assess presence (and strength) or absence of a demographic Allee effect in a population, how can two or more component Allee effects interact when affecting the same population simultaneously, and what is the meaning of a demographic Allee effect in state-structured, multi-dimensional population models.

2.1. Models of component Allee effects

Models of component Allee effects refer to quantitative descriptions of a positive relationship between a fitness component of individuals in a population and population size or density[9,3,10,11]. They can be completely phenomenological or have a mechanistic underpinning. Even though many of the assumptions underlying mechanistic models are unlikely to hold literally in nature, these models may still prove sufficiently flexible to fit a variety of actual observations[9,7,12].

The component Allee effect due to a reduced success of finding mates in low-density populations is in a sense the flagship of component Allee effects – there is hardly any work on Allee effects that would miss mentioning it among the main Allee effect mechanisms. Almost invariably, modelling the mate-finding Allee effect consists of derivation of a function $P(M, F)$, representing the female mating rate or the probability that a female mates within a unit time; M and F denote male and female population density, respectively. The mate-finding process dictates $P(M, F)$ to satisfy a few generic properties (no mate-finding Allee effect corresponds to $P(M, F) = 1$ for any positive M and F, i.e. to the assumption that even one male suffices to fertilize all females):

- $P(0, F) = 0$ for any F
- $P(M, F)$ is a non-decreasing function of M for any fixed F
- $P(M, F)$ is a non-increasing function of F for any fixed M
- $P(M, F)$ approaches 1 for a sufficiently large M/F ratio

The popularity of the mate-finding Allee effect has resulted in a large number of specific $P(M, F)$ forms[13], of which the functions

$$P(M, F) = 1 - \exp(-M/\theta) \tag{1}$$

$$P(M, F) = M/(M + \theta) \tag{2}$$

have been used most frequently, presumably due to their simplicity and plausibility of underlying assumptions. Actually, model (1) fits well the female mating rate in many invertebrate[9,12,14] and vertebrate[15,16] species. Model (2), on the other hand, is more tractable mathematically. Interestingly, the mate-finding Allee effect with Eq. (2) as its descriptor can also be invoked artificially by flooding a wild population with sterile males and so disrupting fertilization – θ now represents the density of sterile males released and $M/(M + \theta)$ the probability that a randomly encountered male is fertile[17,9,13].

Allee effects were long considered to occur mainly in species with specific types of life history, especially those with particular modes of reproduction. Later on, factors external to populations – exploitation[9,2,18] and predation[3,11] – have been demonstrated to generate Allee effects, too. Given this, Allee effects might be relevant to many more taxa than previously thought. Let $p(N)$ be the overall mortality rate in a population of size or density N due to exploitation or predation, and let $F(N) = 1 - [p(N)/N]\Delta t$ denote the probability that an exploited or preyed individual escapes exploitation or predation in a (small) time interval Δt. For any of these two processes to create a component Allee effect in prey survival, we require $F(N)$ to increase with increasing prey population size or density (i.e. $dF/dN > 0$ for all positive N). This implies

$$\frac{dp(N)}{dN} < \frac{p(N)}{N} \quad \text{for all} \ \ N > 0 \tag{3}$$

Exploitation and predation are in many respects analogous, and exploitation has long been one of the major focuses for Allee effect models[13]. Traditionally, researchers distinguish constant effort exploitation, with a population exploited at a rate proportional to its size or density – $p(N) = EN$ for some positive E, and constant yield exploitation, with a population exploited at a constant rate irrespectively of its size or density – $p(N) = E$ for some positive E. Equation (3) shows that whereas constant yield exploitation generates a component Allee effect in species survival, constant effort exploitation does not[9,2]. Actually, constant effort exploitation decreases the overall individual fitness uniformly over the whole range of population sizes or densities, and thus makes existing demographic Allee effects (if there are any) stronger (consider moving the solid or dashed line in Fig. 1 down by a specific amount E).

Regarding predation, the relationship between Allee effects and full predator-prey dynamics is discussed in Sec. 4.1. Here we assume that the predator population does not respond numerically to changes in prey abundance. This situation best corresponds to a generalist predator which is in a dynamic association with another (main) prey and consumes the target prey only as a secondary resource. We may write $p(N) = Nf(N)$ where $f(N)$ is a functional response, the central concept to studies of predation, describing the consumption rate of prey by each individual predator as a function of prey (and generally also predator) population size or density. The literature abounds in mathematical descriptions of the functional response, of which the linear ($f(N) = aN$, $a > 0$), (Holling) type II (e.g. $f(N) = aN^d/(1 + bN^d)$, $a, b > 0$, $0 < d \le 1$) and (Holling) type III (e.g. $f(N) = aN^d/(1 + bN^d)$, $a, b > 0$, $d > 1$) functional responses are used most often[19]. Of these, only the type II functional response, a hyperbolical curve that rises from zero to an asymptote, satisfies Eq. (3) and hence generates a component Allee effect in prey due to predation (or predation-driven Allee effect). This is because when prey density increases, there are more prey individuals per predator attack, and thus a lower probability that any prey individual will be consumed by a predator. Figure 2 shows $f(N)$ and $F(N)$ for the type II functional response.

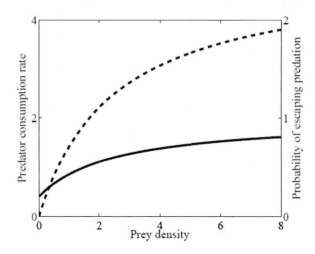

Figure 2. Predator consumption rate (dashed, left axis) and corresponding probability of prey escaping predation (solid, right axis) for the type II functional response. Parameters: $a = 2$, $b = 0.4$, $d = 1$, $\Delta t = 0.4$.

The above discussion of component Allee effect models is by no means complete. Not only that there are plenty of other models describing the mate-finding Allee effect and that a number of formulas can certainly be formulated that satisfy condition (3), but there are also models of other component Allee effects (social Allee effects, inbreeding, etc.), both phenomenological and mechanistic, which are nonetheless hard to present in a unified framework; instead, we suggest consulting Refs. 20, 21, 22, 23 for some examples.

2.2. From component to demographic Allee effects

How can one determine whether there is a demographic Allee effect in a population? Besides a number of empirical approaches[13], it is the one based on modelling population dynamics that we discuss here in more detail. Generally, given a component Allee effect in the population, we may embed its model in a properly structured, dynamical model of this population. Let us start with something simple and consider a continuous-time model

$$dN/dt = Nb(N) - Nd(N) \tag{4}$$

where $b(N)$ and $d(N)$ are (positively or negatively) density-dependent per capita birth and death rates, respectively. Such models can be adequate, e.g., if knowledge about the population is not much detailed or as a raw test of whether an identified component Allee effect is strong enough to generate a demographic Allee effect. Consider a specific version of model (4) which describes dynamics of a population subject to two component Allee effects:

$$b(N) = bN/(N + \theta) \text{ and } d(N) = d(1 + N/K) + E/N \tag{5}$$

Here $b(N)$ represents a positively density-dependent birth rate (mate-finding Allee effect) and $d(N)$ is composed of two terms, a negatively density-dependent survival rate due to factors other than exploitation, and a positively density-dependent survival rate owing to exploitation (Allee effect due to constant yield exploitation). Here $\theta > 0$ defines strength of the mate-finding Allee effect, b is the maximum per capita birth rate, d is the per capita mortality rate at low densities and in absence of exploitation, $K > 0$ scales the carrying capacity of the environment, and $E > 0$ defines the overall exploitation rate (i.e. strength of the exploitation-driven Allee effect). Because setting $\theta = 0$ and $E = 0$ switches off the mate-finding and exploitation-driven Allee effect, respectively, we can assess impacts of both component Allee effects on population dynamics in isolation and also simultaneously (Fig. 3). The most interesting result to see is a kind of synergy between the two component Allee effects: the overall Allee threshold due to the double Allee effect is disproportionately larger than any of the Allee thresholds corresponding to single Allee effects. Interactions between two or more component Allee effects can be intriguing – for example, two component Allee effects each of which results in a weak demographic Allee effect (with no extinction threshold) when alone can give rise to a strong demographic Allee effect (with an extinction threshold) when acting simultaneously[4].

Although simple, "unstructured" models are sufficient to capture and understand dynamics of some populations, in others, population structure must be accounted for, since it often affects population growth rate and size in a significant way. State-structured population models are models in which the term "state" collectively represents structural types of the population as age, sex, body size, developmental stage, local density or any combination of these. In mathematical terms, these are all matrix population

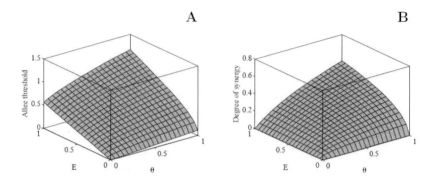

Figure 3. Overall Allee threshold (A) and degree of synergy between two component Allee effects (B) as functions of strengths of the mate-finding Allee effect (θ) and the exploitation-driven Allee effect (E). Not surprisingly, the overall Allee threshold increases with an increase in strength of any of the component Allee effects (A). More interestingly, the degree of synergy, measured as the difference between the overall Allee threshold due to the double Allee effect and the sum of the Allee thresholds corresponding to each single Allee effect increases as well with an increase in strength of any of the component Allee effects (B).

models (whether continuous- or discrete-time) in which component Allee effects appear as positively density-dependent or hump-shaped (ingredients of) matrix elements[7,24]. In principle, any state-structured model can be used to study Allee effects – researchers have already studied Allee effects via population models structured with respect to age[25,26], sex[27,28,29,30], body size[7,31], developmental stage[32,33,24] and local density[34,22].

In state-structured population models we are no longer working with population size or density as a single number, but rather with a whole distribution of sizes or densities across population states (such as densities of males and females in a sex-structured population). This necessarily modifies our understanding of the Allee threshold as compared with that developed for one-equation models discussed above (see also Fig. 1). In multi-dimensional, state-structured models (such as a two-dimensional sex-structured model) it is the combination of sizes or densities of subpopulations corresponding to population states that decides on persistence or extinction of the whole population, not the size or density of the population as a whole. Schreiber[35] showed that for a broad class of state-structured population models with a component Allee effect, one can observe at most three possible dynamical outcomes: (i) population extinction from any initial size or density, (ii) population persistence for any initial size or density, and (iii) bistable dynamics analogous to a strong Allee effect in one-equation models. In particular, regarding (iii), there exists a (continuously differentiable) invariant hypersurface in the space of subpopulation sizes or densities which divides this space into two exclusive parts; for points below the hypersurface the population goes extinct and for points above it the population persists[35]. This is a very important result as it says that the hypersurface, which we refer to as the "extinction boundary" further on, is a generalization of the concept of Allee threshold for state-structured, multi-dimensional population models.

In fact, Schreiber's result does not say anything about actual shape of the extinction boundary other than that it is smooth and bounded. Its specific shape has to be revealed by simulations in each particular instance, as exemplified by Fig. 4A based on the two-dimensional, stage-structured model developed by Gascoigne and Lipcius[24]. The class of models for which such an extinction boundary exists goes beyond Schreiber's assumptions[29,30] (Fig. 4B); this extinction boundary is not bounded.

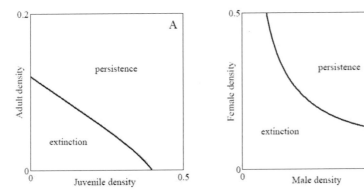

Figure 4. Extinction boundaries in a two-dimensional, age-structured model with a component Allee effect affecting reproduction (A), and in a two-dimensional, sex-structured model with the mate-finding Allee effect (B).

3. Demographic Allee Effects and Metapopulation Dynamics

In this section, we define phenomenological models of demographic Allee effects, show how they can be embedded in spatially explicit metapopulation models, and review the major findings regarding impacts of demographic Allee effects for metapopulation and invasion dynamics.

3.1. *Models of demographic Allee effects*

Majority of continuous-time models of single-species dynamics have the form

$$dN/dt = Ng(N) \qquad (6)$$

while discrete-time models look like

$$N_{t+1} = N_t g(N_t) \qquad (7)$$

In both equations, t represents time, N population size or density, and $g(N)$ the per capita population growth rate. When a population faces a demographic Allee effect, $g(N)$ has a hump-shaped form (Fig. 1). A wide spectrum of functions have been used in the literature to describe such a hump-shaped form, mostly phenomenologically[10,13], of which the following two are used most frequently in continuous-time models:

$$g(N) = r(1 - N/K)(N/K - A/K) \qquad (8)$$
$$g(N) = r(1 - N/K)(1 - (A + C)/(N + C)) \qquad (9)$$

(To be used in discrete-time models, these functions are often transformed as $g(N) \to \exp(g(N))$.) In both models, r is the maximum per capita growth rate in absence of the Allee effect, K the carrying capacity, A the Allee threshold (if A is negative we have a weak Allee effect), and C a non-negative parameter affecting the overall shape of function (9).

3.2. *Allee effects in metapopulations*

Real populations are spatially extended and spatial population models have already become a common tool in population ecology. Some spatial models have been developed to study the effects of spatial variation within (local) populations occupying a relatively homogeneous patch of habitat[29,36]. Scaling one level up, there are models which describe dynamics of metapopulations; that is, of collections of spatially separated, local populations connected by dispersal. Space is usually heterogeneous in this case, composed of a mosaic of inhabitable patches located within a sea of uninhabitable matrix[37]. Metapopulation models are currently of high interest in conservation, since human-induced fragmentation of previously continuous natural habitats is ubiquitous. However, metapopulation models can also be used to describe species in which there is a natural tendency to maintain spatially separated groups in an otherwise homogeneous environment, such as obligately cooperative breeders living in spatially separated breeding groups[6].

Dispersal strategies are extremely important in determining persistence of metapopulations, particularly when local populations are subject to strong Allee effects. The simplest spatial extension of non-spatial models is two-patch metapopulations. With Allee effects and without dispersal, both local populations persist only if they are above its respective Allee threshold. Dispersal modifies this picture – two additional stable equilibria may appear, corresponding to the situation where one population persists below its Allee threshold and the other above it (Fig. 5). The figure is a result of the following model:

$$dN_1/dt = -N_1^3 + 2N_1^2 - dN_1 + \mu(N_2 - N_1)$$
$$dN_2/dt = -N_2^3 + 2N_2^2 - dN_2 + \mu(N_1 - N_2)$$

(10)

in which the cubic terms in local population density N_i define a strong Allee effect in each patch and the exchange term $\mu(N_i - N_j)$ determines the intensity of between-patch migration. Survival below the Allee threshold in one patch is possible because the low-density (sink) population is steadily supplied with individuals from the high-density (source) population[38,39,40]. This basic picture has been further elaborated in many ways[38,39,40,41].

Although these results concern two-patch metapopulations, the underlying mechanisms are general and should operate in multiple-patch systems, too. Only when all local populations fall below the Allee threshold (or, more precisely, below the extinction boundary as delineated, e.g., in Fig. 5) will the whole metapopulation start to decline. Under some circumstances, however, the metapopulation may persist even if all local populations are initially below their respective Allee threshold, provided that dispersal is asymmetric, i.e. some patches receive more immigrants than they lose[42]. The asymmetry in dispersal raises the density of some local populations above their respective Allee threshold and these may persist permanently; some sinks remain maintained through source-sink dynamics, and other local populations go extinct.

Now, consider a metapopulation of a species that lives in spatially separated groups and needs to exceed a minimum size to successfully reproduce, i.e. possesses a strong Allee effect at the group level (such as in the highly endangered African wild dog). Simulations show that the metapopulation is likely to go extinct (the model is stochastic) once the number of groups falls below a critical value[6]. Ecologically, this is because fewer packs generate fewer dispersers which are in turn less likely to colonize a

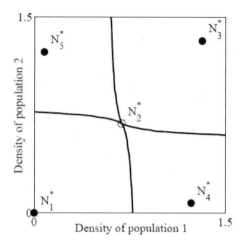

Figure 5. Source-sink dynamics of a two-patch system with strong Allee effects in each patch. There are four locally stable equilibria (full circles) for the adopted parameter values. The point N_2^* locates Allee thresholds corresponding the two (local) populations. Solid lines delimit the areas of attraction of the respective stable equilibria. Parameters of model (10): $d = 0.9$, $\mu = 0.05$.

patch successfully, because they arrive only in small numbers and do not always manage to find other dispersers to successfully reproduce. Natural decreases in group numbers due to group extinctions are thus not balanced by new colonizations in small metapopulations. The threshold behaviour exemplified by this model was termed the Allee-like effect, by analogy with Allee effects operating at the level of (local) populations[43]. In general, the critical number of occupied patches below which the metapopulation as a whole goes extinct is seriously affected by such variables as the Allee effect strength, emigration rate from patches, migration mortality, initial size of local populations and degree of demographic stochasticity[44].

Allee-like effects of this kind may also occur in populations of parasites spreading through hosts – hosts serve as patches for local populations of parasites. If there is a strong Allee effect in local parasite populations due to, e.g., a need to overcome the host's immune system or find a mate for sexual reproduction[45,46], too small a number of infected hosts (i.e. too small a number of occupied patches) can cause the infection to decline rather than spread; see also Sec. 4.2 for Allee effects in host-parasite interactions.

Allee-like effects can also occur in metapopulations that do not suffer from strong Allee effects in local dynamics. When only a few patches are occupied the number of dispersers is likely to be small; these may not succeed in reaching and hence colonizing empty suitable patches and/or rescuing currently occupied patches from extinction, particularly if local population growth rate is low or dispersal mortality is high, even if local populations grow logistically[43]. Since these conditions are apparently more stringent than those given in the previous two paragraphs, metapopulations with Allee effects at the local population level are likely to be more vulnerable to extinction. In any case, all situations of this kind can be grossly described as

$$\frac{dp}{dt} = (cp(1-p) - ep) \frac{p-a}{1-e/c} \tag{11}$$

where p is the fraction of suitable patches that are occupied, c and e are the (maximum) patch colonization rate and the patch extinction rate, respectively, and $0 < a < 1 - e/c$ is the threshold fraction of occupied patches below which the metapopulation as a whole goes extinct and above which it approaches the equilibrium fraction $1 - e/c$ of occupied patches.

Table 1 summarizes the major implications of Allee effects for local populations as well as the metapopulation as a whole along a continuum of fragmentation, a picture stemming from models roughly reviewed in this section.

3.3. Allee effects and invasion dynamics

Invasion dynamics and their interaction with Allee effects has been a topic of wide interest and much work has already been done in this direction[47]. Some of the established model predictions hold equally for strong and weak Allee effects, some are limited only to strong Allee effects. Invasion is a spatial phenomenon which usually consists of a localized appearance of a small number of invading individuals, establishment of an initial population, and spatial expansion out of its initially small area of occurrence. Allee effects, together with demographic and environmental stochasticity, hamper successful establishment of invaders (or (re)introduced species or biological control agents). Small invading populations have a disproportionately reduced chance of establishment; whereas strong Allee effects create an Allee threshold, weak Allee effects lower growth rate of small or sparse populations and hence intensify impacts of stochastic effects[47,13].

Models predominantly used to explore implications of Allee effects for dynamics of invasive species treat space as a continuous entity, although discrete-space models are also occasionally used[48]. Continuous-

Table 1. The major implications of Allee effects for local populations as well as the metapopulation as a whole along a continuum of fragmentation, as implied by mathematical models.

Degree of fragmentation	Modelled consequences for local populations	Modelled consequences for the metapopulation as a whole
No fragmentation – one large continuous population	Likely large or dense enough to exceed any potential Allee threshold	Likely large or dense enough to exceed any potential Allee threshold
Weak fragmentation – metapopulation (high dispersal)	Local populations below the Allee threshold can be rescued from extinction through source-sink dynamics if a neighbour local population is well above it	Metapopulation as a whole can persist thanks to source-sink dynamics if dispersal rates are not too high (in which case we effectively observe one large continuous population)
Medium fragmentation – metapopulation (low dispersal)	Small or sparse populations will not be rescued and vacant patches will have difficulties to be colonised if there is a strong Allee effect in local populations	Low dispersal rates or high dispersal mortalities increase the risk of metapopulation extinction even if local populations do not demonstrate a strong Allee effect
Strong fragmentation – isolated local populations	Any of them likely lives on their own (i.e. there is negligible between-patch dispersal) and is small or sparse enough and thus highly threatened if there is a strong Allee effect	Local populations behave independently, i.e. probability of metapopulation extinction is a product of extinction probabilities of individual local populations. If all fall below the Allee threshold the whole metapopulation goes extinct.

space framework includes discrete-time, integro-difference models[49,50,51] and continuous-time, reaction-diffusion or reaction-diffusion-advection models[52,53,54,55]. In one spatial dimension, integro-difference models evaluate population density at time $t + 1$ and spatial location x as a common contribution of population growth from time t to $t + 1$ at any spatial location y and a chance of dispersal of a fraction of individuals from location y to x:

$$N_{t+1}(x) = \int_{-\infty}^{\infty} N_t(y)g(N_t(y))k(x,y)dy \qquad (12)$$

where $k(x,y)$ specifies a dispersal kernel – the probability that an individual disperses from x to y within a time step. Reaction-diffusion(-advection) models, on the other hand, are defined by means of partial differential equations. In the simplest case of one spatial dimension and random dispersal (i.e. diffusion), they take the form

$$\frac{\partial N(x,t)}{\partial t} = N(x,t)g(N(x,t)) + D\frac{\partial^2 N(x,t)}{\partial x^2} \tag{13}$$

where D is a diffusion constant. In both cases, Allee effects enter the equations as hump-shaped forms of $g(N)$, e.g. as models (8) or (9).

For random dispersers, dispersal works to dilute the population at any given location, thereby requiring higher initial densities to overcome the Allee effect than in comparative non-spatial models[47]. In other words, a founder population subject to a strong Allee effect may fail to establish itself, even when it is initially above its Allee threshold, because its growth may not be sufficient to offset the decline in local population density through diffusion. The success of a founder population in invasion will thus depend not only on the initial population density, but also on the shape and size of the area the population initially occupies – the larger is the initially occupied area, the lower the initial density is allowed to be for the population to invade[52,49,56]. The ability of a founder population to grow and expand will also depend on the overall habitat size[52,57,58,59] and the intensity of advection to which the population is exposed[55,58].

Once a founder population starts to grow, it also starts to spread geographically. Allee effects can adversely affect rates of spread of invading populations. In particular, following an initial period of slower expansion, the rate of spread accelerates, a pattern found in a variety of invading species[52,49,50]. Using an integro-difference model with a strong Allee effect in local dynamics, structured with respect to developmental stage (juveniles and adults) and largely parameterized by data collected on the house finch, Veit and Lewis[50] successfully recreated the patterns of temporal change in the rate of spread of the invading bird population and in its density near the point of initial invasion – both data and model predictions showed an abrupt acceleration to a constant rate of spatial spread, following an initial period of slower expansion. Models also demonstrate that the (asymptotic) rate of spread is negatively correlated with the Allee effect strength, for both weak and strong Allee effects[52,58].

Sometimes, in theory at least, ecological processes are so finely balanced that "range pinning" occurs. Range pinning means that even though other suitable habitat is available nearby, the actually occupied area neither expands nor shrinks[42,60,48,61]. Up to now, range pinning has only been demonstrated in models with discrete space in which the range boundary is maintained via a source-sink equilibrium between an inner patch and a neighbouring outer patch.

Invasion need not always proceed at a constant rate. For example, the gypsy moth, a pest in the northeastern United States causing extensive defoliation in deciduous forests, spreads via periodic pulsed range expansions interspersed with periods of relative quiescence[8]. Modelling showed that a strong Allee effect is a necessary prerequisite for such cyclical invasion dynamics[8]; a strong Allee effect has indeed been detected in this species, as a consequence of reduced mate-finding efficiency in low-density populations[12,62,8,63]. This study bears a fundamental message: it is extremely important to incorporate Allee effects into models of invasion dynamics if they are suspected to occur, especially if the models are used to assess efficiency of management strategies.

4. Demographic Allee Effects and Community Dynamics

Although single-species models have been very useful for understanding Allee effects, their failure to account for interspecific interactions is in many instances an oversimplification. Any of these interactions – be it predation, competition, parasitism, mutualism, or any combination of these – can be affected by Allee effects and demonstrate dynamics which are different from dynamics where Allee effects are absent. Below, we focus on predation and parasitism, and refer to Ref. 13 for a review of the remaining interactions.

4.1. Allee effects and predator-prey dynamics

Here, contrary to the treatment in Sec. 2.1, we consider full predator-prey dynamics, i.e., predators are assumed to numerically respond to their main prey. Arguably the most common modelling framework used to explore predator-prey dynamics involves one equation for the prey and one for the predator:

$$dN/dt = Ng(N) - f(N,P)P$$
$$dP/dt = ef(N,P)P - mP$$

$$(14)$$

where $g(N)$ is the per capita growth rate of prey, $f(N,P)$ is the predator functional response, m is the density-independent per capita mortality rate of predators and e denotes efficiency with which energy obtained from consuming prey is transformed into predator offspring. Although the literature varies on how to include Allee effects in prey, the most straightforward way is to use any (phenomenological) model of demographic Allee effects discussed in Sec. 3.1 for the per capita growth rate of prey $g(N)$. Functions

(8) and (9) are the two most frequent examples[64,65,66,67]. An alternative is to multiply the "growth" term $Ng(N)$ in the prey equation of model (14) by $N/(N + \theta)$, where θ determines the Allee effect strength[68]. In the only study we know of which explored simple predator-prey models with an Allee effect in predators, this Allee effect was modelled via multiplying the "birth" term $ef(N, P)P$ in the predator equation of model (14) by the term $P/(P + \theta)$, where θ determines the Allee effect strength[68].

Allee effects in prey generally destabilize predator-prey dynamics. This destabilization can be "quantified" in a number of different ways:

- Strong Allee effects in prey can prevent predator-prey systems from exhibiting sustained cycles. Indeed, if troughs in cycles observed in the comparative models without Allee effects extend below the Allee threshold in prey density, both prey and predators go extinct[66,69,68,67].

- Strong Allee effects in prey may cause the coexistence equilibrium to change from stable to unstable or may extend the time needed to reach the stable equilibrium[68]. This outcome has also been predicted by the model with the strong Allee effect in predators[68].

- Predators reduce the equilibrium population size of prey with Allee effects more than that of prey without Allee effects and also enlarge the range of model parameters for which both prey and predators go extinct, thus increasing system vulnerability to collapse[66]. This effect also extends to metapopulations (or rather metacommunities): Allee effects at the local population level generate lower metapopulation sizes and higher risks of metapopulation extinction; Allee effects above a certain strength cause metacommunity extinction[66]. Contrary to that, Allee effects in predators increase the equilibrium prey density as compared with predator-prey systems where predators have no Allee effect[68].

- Weak Allee effects in prey cause the predator-prey systems to cycle for a wider range of model parameters than systems without Allee effects, provided that predators have a type II or a weakly sigmoidal type III functional response[67].

Allee effects may also alter spatial dynamics of predator-prey systems.

Consider the following model:

$$\frac{\partial N(\mathbf{x}, t)}{\partial t} = Ng(N) - f(N, P)P + D_1\nabla^2 N(\mathbf{x}, t)$$
$$\frac{\partial P(\mathbf{x}, t)}{\partial t} = ef(N, P)P - mP + D_2\nabla^2 P(\mathbf{x}, t)$$
(15)

in which $g(N)$ has the form (9). Allee effects in prey significantly increase spatio-temporal complexity of predator-prey dynamics, especially as regards the ability of both species to co-invade initially empty habitats[70]. This corresponds to a biological control scenario, where, soon after an alien species invades an empty habitat, its natural enemy is introduced to try and slow down, prevent, or even reverse the invasion. There is a wide set of model parameters for which temporal oscillations in both total and local population densities are chaotic while the spatial distribution of both species remains localized and regular, contrary to predator-prey systems with a weak or no Allee effect in prey where the most complicated behaviour is regular population fluctuations[65]. For some other parameter values, the system demonstrates a patchy invasion, i.e., populations spread via formation, interaction and movement of separate patches, another dynamical regime that is absent when there is no Allee effect in prey[64,71]. Introducing predators quickly after a successful prey invasion is vital to slowing down or preventing further spread, or even to reversing the spread and bringing about prey extinction – the chance of reversing the prey invasion decreases the larger is the delay following prey arrival[72].

It is generally assumed that predators always negatively impact their prey, but there is some evidence that they can also affect prey in a positive way, such as through the ability of predators to mineralize nutrients which limit prey or prey resources or to "transport" prey to places where intraspecific prey competition is lower (e.g. granivores dispersing seeds)[73]. Brown et al.[73] modelled this interaction and showed that an Allee effect can emerge in predators due to this mechanism because low-density populations of predators reduce the availability of their prey. We speak of emergent Allee effects whenever a population model predicts a demographic Allee effect without considering any underlying component Allee effect explicitly, i.e. without using any component Allee effect model.

Positive effects of predators on their prey may also be less direct (and actually not so positive from the prey point of view). In particular, size- or stage-selective predators induce changes in the size or stage distribution of their prey which in turn has a feedback on predator performance. If, in absence of predators, the prey population is regulated by negative density dependence in development through one of its size or stage classes, and there is overcompensation in this regulation such that a decrease in density of the regulating size or stage class will increase its total development rate, predators feeding on a size or stage class other than the regulating class can actually increase density of the size or stage class on which they feed. This positive relationship between predator-induced mortality and density of the consumed prey class occurs because by eating more, predators reduce competition within the regulating class and cause a higher inflow of individuals into the class they consume[74,31,75]. As a consequence, an Allee effect can emerge in predators – small predator populations cannot invade the prey population since they are not able to induce changes in the prey size or stage distribution which are necessary for their own persistence. Contrary to that, predators feeding on the regulating class will always have a negative effect on density of the consumed prey class and thus will never exhibit an (emergent) Allee effect[31]. Some parameter combinations in the underlying model may even give rise to a regime in which there are two alternative, stable predator-prey equilibria[75].

4.2. Allee effects and host-parasite dynamics

Predictions of host-parasite (or disease or infection) dynamics as well as explorations of various infection-based pest control and vaccination strategies now depend heavily on mathematical models[78,76,77]. Parasites are traditionally categorized as macroparasites (helminths and arthropods) or microparasites (bacteria, viruses, protozoans and fungi), and models used to predict dynamics of their interaction with their hosts are fundamentally different for these two broad categories. Host-microparasite systems are usually modelled as a set of differential equations representing dynamics of and flows between different epidemiological stages of the host (such as susceptible, infectious and recovered). Host-macroparasite consider distributions of the numbers of parasites per individual host and describe dynamics of and flows between the host and different life-history stages of the macroparasite.

The simplest model describing dynamics of a microparasitic infection is arguably the so-called SI model

$$dS/dt = b(N)N - m(N)S - T(S, I)$$
$$dI/dt = T(S, I) - m(N)I - \mu I$$
(16)

where $b(N)$ is the host's per capita birth rate, $m(N)$ the per capita mortality rate due to reasons other than infection, μ the disease-induced mortality rate, and $T(S, I)$ the transmission rate of the infection between infectious (I) and susceptible (S) individuals; $N = S + I$ is the overall host population size or density. Two most common forms for the disease transmission rate are the mass action term $T(S, I) = \beta SI$ (often used for air-borne diseases) and the proportionate mixing term $T(S, I) = \beta SI/N$ (commonly used for sexually transmitted or vector-borne diseases), for a positive transmission parameter β, although other forms have also been proposed[78,79,80,81]. Model (16) has seen many modifications (only susceptibles can be allowed to reproduce, pathogens can be transmitted vertically between parents and offspring, there may be a latent stage of the disease, a recovery stage from the infection, etc.)[82,76] and has turned out to be a good starting point for exploring interactions between Allee effects and host-parasite dynamics.

Allee effects which affect hosts in absence of parasites are often treated as a hump-shaped form of the per capita birth rate $b(N)$. Strong Allee effects in hosts give rise to an Allee threshold N_A below which the host population goes extinct. Many infections, on the other hand, face an invasion threshold N_I – there is a minimum number or density of susceptibles for the infection to spread from an infectious individual[76,81]. The relative value of these two thresholds is important: in epidemiology, we require $N_A < N_I$ to eradicate the parasite population without simultaneously bringing the host population below its Allee threshold; in pest control, conversely, $N_A > N_I$ facilitates pest eradication by an introduced infection[81].

In general, Allee effects among hosts reduce the range of host densities for which an initially rare infection can spread and decrease the disease prevalence where it spreads[83]. Hence, host populations affected by Allee effects are better protected or suffer less from the infection than those with no Allee effect. On the other hand, once the infection is established, the impact of Allee effects on the host population can also be negative, as they usually cause the infection to reduce the overall host population density more than in the Allee effect absence and widens the range of parasite species that can lead the host to extinction[83].

As with predation, Allee effects have also been shown to affect spatio-temporal patterns of host-parasite dynamics. Consider an invasive species that succeeds in establishing in a local area and starts to spread spatially. A parasite introduced to control the host spread can slow down or even reverse its invasion, depending on its virulence[71,84]. The reversal occurs when the disease is able to catch up spatially to the expanding host population front. At the edge of this front, when disease-induced mortality lies within a certain range, the infection can both be prevented from fading out and also outweigh the host population growth (which is low due to the Allee effect and low host density at the front edge). As with predation, spatio-temporal patterns can be rather complex[71].

Allee effects may also be inherent in parasites. Many (asexually reproducing) macroparasites and many microparasites need to exceed a threshold in load to overcome any defence of their hosts or to spread effectively through the host population (this kind of threshold behaviour is often referred to as sigmoidal dose-dependent response in parasitology)[85,86]. Regoes et al.[80] analyzed the model

$$dS/dt = \lambda - dS - S\beta(W)$$
$$dI/dt = S\beta(W) - dI - \mu I \qquad (17)$$
$$dW/dt = cI - uW$$

to study the case where the per capita transmission rate $\beta(W)$ of the infection responds sigmoidally to the density W of free parasites in the environment; d and u are the per capita natural mortality of hosts and free parasites, respectively, and c is the rate at which free parasits are produced per infected host. The parasite extinction is always a stable event. Where the stable host-parasite equilibrium exists, an initial concentration of the parasite has to exceed a critical value (Allee threshold) to invade a host population (Fig. 6). This is in sharp contrast to the system in which the infection rate is linear ($\beta(W) = \beta W$, the mass action transmission rate) where the parasite can invade the host population from any initial concentration provided that the host-parasite equilibrium exists[80] (Fig. 6).

A similar need to cross two thresholds in order to invade a host population has been predicted for *Wolbachia*, a bacterium which infects arthropod species, including a high proportion of all insects, and is transmitted vertically from mother to offspring[87,88,89]. These are: (i) the transmission efficiency from infected mother to her offspring has to exceed a threshold level (parasite invasion threshold) and (ii) given condition (i), the initial fraction of infected individuals in the host population must exceed another threshold value (Allee threshold)[87,88,90,89]. In a spatial context, the Allee threshold translates into an initial inoculum size and an area the infected individuals initially occupy; the minimum levels of both for the bacterium to invade depend on the dispersal kernel adopted, i.e. on details of how infected individuals disperse within the host population[90].

Finally, once entering a host, sexually reproducing macroparasites often need to mate there, and hence may suffer from a mate-finding Allee effect[91,45,46]. Focusing their model on parasitic nematodes affecting farmed

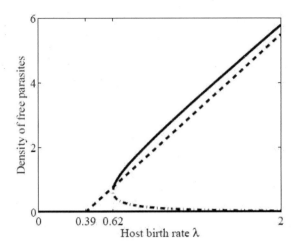

Figure 6. Equilibrium density of free parasites as a function of host birth rate λ for model (17). The dashed line denotes the globally stable equilibrium corresponding to the linear transmission rate $\beta(W) = \beta W$, i.e. the case of no Allee effect (this line coincides with the solid line below $\lambda < 0.39$). The remaining lines are for the model version with the Allee effect, i.e. with sigmoidal $\beta(W)$: solid line is for the stable equilibria (including the zero density of free parasites) and dash-dot line for the unstable equilibria (Allee threshold).

ruminants, Cornell et al.[92] showed that the mean parasite load per host individual varied over the grazing season and depended heavily on both the herd size and the patchiness of the infection (i.e. spatial clumping of parasite larvae in the pasture, which affected the number of larvae ingested simultaneously). In order to invade the host population, there is always a need for the parasite to exceed a threshold density. In addition, small, strongly clumped worm populations were found most vulnerable to extinction if the ruminant herd was small as well.

5. Concluding Remarks

This essay focused on what models used to explore Allee effects look like, how and where they can be used, and, most importantly, how they have contributed to our understanding of the impacts of Allee effects on the structure and dynamics of both single-species populations and multiple-species communities. Allee effects have been examined in the context of most existing population model structures, and have significantly altered our picture of population dynamics based on assuming negative density dependence only. Strong demographic Allee effects are of particular importance in this respect because they introduce thresholds in population size or density below which population vulnerability to extinction disproportionately increases. It is important to realize that these thresholds take on different forms in different modelling frameworks. They are single numbers in one-equation population models, hypersurfaces or extinction boundaries in state-structured, multi-dimensional population models, combinations of numbers or extinction boundaries and spatial distributions in spatially-structured models, and inflection points of the extinction probability as a function of initial population size or density in one-equation stochastic models (not considered in this essay; see Refs. 13, 9, 93, 94).

The impacts of weak demographic Allee effects, which do not result in negative per capita population growth rates in small or sparse populations, have been rather underexplored in studies of population and community dynamics. This is perhaps because systems with weak Allee effects are generally perceived as behaving analogously to systems with no (demographic) Allee effects. This is not always the case, however. Weak Allee effects can prevent successful invasion if the invaded habitat is finite, in much the same way as strong Allee effects in an infinite habitat[52,57,58,59]. Also, they are important to consider when combined with other (weak or strong) Allee effects, as the resulting double Allee effect can be disproportionately stronger than any of the single ones[4].

In this essay, we have indicated (and refer to[13] for a more comprehensive account), among other things, that:

- There is a great variety of models of component and demographic Allee effects that can be used either alone, as statistical models for data fitting, or as building blocks of more complicated models, to predict broader implications of Allee effects for population and community dynamics;
- Allee effects at a local population level can generate thresholds in the number of occupied patches below which the metapopulation is likely to go extinct – Allee-like effects;
- Allee effects can significantly alter patterns of invasion as compared with models without Allee effects, such as lowering the rate of invasion, accelerating the geographical spread after a period of relative quiescence, or allowing cyclical invasion or range pinning;
- Allee effects destabilize predator-prey dynamics;
- Predators or parasites can slow down or even reverse the spread of

prey or host populations if the latter are subject to Allee effects;

- Allee effects may oppositely impact host populations threatened by infections: protecting populations against many infections, but increasing their damage if an infection manages to establish;

The limited length of this essay did not allow us to discuss all the interesting and important issues concerning Allee effects. In particular, we had to omit discussions of how implications of Allee effects are modified by an operation of stochasticity (whether demographic, environmental or both), how Allee effects affect some widely used biodiversity models and how they impact more complex food webs beyond simple two-species relationships (actually, very little is done in this respect), how commonly used empirical Allee effect models look and how they are used to estimate presence and strength of component and demographic Allee effects, what are Allee effects at intermediate population sizes or densities and how they contribute to Allee effects theory (also here, very little is done as yet), and, last but not least, how Allee effects impact and are impacted by predictions of evolutionary models.

The wealth of predictions of population models with demographic Allee effects does not mean that nothing remains to be explored. On the contrary, we expect that future studies will supply us with further novel and interesting insights into many aspects of the interplay of Allee effects and population and community dynamics, especially as regards multiple Allee effects, Allee effects at intermediate population sizes or densities, and the role of Allee effects in multiple-species interactions.

Acknowledgments

I would like to thank Franck Courchamp and Joanna Gascoigne for friendly and stimulating collaboration on our common book on Allee effects from which this essay is a modified extract, and the Oxford University Press for giving me the permission to publish this essay. I also acknowledge funding from the Institute of Entomology (Z50070508) and the Grant Agency of the Academy of Sciences of the Czech Republic (KJB600070602).

References

1. Courchamp F, Clutton-Brock TH and Grenfell BT (1999a). Inverse density dependence and the Allee effect. Trends in Ecology & Evolution, 14, 405-410.
2. Stephens PA and Sutherland WJ (1999). Consequences of the Allee effect

for behaviour, ecology and conservation. Trends in Ecology & Evolution, 14, 401-405.

3. Liermann M and Hilborn R (2001). Depensation: evidence, models and implications. Fish and Fisheries, 2, 33-58.

4. Berec L, Angulo E and Courchamp F (2007). Multiple Allee effects and population management. Trends in Ecology & Evolution, 22, 185-191.

5. Stephens PA, Sutherland WJ and Freckleton R (1999). What is the Allee effect? Oikos, 87, 185-190.

6. Courchamp F, Clutton-Brock TH and Grenfell BT (2000a). Multipack dynamics and the Allee effect in the African wild dog, Lycaon pictus. Animal Conservation, 3, 277-285.

7. Pfister CA and Bradbury A (1996). Harvesting red sea urchins: recent effects and future predictions. Ecological Applications, 6, 298-310.

8. Jonhson DM, Liebhold AM, Tobin PC and Bjornstad ON (2006). Allee effects and pulsed invasion by the gypsy moth. Nature, 444, 361-363.

9. Dennis B (1989). Allee effects: population growth, critical density and the chance of extinction. Natural Resource Modeling, 3, 481-538.

10. Boukal DS and Berec L (2002). Single-species models of the Allee effect: extinction boundaries, sex ratios and mate encounters. Journal of Theoretical Biology, 218, 375-394.

11. Gascoigne J and Lipcius R (2004a). Allee effects driven by predation. Journal of Applied Ecology, 41, 801-810.

12. Tcheslavskaia K, Brewster CC and Sharov AA (2002). Mating success of gypsy moth (Lepidoptera: Lymantriidae) females in Southern Wisconsin. Great Lakes Entomologist, 35, 1-7.

13. Courchamp F, Berec L, Gascoigne J (2008). Allee effects in ecology and conservation. Oxford University Press, Oxford.

14. Vogel H, Czihak G, Chang P and Wolf W (1982). Fertilization kinetics of sea urchin eggs. Mathematical Biosciences, 58, 189-216.

15. Mosimann JE (1958). The evolutionary significance of rare matings in animal populations. Evolution, 12, 246-261.

16. Rowe S, Hutchings JA, Bekkevold D and Rakitin A (2004). Depensation, probability of fertilization, and the mating system of Atlantic cod (Gadus morhua L.). ICES Journal of Marine Science, 61, 1144-1150.

17. Barclay H and Mackauer M (1980a). The sterile insect release method for pest control: a density-dependent model. Environmental Entomology, 9, 810-817.

18. Courchamp F, Angulo E, Rivalan P et al. (2006). Rarity value and species extinction: the anthropogenic Allee effect. Plos Biology, 4, 2405-2410.

19. Skalski GT and Gilliam JF (2001). Functional responses with predator interference: viable alternatives to the Holling type II model. Ecology, 82, 3083-3092.

20. Stephens PA, Frey-Roos F, Arnold W and Sutherland WJ (2002a). Model complexity and population predictions. The alpine marmot as a case study. Journal of Animal Ecology, 71, 343-361.

21. Wertheim B, Marchais J, Vet LEM and Dicke M (2002). Allee effect in larval resource exploitation in Drosophila: an interaction among density of adults,

larvae, and micro-organisms. Ecological Entomology, 27, 608-617.

22. Taylor CM, Davis HG, Civille JC, Grevstad FS and Hastings A (2004). Consequences of an Allee effect in the invasion of a pacific estuary by *Spartina alterniflora*. Ecology, 85, 3254-3266.

23. Rohlfs M and Hoffmeister TS (2003). An evolutionary explanation of the aggregation model of species coexistence. Proceedings of the Royal Society of London Series B-Biological Sciences, 270, S33-S35.

24. Gascoigne J and Lipcius RN (2005). Periodic dynamics in a two-stage Allee effect model are driven by tension between stage equilibria. Theoretical Population Biology, 68, 237-241.

25. Cushing JM (1994). Oscillations in age-structured population models with an Allee effect. Journal of Computational and Applied Mathematics, 52, 71-80.

26. Kulenovic MRS and Yakubu AA (2004). Compensatory versus overcompensatory dynamics in density-dependent Leslie models. Journal of Difference Equations and Applications, 10, 1251-1265.

27. Hopper KR and Roush RT (1993). Mate finding, dispersal, number released, and the success of biological-control introductions. Ecological Entomology, 18, 321-331.

28. Ashih AC and Wilson WG (2001). Two-sex population dynamics in space: effects of gestation time on persistence. Theoretical Population Biology, 60, 93-106.

29. Berec L, Boukal DS and Berec M (2001). Linking the Allee effect, sexual reproduction, and temperature-dependent sex determination via spatial dynamics. American Naturalist, 157, 217-230.

30. Berec L and Boukal DS (2004). Implications of mate search, mate choice and divorce rate for population dynamics of sexually reproducing species. Oikos, 104, 122-132.

31. de Roos AM, Persson L and Thieme HR (2003). Emergent Allee effects in top predators feeding on structured prey populations. Proceedings of the Royal Society of London Series B-Biological Sciences, 270, 611-618.

32. Gascoigne J and Lipcius R (2004b). Allee effects in marine systems. Marine Ecology Progress Series, 269, 49-59.

33. Gascoigne J and Lipcius RN (2004c). Conserving populations at low abundance: delayed functional maturity and Allee effects in reproductive behaviour of the queen conch *Strombus gigas*. Marine Ecology-Progress Series, 284, 185-194.

34. Taylor CM and Hastings A (2004). Finding optimal control strategies for invasive species: a density-structured model for *Spartina alterniflora*. Journal of Applied Ecology, 41, 1049-1057.

35. Schreiber SJ (2004). On Allee effects in structured populations. Proceedings of the American Mathematical Society, 132, 3047-3053.

36. South AB and Kenward RE (2001). Mate finding, dispersal distances and population growth in invading species: a spatially explicit model. Oikos, 95, 53-58.

37. Hanski I and Gaggiotti OE (2004). Ecology, genetics, and evolution of metapopulations. Elsevier Academic Press.

38. Gruntfest Y, Arditi R and Dombrovsky Y (1997). A fragmented population in a varying environment. Journal of Theoretical Biology, 185, 539-547.

39. Amarasekare P (1998b). Interactions between local dynamics and dispersal: insights from single species models. Theoretical Population Biology, 53, 44-59.

40. Gyllenberg M, Hemminki J and Tammaru T (1999). Allee effects can both conserve and create spatial heterogeneity in population densities. Theoretical Population Biology, 56, 231-242.

41. Greene CM (2003). Habitat selection reduces extinction of populations subject to Allee effects. Theoretical Population Biology, 64, 1-10.

42. Padrón V and Trevisan MC (2000). Effect of aggregating behavior on population recovery on a set of habitat islands. Mathematical Biosciences, 165, 63-78.

43. Amarasekare P (1998a). Allee effects in metapopulation dynamics. American Naturalist, 152, 298-302.

44. Zhou SR and Wang G (2004). Allee-like effects in metapopulation dynamics. Mathematical Biosciences, 189, 103-113.

45. May RM (1977a) Togetherness among Schistosomes: its effects on the dynamics of the infection. Mathematical Biosciences, 35, 301-343.

46. Garrett KA and Bowden RL (2002). An Allee effect reduces the invasive potential of Tilletia indica. Phytopathology, 92, 1152-1159.

47. Taylor CM and Hastings A (2005). Allee effects in biological invasions. Ecology Letters, 8, 895-908.

48. Hadjiavgousti D and Ichtiaroglou S (2004). Existence of stable localized structures in population dynamics through the Allee effect. Chaos, Solitons & Fractals, 21, 119-131.

49. Kot M, Lewis MA and van den Driessche P (1996). Dispersal data and the spread of invading organisms. Ecology, 77, 2027-2042.

50. Veit RR and Lewis MA (1996). Dispersal, population growth, and the Allee effect: dynamics of the house finch invasion of eastern North America. American Naturalist, 148, 255-274.

51. Wang MH, Kot M and Neubert MG (2002). Integrodifference equations, Allee effects, and invasions. Journal of Mathematical Biology, 44, 150-168.

52. Lewis MA and Kareiva P (1993). Allee dynamics and the spread of invading organisms. Theoretical Population Biology, 43, 141-158.

53. Lewis MA and van den Driessche P (1993). Waves of extinction from sterile insect release. Mathematical Biosciences, 116, 221-247.

54. Wang MH and Kot M (2001). Speeds of invasion in a model with strong or weak Allee effects. Mathematical Biosciences, 171, 83-97.

55. Petrovskii SV and Li BL (2003). An exactly solvable model of population dynamics with density- dependent migrations and the Allee effect. Mathematical Biosciences, 186, 79-91.

56. Soboleva TK, Shorten PR, Pleasants AB and Rae AL (2003). Qualitative theory of the spread of a new gene into a resident population. Ecological Modelling, 163, 33-44.

57. Hastings A (1996). Models of spatial spread: a synthesis. Biological Conser-

vation, 78, 143-148.

58. Almeida RC, Delphim SA and da S Costa MI (2006). A numerical model to solve single-species invasion problems with Allee effects. Ecological Modelling, 192, 601-617.

59. Shi JP and Shivaji R (2006). Persistence in reaction diffusion models with weak Allee effect. Journal of Mathematical Biology, 52, 807-829.

60. Keitt TH, Lewis MA and Holt RD (2001). Allee effects, invasion pinning, and species' borders. American Naturalist, 157, 203-216.

61. Hadjiavgousti D and Ichtiaroglou S (2006). Allee effect in population dynamics: existence of breather-like behavior and control of chaos through dispersal. International Journal of Bifurcation and Chaos, 16, 2001-2012.

62. Liebhold A and Bascompte J (2003). The Allee effect, stochastic dynamics and the eradication of alien species. Ecology Letters, 6, 133-140.

63. Tobin PC, Whitmire SL, Johnson DM, Bjornstad ON and Liebhold AM (2007). Invasion speed is affected by geographical variation in the strength of Allee effects. Ecology Letters, 10, 36-43.

64. Petrovskii SV, Morozov AY and Venturino E (2002). Allee effect makes possible patchy invasion in a predator-prey system. Ecology Letters, 5, 345-352.

65. Morozov A, Petrovskii S and Li BL (2004). Bifurcations and chaos in a predator-prey system with the Allee effect. Proceedings of the Royal Society of London Series B-Biological Sciences, 271, 1407-1414.

66. Courchamp F, Grenfell BT and Clutton-Brock TH (2000b). Impact of natural enemies on obligately cooperative breeders. Oikos, 91, 311-322.

67. Boukal DS, Sabelis MW and Berec L (2007). How predator functional responses and Allee effects in prey affect the paradox of enrichment and population collapses. Theoretical Population Biology, in press.

68. Zhou SR, Liu YF and Wang G (2005). The stability of predator-prey systems subject to the Allee effects. Theoretical Population Biology, 67, 23-31.

69. Kent A, Doncaster CP and Sluckin T (2003). Consequences for predators of rescue and Allee effects on prey. Ecological Modelling, 162, 233-245.

70. Petrovskii SV, Morozov AY and Li BL (2005a). Regimes of biological invasion in a predator-prey system with the Allee effect. Bulletin of Mathematical Biology, 67, 637-661.

71. Petrovskii SV, Malchow H, Hilker FM and Venturino E (2005b). Patterns of patchy spread in deterministic and stochastic models of biological invasion and biological control. Biological Invasions, 7, 771-793.

72. Fagan W, Lewis M, Neubert M and van den Driessche P (2002). Invasion theory and biological control. Ecology Letters, 5, 148-157.

73. Brown DH, Ferris H, Fu S and Plant R (2004). Modeling direct positive feedback between predators and prey. Theoretical Population Biology, 65, 143-152.

74. de Roos AM and Persson L (2002). Size-dependent life-history traits promote catastrophic collapses of top predators. Proceedings of the National Academy of Sciences of the United States of America, 99, 12907-12912.

75. van Kooten T, de Roos AM and Persson M (2005). Bistability and an Allee effect as emergent consequences of stage-specific predation. Journal of The-

oretical Biology, 237, 67-74.

76. Hethcote HW (2000). The mathematics of infectious diseases. SIAM Review, 42, 599-653.

77. Shea K, Thrall PH and Burdon JJ (2000). An integrated approach to management in epidemiology and pest control. Ecology Letters, 3, 150-158.

78. Courchamp F and Cornell SJ (2000). Virus-vectored immunocontraception to control feral cats on islands: a mathematical model. Journal of Applied Ecology, 37, 903-913.

79. McCallum H, Barlow N and Hone J (2001). How should pathogen transmission be modelled? Trends in Ecology & Evolution, 16, 295-300.

80. Regoes RR, Ebert D and Bonhoeffer S (2002). Dose-dependent infection rates of parasites produce the Allee effect in epidemiology. Proceedings of the Royal Society of London Series B-Biological Sciences, 269, 271-279.

81. Deredec A and Courchamp F (2003). Extinction thresholds in host-parasite dynamics. Annales Zoologici Fennici, 40, 115-130.

82. May RM and Anderson RM (1979). Population biology of infectious diseases: part I. Nature, 280, 361-367.

83. Deredec A and Courchamp F (2006). Combined impacts of Allee effects and parasitism. Oikos, 112, 667-679.

84. Hilker FM, Lewis MA, Seno H, Langlais M and Malchow H (2005). Pathogens can slow down or reverse invasion fronts of their hosts. Biological Invasions, 7, 817-832.

85. Devi KU and Rao CUM (2006). Allee effect in the infection dynamics of the entomopathogenic fungus *Beauveria bassiana* (Bals) Vuill. on the beetle, *Mylabris pustulata*. Mycopathologia, 161, 385-394.

86. Ebert D, Zschokke-Rohringer CD and Carius HJ (2000). Dose effects and density-dependent regulation of two microparasites of *Daphnia magna*. Oecologia, 122, 200-209.

87. Turelli M and Hoffmann AA (1991). Rapid spread of an inherited incompatibility factor in California *Drosophila*. Nature, 353, 440-442.

88. Turelli M and Hoffmann AA (1995). Cytoplasmic incompatibility in *Drosophila simulans*: dynamics and parameter estimates from natural populations. Genetics, 140, 1319-1338.

89. Keeling MJ, Jiggins FM and Read JM (2003). The invasion and coexistence of competing Wolbachia strains. Heredity, 91, 382-388.

90. Schofield P (2002). Spatially explicit models of Turelli-Hoffmann *Wolbachia* invasive wave fronts. Journal of Theoretical Biology, 215, 121-131.

91. Rohlf FJ (1969). The effect of clumped distributions in sparse populations. Ecology, 50, 716-721.

92. Cornell SJ, Isham VS and Grenfell BT (2004). Stochastic and spatial dynamics of nematode parasites in farmed ruminants. Proceedings of the Royal Society of London Series B-Biological Sciences, 271, 1243-1250.

93. Dennis B (2002). Allee effects in stochastic populations. Oikos, 96, 389-401.

94. Allen LJS, Fagan JF, Högnäs G and Fagerholm H (2005). Population extinction in discrete-time stochastic population models with an Allee effect. Journal of Difference Equations and Applications, 11, 273-293.

THE DECOUPLING & SOLUTION OF LOGISTIC & CLASSICAL TWO-SPECIES LOTKA-VOLTERRA DYNAMICS WITH VARIABLE PRODUCTION RATES

CHARLES E. M. PEARCE

School of Mathematical Sciences,
The University of Adelaide,
Adelaide SA 5005, Australia
E-mail: charles.pearce@adelaide.edu.au

ROY B. LEIPNIK*

Department of Mathematics,
University of California at Santa Barbara,
California 93106-3080, USA

Central to the dynamics of population biology are various versions of the Lotka-Volterra equations. Particular cases may be used to model competitive, commensal, predatory and other behaviour. Similar equations describe macro-economic interactions, epidemics and other processes of mass action. Refinements of many versions of these equations have been exhibited in the BIOMAT meetings to describe new biological features. The solutions to such equations may display a variety of forms. Broadly speaking, quite a lot of qualitative information may often be obtained about the solutions. In many cases it is convenient to eliminate time from the equations, so obtaining equations for the joint values of population sizes. By contrast, determining explicit expressions for the evolution with time of the component population sizes frequently appears to be infeasible, although numerical procedures may be available. This is less than fully satisfactory, as numerical work with many joint choices of the driving parameters, involving a number of dimensions, may be needed to obtain any sort of collective overview of the population dynamics. Perhaps surprisingly, in a number of general situations it is in fact possible to derive explicit solutions for the population sizes as functions of time. We may indeed decouple the basic equations and perform explicit quadratures. Several techniques are available for our purposes, such as Euler substitution and separation and the methods of Gambier and Painlevé. The working is sometimes heavy but often manageable. We illustrate these techniques by considering a number of well-known classical biological models from the literature.

---

*Deceased October 10, 2006.

1. Background and Introduction

Lotka-Volterra dynamics occur in a variety of applications, such as with population biology (see, for example, May[1,2] and Hassell and May[3]) and macro-economic interactions (see Goodwin[4]). A simple version is the logistic two-species equations

$$\dot{x}_1 = a_1(t)x_1 - b_1 x_1^2 + c_1 x_1 x_2,$$
$$\dot{x}_2 = a_2(t)x_2 - b_2 x_2^2 + c_2 x_1 x_2 \tag{1}$$

with interaction. Here b_1, $b_2 \geq 0$, c_1, c_2 are real and $a_1(t)$, $a_2(t)$ are integrable over bounded sets. These equations are often considered to be analytically inaccessible, though see the limit-cycle results of Bautin and Leontovich[5], referenced also in the general treatment of planar limit cycles in Andronov et al.[6,7].

The Euler substitution $x = \dot{u}/(bu)$ resolves the classic Riccati equation

$$\dot{x} = a(t)x - bx^2$$

by reducing it to

$$\ddot{u} = a(t)\dot{u}.$$

In this chapter we show how this idea may be extended to treat (1). Examination of four special cases for $a_j(t)$ variable and constant and b_j positive and zero culminates in reducing the classic Lotka-Volterra system to quadrature for each species for nearly all parametric choices. We cover systems which are competitive ($a_j > 0$, $c_j < 0$), commensal ($a_j > 0$, $c_j > 0$), predatory ($a_1, c_2 > 0$, $a_2, c_1 < 0$), suicidal ($a_j < 0$, $c_j < 0$), paradoxical ($a_1, c_1 > 0$, $a_2, c_2 < 0$), etc. The critical quantities are a_1/c_1, a_2/c_2, $a_1/a_2 - 1$, $a_2/a_1 - 1$. The quadrature details are difficult, as they involve functions such as $x \cdot \exp(\pm a/x)$ combined with $\exp(\pm c x^{\pm \ell})$.

2. Euler Substitution and Separation

We begin by making the Euler substitution

$$x_i = \dot{u}_i/(b_i u_i) \quad \text{for } i = 1, 2$$

in (1). This yields the preliminary separation

$$\ddot{u}_1/\dot{u}_1 = a_1(t) + (c_1/b_2)\dot{u}_2/u_2,$$
$$\ddot{u}_2/\dot{u}_2 = a_2(t) + (c_2/b_1)\dot{u}_1/u_1,$$

which integrates to give

$$\ln |\dot{u}_1(t)/\dot{u}_1(t_0)| = A_1(t) + (c_1/b_2) \ln |u_2(t)/u_2(t_0)|,$$
$$\ln |\dot{u}_2(t)/\dot{u}_2(t_0)| = A_2(t) + (c_2/b_1) \ln |u_1(t)/u_1(t_0)|,$$

where

$$A_j(t) = \int_{t_0}^{t} a_j(t_j) \, dt_j \quad \text{for } t \geq t_0 \text{ and } j = 1, 2.$$

In the usual applications $x_1, x_2 \geq 0$, so the variables u_i and $\dot{u}_i$ $(i = 1, 2)$ are taken as positive. This permits the power substitutions

$$u_1 = v_1^{\alpha}, \quad u_2 = v_2^{\beta}$$

for $v_1, v_2 > 0$ and α, β real and nonzero.

The new equations are

$$\ln \left| \alpha v_1^{\alpha-1} \dot{v}_1 / \dot{u}_1(t_0) \right| = A_1(t) + (c_1/b_2) \ln \left(v_2^{\beta} / u_2(t_0) \right),$$
$$\ln \left| \beta v_2^{\beta-1} \dot{v}_2 / \dot{u}_2(t_0) \right| = A_2(t) + (c_2/b_1) \ln \left(v_1^{\alpha} / u_1(t_0) \right),$$

so

$$\begin{aligned}
\ln \dot{v}_1 = {}& A_1(t) + (1 - \alpha) \ln v_1 + (c_1/b_2)\beta \ln v_2 \\
& + \ln \dot{u}_1(t_0) - \ln \alpha - (c_1/b_2) \ln u_2(t_0), \\
\ln \dot{v}_2 = {}& A_2(t) + (1 - \beta) \ln v_2 + (c_2/b_1)\alpha \ln v_1 \\
& + \ln \dot{u}_2(t_0) - \ln \beta - (c_2/b_1) \ln u_1(t_0).
\end{aligned}$$

The choices $1 - \alpha = c_2\alpha/b_1$ and $1 - \beta = c_1\beta/b_2$ yield

$$\alpha = b_1/(b_1 + c_2), \quad \beta = b_2/(b_2 + c_1)$$

and

$$\ln \dot{v}_1 = A_1(t) + k_1 + \ln v,$$
$$\ln \dot{v}_2 = A_2(t) + k_2 + \ln v,$$

where

$$v = v_1^{1-\alpha} v_2^{1-\beta},$$
$$k_1 = \ln \left(\dot{u}_1(t_0)/\alpha [u_2(t_0)]^{c_1/b_2} \right),$$
$$k_2 = \ln \left(\dot{u}_2(t_0)/\beta [u_1(t_0)]^{c_2/b_1} \right).$$

Hence

$$\dot{v}_i = v \exp(A_i + k_i) \quad \text{for } i = 1, 2$$

and

$$\dot{v}_1/\dot{v}_2 = \Phi_1/\Phi_2, \tag{2}$$

where

$$\Phi_j(t) = \exp(A_j + k_j) \quad \text{for } j = 1, 2.$$

The major result is the pair of separations

$$v_1^{\alpha-1}\dot{v}_1 = v_2^{1-\beta}\Phi_1, \quad v_2^{\beta-1}\dot{v}_2 = v_1^{1-\alpha}\Phi_2. \tag{3}$$

3. Equal Production Rates

In the simple case where (1) holds with $a_1(t) = a_2(t) = a(t)$, we have

$$A_1(t) = A_2(t) = A(t).$$

Thus (2) reduces to $\dot{v}_1/\dot{v}_2 = \exp(k_1 - k_2) = K$, constant, and

$$v_1 - v_1(0) = K[v_2 - v_2(0)]. \tag{4}$$

From (3),

$$v_1^{\alpha-1}\dot{v}_1 = [(v_1 - v_1(0) + Kv_2(0))'K]^{1-\beta}\Phi_1,$$

which integrates to

$$\int_{v_1(t_0)}^{v_1(t)} w^{\alpha-1}[w/K + L]^{\beta-1}dw = \int_{t_0}^{t} \Phi_1(t_1)dt_1.$$

The left–hand side is a semi-elementary Eulerian integral function of $v_1(t)$ and $\Phi_1(t) = \exp(A(t) + k_1)$, where $L = v_2(0) - v_1(0)/K$. Thus $x_1(t)$ can be calculated in a straightforward way in terms of $a(t)$ and the other parameters. From (4), $v_2(t)$ and hence $x_2(t)$ are determined also.

A modified version of some economic interest occurs when $a_1(t) = \mu a_2(t)$, so $A_1(t) = \mu A_2(t)$ and

$$\dot{v}_1/\dot{v}_2 = \exp[(\mu - 1)A_2 + k_1 - k_2].$$

This permits some inferences about the relative production x_1/x_2.

4. Separated Dynamics

Differentiation of (3) produces

$$(\alpha - 1)v_1^{\alpha-2}\dot{v}_1^2 + v_1^{\alpha-1}\ddot{v}_1 = \dot{\Phi}_1 v_2^{1-\beta} + (1-\beta)\Phi_1 v_2^{-\beta}\dot{v}_2$$
$$= \dot{\Phi}_1 v_1^{\alpha-1}\dot{v}_1/\Phi_1 + (1-\beta)\Phi_1\Phi_2 v_1^{1-\alpha}v_2^{1-2\beta}$$
$$= \frac{\dot{\Phi}_1}{\Phi_1}v_1^{\alpha-1}\dot{v}_1 + (1-\beta)\Phi_1\Phi_2^{(1-2\beta)/(1-\beta)},$$

on repeated use of both parts of (3). Division by $\dot{v}_1 v_1^{\alpha-1}$ (for $\dot{v}_1 \neq 0$) leads to

$$\frac{\ddot{v}_1}{\dot{v}_1} + (\alpha - 1)\frac{\dot{v}_1}{v_1} = \frac{\dot{\Phi}_1}{\Phi_1} + (1-\beta)\Phi_1^{\beta/(1-\beta)}\Phi_2 v_1^{(\alpha-1)/(\beta-1)}\dot{v}_1^{\beta/(\beta-1)}. \tag{5}$$

Similarly

$$\frac{\ddot{v}_2}{\dot{v}_2} + (\beta - 1)\frac{\dot{v}_2}{v_2} = \frac{\dot{\Phi}_2}{\Phi_2} + (1-\alpha)\Phi_2^{\alpha/(1-\alpha)}\Phi_1 v_2^{(\beta-1)/(\alpha-1)}\dot{v}_2^{\alpha(\alpha-1)}. \tag{6}$$

Together (5) and (6) give the dynamics for the decoupled variables v_1 and v_2 when $b_1, b_2 > 0$. These equations can be viewed as Newtonian equations of motion with three force types:

(a) pseudo-centrifugal (or centripetal) forces of the form $\dot{v}_i^2/v_i$, with constant coefficients for $c_1 + b_2$, $c_2 + b_1 \neq 0$,

(b) a negative or positive friction with variable coefficient and

(c) multiplicative terms $v_i^{\lambda_i}\dot{v}_i^{\mu_i}$ with variable coefficients and different exponents.

The third force type has been studied extensively (with a power time dependence) in astrophysics in the Emden-Fowler equations.

As might be expected, the combination of logistics and variable coefficients yields formidable decoupled dynamics. Four obvious special cases are of interest:

I – equal production rates, as treated in Section 3;

II – logistics are retained but a_1 and a_2 are constant;

III – a_1 and a_2 are variable but the logistics are dropped by sending b_1 (and so α) to zero and b_2 (and so β) to zero;

IV – the logistics are dropped and a_1, a_2 are constant (the classical Lotka-Volterra case).

We consider cases II, III, IV in turn in the following sections. For III and IV the original variables are kept to bypass singularities. Case IV is integrated explicitly to semi-elementary parametric integrals, thus solving the simplest Lotka-Volterra system.

5. The Painlevé Property

The Painlevé property is defined in Painlevé[8] for a differential equation

$$E\left(x, u, u', \ldots, u^{(N)}\right) = 0$$

as the absence of movable critical singularities in the general solution of that differential equation. Here a *movable* singularity is one whose location in the complex x-plane depends on the initial conditions. A singularity is *critical* if the solution is multivalued around it. Bureau[9] applied the term *stability* for a differential equation possessing the Painlevé property. This may be rephrased as saying that any movable singularities must be poles. The Painlevé property is defined in the same way for a system of ordinary differential equations.

A homely example of nonstability is provided by the equation $du/dx = 1/(2u)$, which has solution $u = \sqrt{x - c}$ for any constant c. This solution has a branchpoint at $x = c$. This branchpoint depends on the choice of solution and so is movable.

UIn practice one finds that a dynamical system is completely integrable only for parameters for which it has the Painlevé property (see, for example, Tabor[17]. However the integrability of an arbitrary system with the Painlevé property is yet to be proved. The Painlevé property has been quintessential in the domain of integrable systems.

The Painlevé property is invariant under the group of birational transformations

$$(u, x) \rightarrow (U, X): \ u = r(x, U, dU/dX, \ldots, d^{N-1}U/dX^{N-1}), \ x = \Xi(X),$$

$$(U, X) \rightarrow (u, x): \ U = R(X, u, du/dx, \ldots, d^{N-1}u/dx^{N-1}), \ X = \xi(x),$$

where r and R are rational in U, u and their derivatives and analytic in x, X. This has a more manageable subgroup consisting of the homographic transformations

$$(u, x) \rightarrow (U, X): \ u = \frac{aU + b}{cU + d}, \quad X = \xi(x), \ ad - bc \neq 0,$$

where a, b, c, d, ξ are arbitrary analytic functions of x. Gambier[10] found equivalence classes for these two groups, with later corrections by Bureau[11] and Cosgrove and Scoufis[13].

Necessary conditions for the Painlevé have been given by the test of Kowalevski and Gambier[10]. There are further tests for when that of gambier and Kowalevski fails to detect a movable multivaluedness in the general solution. For details see Conte[16].

6. Constant Production Rates

The case of constant production rates is quite common in the modelling literature and permits some further simplification. If we define

$$\Psi_1(t) = \exp\left[\left\{\frac{a_1\beta}{1-\beta} + a_2\right\}(t-t_0) + \frac{k_1\beta}{1-\beta} + k_2\right],$$

then (5) becomes

$$\frac{\ddot{v}_1}{\dot{v}_1} + (\alpha - 1)\frac{\dot{v}_1}{v_1} = a_1 + (1-\beta)\Psi_1 v_1^{(\alpha-1)/(\beta-1)}\dot{v}_1^{\beta/(\beta-1)}. \tag{7}$$

If $s_1 = \dot{v}_1/v_1$, then $v_1 = \exp(\int s_1 dt)$ and (7) yields

$$\frac{\dot{s}_1}{s_1} + \alpha s_1 = a_1 + (1-\beta)\Psi_1 s_1^{\beta/(\beta-1)}\exp\left[\frac{\beta+\alpha-1}{\beta-1}\int s_1 dt\right].$$

Redifferentiation then leads to the second-order equation

$$s_1\ddot{s}_1 - \dot{s}_1^2 + \alpha s_1^2 \dot{s}_1 = (\dot{s}_1 + \alpha s_1^2 - a_1 s_1)$$
$$\times\left[\frac{\beta}{\beta-1}\dot{s}_1 + \frac{\alpha+\beta-1}{\beta-1}s_1^2 + \left(\frac{a_1\beta}{1-\beta} + a_2\right)s_1\right] \tag{8}$$

for s_1, which is proportional to x_1, and a similar equation for s_2.

This equation does not appear to have been studied previously. It is generally unstable in the sense of Bureau: its general solution is not uniformizable in the complex plane. In Section 7 we consider it for the classical limit case $\alpha, \beta \to 0$ with α/β finite, in which the seven types of terms above – in $s\ddot{s}$, $\dot{s}^2$, $s^2\dot{s}$, $s\dot{s}$, s^2 s^3 and s^4 – reduce to five – in $s\ddot{s}$, $\dot{s}^2$, $s\dot{s}$, s^2 and s^3. The latter was studied a century ago by Painlevé and essentially solved.

With $\beta = 0$ we obtain an equation of Painlevé-Gambier type, even though $\alpha \neq 0$. Equation (8) then reduces to

$$\ddot{s}_1 - \frac{\dot{s}_1^2}{s_1} = (1-2\alpha)s_1\dot{s}_1 + a_2\dot{s}_1 + (\alpha s_1 - a_1)s_1[(1-\alpha)s_1 + a_2].$$

The cubic in s_1 allows integration by Painlevé methods.

7. Classical Case, Variable Rates

When $b_1, b_2 \to 0$ with $b_1/b_2 \to$ constant and a_1, a_2 are variable, (3) simplifies to

$$\dot{v}_1 = v_1 v_2 \overline{\Phi}_1, \quad \dot{v}_2 = v_1 v_2 \overline{\Phi}_2,$$

where the bars indicate the limiting values of the coefficients Φ_1, Φ_2. Thus

$$
\begin{aligned}
\ddot{v}_1 &= \dot{v}_1 v_2 \bar{\Phi}_1 + v_1 \dot{v}_2 \bar{\bar{\Phi}} + v_1 v_2 \dot{\bar{\Phi}}_1 \\
&= v_1 v_2^2 \bar{\Phi}_1 + v_1^2 v_2 \bar{\Phi}_1 \bar{\Phi}_2 + \left(\dot{v}_1 / \bar{\Phi}_1 \right) \dot{\bar{\Phi}}_1 \\
&= v_1 \left(\frac{\dot{v}_1}{v_1 \bar{\Phi}_1} \right)^2 \bar{\Phi}_1 + v_1^2 \left(\frac{\dot{v}_1}{v_1 \bar{\Phi}_1} \right) \bar{\Phi}_1 \bar{\Phi}_2 + \left(\dot{v}_1 / \bar{\Phi}_1 \right) \dot{\bar{\Phi}}_1 \\
&= \frac{\dot{v}_1^2}{v_1 \bar{\Phi}_1} + v_1 \dot{v}_1 \bar{\Phi}_2 + \dot{v}_1 \frac{\dot{\bar{\Phi}}_1}{\bar{\Phi}_1} .
\end{aligned}
\tag{9}
$$

This belongs to the Picard family

$$
\ddot{w} = L(w,t)\dot{w}^2 + M(w,t)\dot{w} + N(w,t), \tag{10}
$$

which is necessary but not sufficient form for a second-order equation with no movable critical points. Another necessary condition is that the degenerate case $\ddot{w} = L(w,t)\dot{w}^2$ of (10) also have no movable critical points. Since $L = 1/(w\bar{\Phi}_1(t))$, the latter equation reads

$$
\frac{\ddot{w}}{\dot{w}} = \frac{\dot{w}}{w\bar{\Phi}_1(t)} . \tag{11}
$$

If $\dot{w} = qw$, then $\ddot{w} = \dot{q}w + q^2 w$ and

$$
\frac{\dot{q}}{q^2} = \frac{1}{\bar{\Phi}_1 - 1}, \qquad \frac{1}{q} = \frac{1}{q_0} - \int_{t_0}^{t} [\bar{\Phi}_1(t_1) - 1]^{-1} dt_1 .
$$

Thus

$$
q = \left\{ \frac{1}{q_0} - \int_{t_0}^{t} [\bar{\Phi}_1(t_1) - 1]^{-1} dt_1 \right\}^{-1} .
$$

The necessary conditions referred to are then that

$$
w_j = w_0 \exp \left\{ \int_{t_0}^{t} \left[\frac{1}{q_0} - \int_{t_0}^{t_3} [\bar{\Phi}_j(t_j) - 1]^{-1} dt_j \right]^{-1} dt_3 \right\}
$$

for $j = 1, 2$ not to have movable critical points. Recall that

$$
\Phi_j(t) = \exp \left(\int_{t_0}^{t} a_j(t') dt' + k_j \right) ,
$$

so a chain of two exponential integrals appears. Examination of Ince shows that few choices of the a_j will produce fixed critical points, so that perturbation methods for such nonlinear problems are likely to be unsuccessful. Fortunately constant a_j can be managed by a different but related approach. Note that if $\bar{\Phi}_1 = 1$, (9) is of the Painlevé-Gambier type considered in the next section.

8. Classical Lotka-Volterra System

We now return to the classical system

$$\dot{x}_1 = x_1(a_1 + c_1 x_2), \quad \dot{x}_2 = x_2(a_2 + c_2 x_1), \tag{12}$$

noting the steady state $x_1 = -a_2/c_2$, $x_2 = -a_1/c_1$ and the first integral

$$F_0 = -a_2 \ln|x_1| + a_1 \ln|x_2| - c_2 x_1 + c_1 x_2.$$

Taking limits of (8) as $b_1, b_2 \to 0$ results in the separated equations

$$\ddot{x}_1 - \frac{\dot{x}_1^2}{x_1} = a_2 \dot{x}_1 - a_1 a_2 x_1 - a_1 c_2 x_1^2 + c_2 x_1 \dot{x}_1 + \epsilon g_1(t), \tag{13}$$

$$\ddot{x}_2 - \frac{\dot{x}_2^2}{x_2} = a_1 \dot{x}_2 - a_1 a_2 x_2 - a_2 c_1 x_2^2 + c_1 x_2 \dot{x}_2 + \epsilon g_2(t).$$

The added "gravitational" terms in ϵ are to prevent singularity and are part of standard Painlevé–Gambier methodology. They will be relaxed later after partial solution of the equations.

We now outline the methods of Gambier and Painlevé for the generalized form

$$
\begin{aligned}
0 = N(U, w) \\
= \ddot{w} - \dot{w}^2/w + [A(t)w + B(t) + C(t)/w]\dot{w} \\
+ D(t)w^3 + E(t)w^2 + F(t) + G(t) + H(t)/w,
\end{aligned}
$$

where $U = (A, B, C, D, E, F, G, H)$. This form depends on t for $w = w(t)$. Equation (13) is of the form $N(U, x_1) = 0$, with

$$U = U^{(\epsilon)} = (-c_2, -a_2, 0, 0, a_1 c_2, a_1 a_2, -\epsilon g_1(t), 0).$$

If $\tilde{U} = (C, B, A, -H, -G, -F, -E, -D)$, then

$$N(\tilde{U}, w^{-1}) = -w^{-2} N(U, w) \quad \text{for } w \neq 0.$$

Thus w is a solution of $N(U, w) = 0$ if and only if $N(\tilde{U}, w^{-1}) = 0$. Here

$$\tilde{U} = (0, -a_2, -c_2, 0, \epsilon g_1(t), -a_1 a_2, -a_1 c_2, 0)$$

and

$$M^{(\epsilon)}(y) = N(\tilde{U}, x_1^{-1}) = N(\tilde{U}, \lambda[y]),$$

where

$$\lambda[y] = \frac{\alpha(t)y(t) + \beta(t)}{\gamma(t)y(t) + \delta(t)}$$

is a Möbius (homographic) mapping and

$$z = \mu(t), \quad \lambda[y](z) = x_1^{-1}$$

is the mapping to produce a canonical form in (y, z) replacing (x_1^{-1}, t).
Explicitly, if $\lambda[y]$ reduces to $\lambda(t)y(t)$, then

$$M^{(\epsilon)}(y) = \left(\ddot{\lambda} - \frac{\dot{\lambda}^2}{\lambda} - a_1 a_2 \lambda - a_2 \dot{\lambda} \right) y + (\lambda \ddot{\mu} - \lambda \dot{\mu} a_2)y' + \epsilon \lambda^2 g_1(t) y^2$$

$$+ \lambda \dot{\mu}^2 \left(y'' - \frac{(y')^2}{y} \right) - c_2 \dot{\mu} \frac{y'}{y} - a_1 c_2 - c_2 \frac{\dot{\lambda}}{\lambda}.$$

The normalization and reduction are performed by taking $\lambda \dot{\mu}^2 = c_2 \dot{\mu}$
and $\ddot{\mu} = \dot{\mu} a_2$. The result $M^{(\epsilon)}(y) = 0$ is of the form

$$y'' - \frac{(y')^2}{y} - \frac{y'}{y} = r(z)y^2 - \left(\frac{r'(z)}{r(z)} \right)' y. \tag{14}$$

This is a Painlevé-Gambier equation of type II class 3^0 for each $\epsilon \neq 0$,
as exhibited on pp. 334–335 of the catalogue of Ince[14]. Here $r(z)$ depends
on λ, μ, ϵ, g_1 and a_1, a_2, c_2. Thus $x_1^{-1} = \lambda y$. A similar result holds for
x_2^{-1} with similar, but different, choices depending on a_1, a_2, c_1, ϵg_2. If
$g_1(t) = g_1$ independent of t, then $r(z)$ reduces to a constant multiple of
$\epsilon g_1 z^{-3}$.

Equation (14) has explicit first integrals of the form

$$y' = -1 - y\frac{r'(z)}{r(z)} \pm \sqrt{2r(z)y^3 + \left(2\int_{z_0}^z r(z_1)dz_1 + k \right) y^2} \tag{15}$$

with k arbitrary (see Ince). The square root of a cubic suggests periodicity.
These are Abelian integrals when $r(z)$ and $\int_{z_0}^z r(z_1)dz_1$ are rational. Even
when non-Abelian, the solutions for $\epsilon \neq 0$ have fixed critical points.

As $\epsilon \to 0$ the second-order equation (14) has a degenerate form, $r(z)$
simplifies and both parts of (15) change significantly. However another tack
is available. Put (13) with $\epsilon = 0$ directly in a canonical form by the mapping
$x_1 = \lambda(t)y$, $z = \phi(t)$ and express $y(z)$ through a differential equation as
before. The function y is, of course, different from before. Thus

$$(\lambda y)\frac{d^2}{dt^2}(\lambda y) - \left[\frac{d}{dt}(\lambda y) \right]^2 = a_2(\lambda y)\frac{d}{dt}(\lambda y) - a_1 a_2(\lambda y)^2$$

$$-a_1 c_2(\lambda y)^3 + c_2(\lambda y)^2 \frac{d}{dt}(\lambda y). \tag{16}$$

Then

$$\frac{d}{dt}(\lambda y) = \dot{y}y + \lambda\dot{\phi}y',$$

$$\frac{d^2}{dt^2}(\lambda y) = \ddot{y}y + 2\dot{\lambda}\dot{\phi}y' + \lambda\ddot{\phi}y' + \lambda\dot{\phi}^2 y''$$

and substitution into (16) and grouping terms provides

$$0 = y^2\left[\lambda\ddot{y} - \dot{\lambda}^2 - a_2\lambda\dot{\lambda} + a_1 a_\lambda^2\right] + yy'\lambda^2\left[\ddot{\phi} - a_2\dot{\phi}\right]$$
$$+ y^3\lambda^2 c_2\left[a_1\lambda - \dot{\lambda}\right] - c_2\lambda^3\dot{\phi}y^2 y' + \lambda^2\dot{\phi}^2(yy'' - y'^2).$$

The choices $z = \phi = \exp(a_2 t)$, $\lambda = \exp(a_1 t)$ eliminate the terms in yy' and y^3, respectively and also the term in y^2, leaving the three-term relation

$$yy'' - y'^2 = \frac{c_2}{a_2}y^2 y' e^{(a_1-a_2)t} = \frac{c_2}{a_2}y^2 y' z^{(a_1/a_2)-1}. \tag{17}$$

This is quite different from (14) when $\epsilon = 0$. It is rational for selected values of a_1/a_2. The extension of rationality to general values of a_1/a_2 is found in the review paper of Conte[16]. The degenerate form (17) does not appear in the Ince catalogue and possibly not in the original Gambier papers, but it is reducible, that is, its order can be lowered.

Let θ_1 be defined for $y > 0$ by

$$y(z) = \exp(\theta_1(z)). \tag{18}$$

From (17) and (18) we obtain the nonlinear differential equation

$$\theta_1'' = d_1\theta_1 e^{\theta_1} z^{\sigma_1} \tag{19}$$

for $\theta_1(z)$, where $d_1 = c_2/a_2$ and $\sigma_1 = a_1/a_2 - 1$.

Introduce a new independent variable (parameter) p_1 by

$$z = p_1(z)\exp\left[-\frac{\hat{\theta}_1(p-1)}{\sigma_1 + 1}\right] \quad \text{with } \hat{\theta}_1(p_1) = \theta_1(z) \tag{20}$$

and χ_1 defined by

$$\hat{\theta}_1(p_1) = \int_{p_o}^{p}\frac{dp}{\chi_1(p)} + k_1.$$

If the operator d/dp_1 is denoted by $*$, then

$$\hat{\theta}_1^*(p-1) = 1/\chi_1(p_1). \tag{21}$$

Also

$$\theta_1'(z) = \hat{\theta}_1^*(p_1)p_1'(z),$$
$$\theta_1''(z) = \hat{\theta}_1^{**}(p_1)\left[p_1'(z)\right]^2 + \hat{\theta}_1^*(p_1)p_1''(z)$$

and from (19)

$$\frac{\theta_1''(z)}{\theta_1'(z)} = d_1\theta_1 e^{\theta_1(z)}z^{\sigma_1} = d_1\theta_1 e^{\hat{\theta}_1(p_1)}z^{\sigma_1} = \frac{p_1''(z)}{p_1'(z)} + \frac{\hat{\theta}_1^{**}(p_1)}{\hat{\theta}_1^*(p_1)}p_1'(z). \qquad (22)$$

The result of differentiating (21) once and (20) twice with respect to z, eliminating z through (20) and combination with (22) is the nonlinear differential equation

$$\chi_1(\chi_1^* - 1/\ell_1) = (1/\ell_1)(\chi_1 - p/\ell_1) - d_1 p^{\ell_1-1}(\chi_1 - p/\ell_1)^2,$$

which integrates to

$$\ln\left|\chi_1 - \frac{p_1}{\ell_1}\right| = \frac{p_1}{\ell_1\chi_1 - p_1} - \frac{d_1}{\ell_1}p_1^{\ell_1} + L_1, \qquad (23)$$

where L_1 is a constant and $\ell_1 = \sigma_1 + 1$. This determines $\chi_1 p_1$). Note that $\chi_1(0) = \exp(L_1)$ if $\ell_1 > 0$. Formula (23), without explanations, is given essentially in Section 2.3.7, no. 7, p. 383 of Polyanin and Zaitsev[15].

This result must be combined with

$$\chi_1(p_1) = 1/\hat{\theta}_1^*(p_1) \quad \text{and} \quad z(p_1) = p_1\exp\left(-\hat{\theta}_1(p_1)/(\sigma_1+1)\right)$$

to determine $y(z) = \exp\hat{\theta}_1(p_1) = \exp\theta_1(z)$ and $x_1(t) = e^{a_1 t}y(e^{a_2 t})$. Thus it is the independent variable z that requires a quadrature.

Another approach is to assume the elegant Ansatz

$$\frac{d\chi_1}{dp} = \chi_1^*(p) = A_1(p)/\chi_1 + B_1(p) + C_1(p)\chi_1, \qquad (24)$$

and find that

$$A_1(p) = -p/\ell_1^2 - d_1 p^{\ell_1+1}/\ell_1^2,$$
$$B_1(p) = 2/\ell_1 + 2d_1 p^{\ell_1}/\ell_1,$$
$$C_1(p) = d_1 p^{\ell_1-1}$$

are required to satisfy the various conditions. If $\ell_1 = 0$ and $d_1 = -1$, then $A_1 = 0$ and χ_1 satisfies a linear differential equation. Equation (24) is a first-order differential equation that can be reduced to an Abelian equation of the form

$$ww' = f_1(x)w + f_2(x).$$

Unfortunately the resulting $f_1(x)$ and $f_2(x)$ do not place the equation in the short list of solved Abel forms unless $\ell_1 = \sigma_1 + 1 = 0$ or 1.

The problem reduces to quadrature of a rather peculiar composition of inverse $x \exp(-a/x)$ functions with $\exp(-cx^d)$ functions. Both functions are monotone, with directions dependent on the signs of $\sigma_1 + 1$ and d_1. The relevance is seen by exponentiating (23).

For determining the behaviour of the second species x_2, it is necessary only to replace σ_1 by $\sigma_2 = a_2/a_1 - 1$ and d_1 by $d_2 = c_1/a_1$. This takes the classical Lotka-Volterra system into the list of solved problems.

References

1. R.M. May, *Stability and Complexity in Model Ecosystems*, Princeton University Press, Princeton 1973.
2. R.M. May, Chaos and the dynamics of biological populations, *Proc. Roy. Soc. London Ser. A* **413** (1987), 27–44.
3. M.P. Hassell and R.M. May, The population biology of host–parasite and host–parasitoid associations, in *Perspectives in Ecological Theory*, Eds J. Roughgarden, R.M. May and S.A. levin, Princeton University Press, Princeton 1989, 319–347.
4. R.M. Goodwin, A growth cycle, in *Socialism, Capitalism and Economic Growth*, Ed. C.H. Feinstein, Cambridge University Press, Cambridge 1967, 54–58.
5. N.N. Bautin and E.A. Leontovich, *Methods and Rules for the Qualitative Study of Dynamical Systems on the Plane*, 2nd ed., Mathematical Reference Library, Nauka, Moscow 1990 (in Russian).
6. A.A. Andronov, A.A. Witt and S.E. Chaikin, *The Theory of Oscillations* (corrected reprint of the 1937 original), Nauka, Moscow 1981 (in Russian).
7. A.A. Andronov, E.A. Leontovich, I.I. Gordon and A.G. Maier, *Theory of Bifurcations of Dynamic Systems on a Plane*, Halsted Press, New York 1973.
8. P. Painlevé, *Leçons sur la théorie analytique des équations différentielles* (Leçons de Stockholm, 1895), Hermann, Paris 1897. reprinted, *Oeuvres de Paul Painlevé, vol. I* Éditions du CNRS, Paris 1973.
9. F.J. Bureau, Sur la recherche des équations différentielles du second ordre dont l'intégrale générale est à points fixes, *Bulletin de la Classe des Sciences* **XXV** 1939, 51–68.
10. B. Gambier, Sur les équations différentielles du second ordre et du premier degré dont l'intégrale générale est à points fixes, *Acta Mathematica* **33** 1910, 1–55.
11. F.J. Bureau, Differential equations with fixed critical points, *Annali di Matematica pura ed applicata* **64** 1964, 229–364.
12. F.J. Bureau, Differential equations with fixed critical points, *Annali di Matematica pura ed applicata* **66** 1964, 1–116.
13. C.M. Cosgrove and G. Scoufis, Painlevé classification of a class of differ-

ential equations of the second order and second degree, *Studies in Applied Mathematics* **88** 1993, 25–87.

14. E.L. Ince, *Ordinary Differential Equations*, Longmans, Green and Co., London 1927.
15. A.D. Polyanin and V.F. Zaitsev, *Handbook of Exact Solutions for Ordinary Differential Equations*, 2nd ed. Chapman and Hall/CRC, Boca Raton 2003.
16. R. Conte, *The Painlevé Property. One Century Later*. Ed. R. Conte. CRM Series in Mathematical Physics. Springer-Verlag, New York 1999.
17. M. Tabor, *Chaos and Integrability in Nonlinear Dynamics*, Wiley, New York 1989, URSS, Moscow 2000.

ALLEE EFFECT, EMIGRATION AND IMMIGRATION IN A CLASS OF PREDATOR-PREY MODELS *

EDUARDO GONZÁLEZ-OLIVARES , JAIME MENA-LORCA

HÉCTOR MENESES-ALCAY, BETSABÉ GONZÁLEZ-YAÑEZ

Grupo Ecología Matemática, Instituto de Matemáticas,
Pontificia Universidad Católica de Valparaíso.
Casilla 4950, Valparaíso, Chile.
E-mails: `ejgonzal@ucv.cl, jmena@ucv.cl, hmeneses@ucv.cl,`
`betsabe.gonzalez@ucv.cl`

JOSÉ D. FLORES

Department of Mathematical Sciences
The University of South Dakota,
314 East Clark Street,
Vermillion, SD 57069, USA
E-mail: `jflores@usd.edu`

In this work we analyze a predator-prey model proposed by Kent et. al. in[16], in which two aspect of the model are considered: an effect of emigration or immigration on prey population to constant rate and a prey threshold level for predators. We prove that the system when the immigration effect is introduced in the model has a dynamics that is similar to the Rosenzweig-MacArthur model. Also, when emigration is considered in the model, we show that the behavior of the system is strongly dependent on this phenomenon, this due to the fact that trajectories are highly sensitive to the initial conditions, in similar way as when Allee effect is assumed on prey. Furthermore, we determine constraints in the parameters space for which two stable attractor exist, indicating that the extinction of both population is possible in addition with the coexistence of oscillating of populations size in a unique stable limit cycle. We also show that the consideration of a threshold level of prey population for the predator is not essential in the dynamics of the model.

---

*This work is partially supported by DI-PUCV project 124.711/2007, and by the Department of Mathematics at USD.

1. Introduction

In this paper we analyze the predation model proposed by Kent et al.[16] where they model the case of resource subsidy supposing that the resource is a limitation prey to predators and assuming influx (immigration) or outflux (emigration) of prey. They affirm that the system is: (i) stabilized by an influx of prey in the form of a rescue effect, and (ii) destabilized by an outflux of prey in the form of an Allee effect.

In Population Dynamics, any mechanism that can lead to a positive relationship between a component of individual fitness and either the number or density of conspecific can be termed a mechanism of *Allee effect*[5,25], or *depensation*[4], or *negative competition effect*[29]. The outflux of prey to constant rate can be considered as Allee effect because a change on interaction dynamics is provoked, for instance, due to difficult of encountering mate.

Other implicit assumptions for the proposed continuous time model model are; the populations are uniformly distributed in the environment, neither sex and nor age structure are considered, and abiotic phenomenon is not influencing on the growth of both populations.

In this paper we show that when immigration or emigration are not assumed, the model proposed in[16] is topological equivalent to well known Rosenzweig-MacArthur predation model[22,28], for which the effects of entrance or exit of individual prey will be considered by modification to this last model.

In order to make an exhaustive study of the proposed system we consider separately the immigration and emigration effects. The influx (immigration)or outflux (emigration) of prey in the interaction is of constant rate, for which the new model is not of Kolmogorov type system[6]. The lack of these mathematical properties of Kolmogorov type models provoke significant changes in the dynamics of new system with respect to the original Rosenzweig-MacArthur model when the emigration is assumed different to zero.

Moreover, we show that if immigration (the influx) is considered, the new system has a similar behavior to the Rosenzweig-MacArthur model[19,23]; that is, there exists a parameter set for which the unique equilibrium point at interior of the first quadrant is globally asymptotically stable or else the existence of a unique limit cycle is assured.

We prove that there is a separatrix curve determined by an attractor point in the second quadrant, which divides the behavior of the trajectories. This result implies that the proposed model when prey emigrates at constant rate, is highly sensitive to initial conditions, since it is possible to find points close enough in the phase plane whose trajectories can have different ω-*limit*, an equilibrium point at the second quadrant and the unique positive equilibrium point in the first quadrant, moreover this later singularity could be surrounded by a unique limit cycle. Then, in this sense, a unique limit cycle can coexist with the extinction of both species depending only on the initial conditions.

We consider that the extinction of both species is a result inherent to the Allee effect, and in no way is consequence of the combine presence of both, an Allee effect on prey and a predator functional response type II. Similar conclusion has been shown in[13] by using another type of functional response or another predator growth equation[7,10,12].

2. The Model

In this section we analyze the consequences of the emigration (outflux) and the immigration (influx) effect of prey in a predation model. In the analysis of the model it is assumed that prey are consumed by predators at rate per capita that depends on prey abundance according to the classic disk equation $h(x) = \frac{Qx}{Qhx+1}$ proposed by Holling[18]. The deterministic continuous time model is described by the autonomous bi-dimensional differential equations system

$$X_\lambda^\alpha : \begin{cases} \frac{dx}{dt} = r\left(x + \alpha\right)\left(1 - \frac{x}{K}\right) - \frac{Qxy}{Qhx+1} \\ \frac{dy}{dt} = \frac{PQ(x-\beta)}{1+Qhx}y \end{cases} \tag{1}$$

where $\lambda = (r, K, Q, h, P, \beta) \in \mathbb{R}_+^6$ and $\alpha \in \mathbb{R}$, and the parameters have the following meanings:

r is the intrinsic rate, that is the average rate of prey births per capita in a pristine environment,

α is the average size of prey immigrants for positive values or emigrants for negative values, or intrinsic prey flux into or out of a prey population of given carrying capacity,

K is the carrying capacity of prey at which births and migration reduce to zero,

Q is a the predator's average searching rate for prey,

h is the handling time for each encountered prey,

P is a conversion ratio of consumed prey into viable predator offspring, and

β is the size of prey that sustains one predator, and replaces it with a single offspring when it dies or the marginal subsistence demand for prey;

also, it is considered as the threshold prey density below which predator density declines.

We observed that in the system (1) which is defined in the first quadrant,

$$\Omega = \{(x,y) \in \mathbb{R}^2 : x \geq 0, \ y \geq 0\} = \mathbb{R}_0^+ \times \mathbb{R}_0^+$$

the y-axis is not an invariant set, that is, the system is not of Kolmogorov type[6], except for the case $\alpha = 0$.

The equilibrium points of system (1) (or singularities of vector field X_μ^α) are $P_\alpha = (-\alpha, 0)$, $P_{Qh} = (-\frac{1}{Qh}, 0)$, $P_K = (K, 0)$ and $P_e = (x_e, y_e)$, which is the positive(interior) equilibrium point with

$$x_e = \beta \quad \text{and} \quad y_e = \frac{r}{Q\beta}\left(1 - \frac{\beta}{K}\right)(\alpha + \beta)(1 + Qh\beta).$$

Then, $y_e > 0$, if and only if $\beta < K$.

In the next two subsections we will analyze separately the two cases; (a) *emigration* of the prey when $\alpha \in \mathbb{R}^+$, and (b) *immigration* of prey when $\alpha \in \mathbb{R}^-$ both cases in the framework of the Rosenzweig-MacArthur predator-prey model[22,28].

2.1. *Immigration*

Here we consider the effect of immigration (influx) of prey in model (1). That is, assuming the case $\alpha \in \mathbb{R}^+$ we have the following result.

Lemma 2.1. *The system (1) is topologically equivalent to the following system*

$$X_\mu^\alpha : \begin{cases} \frac{dx}{dt} = r(x+\alpha)\left(1 - \frac{x}{K}\right) - \frac{qxy}{x+a} \\ \frac{dy}{dt} = \left(\frac{px}{x+a} - c\right)y \end{cases} \tag{2}$$

with $\mu = (r, K, q, a, p, c, \alpha) \in \mathbb{R}_+^7$.

Proof.

First we reparametrize the function $h(x) = \frac{Qx}{Qhx+1}$ by using the substitutions $a = \frac{1}{Qh}$ and $q = \frac{1}{h}$, hence $h(x)$ has the form $h(x) = \frac{qx}{a+x}$.

a) We can see clearly that in the second equation of system (2),

$$\frac{px}{x+a} - c = \frac{(p-c)\left(x - \frac{ac}{p-c}\right)}{x+a},$$

and using the second equation from system (1) we have

$$\frac{PQ(x-\beta)}{1+Qhx} = \frac{Pq(x-\beta)}{a+x}.$$

Then by making the substitution $\beta = \frac{ac}{p-c}$ and $p - c = Pq$, we have that $c = \beta PQ$ and $p = P\left(\frac{1}{h} + Q\beta\right)$, obtaining one part of the result.

b) Reciprocally, the function $\frac{PQ(x-\beta)}{1+Qhx}$ can be written as $\frac{px}{x+a} - c$ to complete the proof. $\qquad\square$

It is worth to note that the parameter $a = \frac{1}{Qh}$ is a measure of abruptness of the functional response[8]. Clearly, if $a \to 0$ ($Qh \to \infty$), the curve increase steepest, whereas if $a \to K$ ($Qh \to \frac{1}{K}$), the curve increase slowly, that is, a larger quantity of prey is necessary to attain $\frac{q}{2} = \frac{1}{2h}$.

Corollary 2.1. *In the particular case when $\alpha = 0$, the system (1) is topologically equivalent to the Rosenzweig-MacArthur model[22,28].*

$$X_\mu : \begin{cases} \frac{dx}{dt} = \left(r\left(1 - \frac{x}{K}\right) - \frac{qy}{x+a}\right)x \\ \frac{dy}{dt} = \left(\frac{px}{x+a} - c\right)y \end{cases} \tag{3}$$

Using the results from[11], we have shown that the system (3) has a unique equilibrium point (x_e, y_e) in the interior of the first quadrant , with

$$x_e = \frac{ac}{p-c} \quad \text{and} \quad y_e = \frac{r}{q}\left(1 - \frac{1}{K}x_e\right)(x_e + a),$$

which is

a) a stable equilibrium point if and only if $0 < \frac{p-c}{p+c} < \frac{a}{K} < \frac{p-c}{c} < 1$, and

b) an unstable equilibrium point surrounded by a unique limit cycle if and only if, $0 < \frac{a}{K} < \frac{p-c}{p+c} < 1$.

Furthermore, the point $(K, 0)$ is an stable equilibrium point if and only if,

$$0 < \frac{p-c}{c} < \frac{a}{K} < 1,$$

in this case, the ratio $\frac{c}{p} = \frac{\beta PQ^2 h}{PQ(1+\beta Qh)} = \frac{\beta Qh}{1+\beta Qh} < 1$.

In order to simplify the calculus for further analysis in the next section we make the reparameterization of system (1) using the function

$$\varphi : \bar{\Omega} \times \mathbb{R} \longrightarrow \Omega \times \mathbb{R}$$

such that

$$\varphi(u, v, \tau) = \left(Ku, \frac{rK}{q}v, \frac{u + \frac{a}{K}}{r}\tau\right) = (x, y, t),$$

and we have that

$$\det(D\varphi(u,v,\tau)) = \frac{K}{q}\left(u + \frac{a}{K}\right) > 0,$$

thus, φ is a diffeomorphism, for which the vector field (3) on the new coordinates system is topologically equivalent to the vector field $Y_\nu = \varphi \circ X_\mu{}^{1,24}$, which has the form $Y_\nu(u,v) = P(u,v)\frac{\partial}{\partial u} + Q(u,v)\frac{\partial}{\partial v}$. With this scaling we obtained an associated third degree polynomial differential equations system which is given by

$$Y_\nu : \begin{cases} \frac{du}{d\tau} = (u+I)(1-u)(u+A) - uv \\ \frac{dv}{d\tau} = B(u-C)v \end{cases} \tag{4}$$

with $\nu = (A, C, B, I) \in \mathbb{R}_+^4$, where the new constants are as follows;

$$I = \frac{\alpha}{K}; \quad A = \frac{a}{K} < 1; \quad B = \frac{p}{r}\left(1 - \frac{c}{p}\right) \quad \text{and} \quad C = \frac{\frac{c}{p}\frac{a}{K}}{1 - \frac{c}{p}}.$$

The equilibrium points of system (4) are

$$P_I = (-I, 0), P_A = (-A, 0), P_1 = (1, 0), \quad \text{and} \quad P_e = (u_e, v_e)$$

with

$$u_e = C \quad \text{and} \quad v_e = \frac{(C+I)(1-C)(C+A)}{C}$$

The points P_I and P_A have no biological meanings and they do not exert influence in the behavior of the system.

2.2. *Emigration*

In this subsection we assume that emigration (outflux) of prey is introduced in model (1); in this case the model is described by the following autonomous differential equations system

$$X_\lambda^\delta : \begin{cases} \frac{dx}{dt} = r(x - \delta)\left(1 - \frac{x}{K}\right) - \frac{Qxy}{Qhx+1} \\ \frac{dy}{dt} = \frac{PQ(x-\beta)}{1 + Qhx}y \end{cases} \tag{5}$$

where $\lambda = (r, K, Q, h, P, \beta) \in \mathbb{R}_+^6$ and $\delta \in \mathbb{R}^+$ and the parameters have the same meanings as above. By biological considerations, we have that $\delta < x < K$.

System (5) (or vector field X_λ^δ) is defined in the first and second quadrants, that is

$$\Omega = \{(x, y) \in \mathbb{R}^2 : x \in \mathbb{R}, \; y \geq 0\} = \mathbb{R} \times \mathbb{R}_0^+.$$

Similarly as in system (1) we have observed that the x-axis is not an invariant set and therefore system (5) is not of Kolmogorov type[6] due to this δ-perturbation.

The equilibrium points of system (5) or singularities of the vector field X_μ^δ are $P_{Qh} = (-\frac{1}{Qh}, 0)$, $P_K = (K, 0)$, $P_\delta = (\delta, 0)$ and $P_e = (x_e, y_e)$, which is the positive (interior) equilibrium point, with

$$x_e = \beta \quad \text{and} \quad y_e = \frac{r}{Q\beta}\left(1 - \frac{\beta}{K}\right)(\delta + \beta)(1 + Qh\beta).$$

Hence, $y_e > 0$, if and only if $\beta < K$.

Using similar computations as in system (2) in lemma 2.1, system (5) can be described as follows;

$$X_\mu : \begin{cases} \frac{dx}{dt} = r(x - \delta)\left(1 - \frac{x}{K}\right) - \frac{qxy}{x+a} \\ \frac{dy}{dt} = \left(\frac{px}{x+a} - c\right)y \end{cases} \tag{6}$$

with $\mu = (r, K, q, a, p, c, \delta) \in \mathbb{R}_+^7$, and again for the case when $\delta = 0$, the system (6) is topological equivalent [1,3] to the Rosenzweig-MacArthur model [22,28].

To prepare the system (6) for further computations we make the reparameterization and time rescaling using the function

$$\varphi : \bar\Omega \times \mathbb{R} \longrightarrow \Omega \times \mathbb{R},$$

such that

$$\varphi(u, v, \tau) = \left(Ku, \frac{rK}{q}v, \frac{u + \frac{a}{K}}{r}\tau\right) = (x, y, t),$$

with

$$\bar\Omega = \{(u, v) \in \mathbb{R}^2 : u \in \mathbb{R}, v \geq 0\}$$

and we have that

$$\det(D\varphi(u, v, \tau)) = \frac{K}{q}\left(u + \frac{a}{K}\right) > 0,$$

obtaining that φ is a diffeomorphism[1,3,17] and we obtain the fifth degree polynomial differential equations system which is topologically equivalent to system (1) or (2) and is given by

$$Y_\nu^E : \begin{cases} \frac{du}{d\tau} = (u - E)(1 - u)(u + A) - uv \\ \frac{dv}{d\tau} = B(u - C)v \end{cases} \tag{7}$$

where

$$A = \frac{a}{K}, B = \frac{p}{r}\left(1 - \frac{c}{p}\right), C = \frac{\frac{c}{p}\frac{a}{K}}{1 - \frac{c}{p}} \quad \text{and} \quad E = \frac{\delta}{K},$$

with $\nu = (A, C, E) \in]0, 1[^3$ and $B \in \mathbb{R}^+$.

The equilibrium points of system (7) or vector field Y_ν^E are: $Q_1 = (1, 0)$, $Q_E = (E, 0)$, $Q_A = (-A, 0)$, and $Q_e = (u_e, v_e)$, with

$$u_e = C \quad \text{and} \quad v_e = \frac{1}{C}(C - E)(1 - C)(C + A) > 0,$$

hence, $v_e > 0$ if and only if, $E < C < 1$.

We have noted that the point Q_A has no biological interest, but it will be shown that this equilibrium point has a strongly influence on behavior of the system (7). The point $(-A, 0)$ is equivalent to $(-\frac{1}{Qh}, 0)$ in system (5) and to $(-a, 0)$ in system (6), and neither of these systems are defined these points, because the respective denominators in the functional response vanished.

3. Main Results

In this section we present the main mathematical results for systems of the models described in section 2. We present detailed proves for systems (4) and (7) and we replicate the equivalent results for the other systems in section 2. Starting with system (4) we have the following results;

Lemma 3.1.

   (a) The set $\bar{\Gamma} = \left\{(u, v) \in (\mathbb{R}^+)^2 : 0 \le u \le 1, v \ge 0\right\}$ is a invariant region.

   (b) The solutions of system are bounded.

Proof. (a) Clearly, if $u = 1$, we have that $\frac{du}{dt} = -uv < 0$, therefore, all solutions originated in region $\bar{\Gamma}$ enter there and they remain at $\bar{\Gamma}$, for any sign of $\frac{dv}{dt}$.

Moreover, if $u = 0$, we have that $\frac{du}{dt} = AI > 0$ and $\frac{dv}{d\tau} = -BCv < 0$, then all solutions remains at the first quadrant, since the resultant vector point towards right.

(b) In order to study the boundedness of trajectories, we analyze the behavior at infinity using the Poincaré compactification[3,21] given by the transformation $X = \frac{u}{v}$ and $Y = \frac{1}{v}$, then we have that $\frac{dX}{dt} = \frac{1}{v^2}\left(v\frac{du}{d\tau} - u\frac{dv}{d\tau}\right)$ and $\frac{dY}{dt} = -\frac{1}{v^2}\frac{dv}{d\tau}$. Analyzing the Jacobian matrix for this new system we have that the point $(0, \infty)$ is saddle point. $\qquad \square$

Even though, it could exist trajectories crossing from second to first quadrant, but by biological reason, initial conditions at second quadrant $(x(0) < 0)$, they do not need to be considered.

To simplify the notation of the following results we introduce the function $G(u) = \frac{(u+I)(1-u)(u+A)}{u}$, then the system (4) can be rewritten as;

$$\begin{cases} \frac{du}{d\tau} = (G(u) - v)\,u \\ \frac{dv}{d\tau} = B(u - C)v \end{cases}$$

Then jacobian matrix is given by

$$DY_\nu(u,v) = \begin{pmatrix} uG'(u) + G(u) - v & -u \\ Bv & B(u - C) \end{pmatrix} \tag{8}$$

Lemma 3.2. *The stability of the equilibrium point $P_1 = (1,0)$, is as follows;*

(a) a saddle point if $C < 1$,

(b) an attractor node if $C > 1$ which implies the nonexistence of an interior equilibrium point, and

(c) a non-hyperbolic attractor node if $C = 1$.

Proof.

The jacobian matrix (8) evaluated at the equilibrium point $P_1(1,0)$ is giving by;

$$DY_\nu(1,0) = \begin{pmatrix} -(1 + I)(1 + A) & -1 \\ 0 & B(1 - C) \end{pmatrix}$$

with trace, $\mathrm{tr}(DY_\nu(1,0)) = B(1 - C) - (1 + I)(1 + A)$ and determinant $\det(DY_\nu(1,0)) = B(C - 1)(1 + I)(1 + A)$. Then,

(a) the result follows by analyzing the determinant and the trace the Jacobian matrix at $(1,0)$.

(b) If $C \geq 1$, by applying the Poincaré-Bendixon Theorem, $P_1 = (1,0)$ is a globally asymptotically stable equilibrium point.

(c) The result follows from $DY_\nu(1,0)$ in the case $C = 1$ □

Theorem 3.1. *Assuming $C < 1$, the singularity $P_e = (u_e, v_e)$ is*

(a) an attractor equilibrium point if and only if $I > C^2 \frac{1-A-2C}{C^2+A}$. See Figure 4.

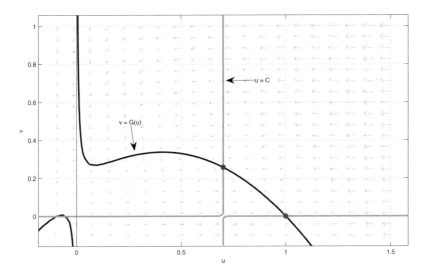

Figure 1.   The isoclines function for $I = 0.1$, $A = 0.05$, $C = 0.7$

(b) a repellor equilibrium point surrounded by an stable limit cycle if and only if, $I < C^2 \frac{1-A-2C}{C^2+A}$. See Figure 5.

(c) a first order weak focus if and only if, $I = C^2 \frac{1-A-2C}{C^2+A}$

Proof. The jacobian matrix (8) evaluated at $P_e = (u_e, v_e)$ with $u_e = C$ and $v_e = G(C)$ is;

$$DY_\nu(u_e, v_e) = \begin{pmatrix} CG'(C) & -C \\ BG(C) & 0 \end{pmatrix}$$

and since the determinant $\det(DY_\nu(u_e, v_e)) = BCG(C) > 0$, then the nature of this critical point depends on the trace of the jacobian matrix which is given by

$$\text{tr}(DY_\nu(u_e, v_e)) = CG'(C) = -\frac{1}{C}\left(2C^3 + C^2(A + I - 1) + IA\right)$$

and therefore the sign of the trace depends on the sign of $G'(C)$, see Figure 1, or which is equivalent, the sign of the trace will be determined by the sign of the term

$$T = 2C^3 + C^2 (A + I - 1) + IA.$$

(a) If $T > 0$, that is, $A > \left(\frac{C^2}{C^2+I}\right)(1 - I - 2C)$ or $I > \left(\frac{C^2}{C^2+A}\right)(1 - A - 2C)$ then (u_e, v_e) is globally asymptotically stable equilibrium point.

(b) If $T < 0$, that is, $I > \frac{C^2(1-A-2C)}{C^2+A}$, thus $P_e = (u_e, v_e)$ is a unstable (repellor) equilibrium point.

(c) Let $T = 0$ and in order to simplify the calculus let assume that $A = I$. Then $A = C - 2C^2$ and the singularity becomes $(C, 4C(1 - C)^3)$.

In this case we have the system with two parameters described by

$$Y_\nu : \begin{cases} \frac{du}{d\tau} = (1 - u)(u + C(1 - 2C))^2 - uv \\ \frac{dv}{d\tau} = B(u - C)v \end{cases}$$

Using the standard substitution

$$u = U + C \text{ and } v = V + 4C(1 - C)^3$$

the vector field Y_ν becomes

$$\bar{Y}_\nu : \begin{cases} \frac{dU}{d\tau} = (1 - U - C)(U + 2C(1 - C))^2 - (U + C)\left(V + 4C(1 - C)^3\right) \\ \frac{dV}{d\tau} = BU\left(V + 4C(1 - C)^3\right) \end{cases}$$

At $(0,0)$ the jacobian matrix of the vector field $\bar{Y}_\nu$ is

$$D\bar{Y}_\nu(0,0) = \begin{pmatrix} 0 & -C \\ 4BC(1 - C)^3 & 0 \end{pmatrix},$$

with $\det(D\bar{Y}(0,0)) = 4BC^2(1 - C)^3 = H^2$, this is, $B = \dfrac{H^2}{4C^2(1 - C)^3}$ and

therefore the jacobian matrix at $(0,0)$ is reduced to

$$D\bar{Y}(0,0) = \begin{pmatrix} 0 & -C \\ \frac{H^2}{C} & 0 \end{pmatrix}.$$

Then making the substitution, $U = -Hy$ and $V = \dfrac{H^2}{C}x$, which is equivalent to, $x = \dfrac{C}{H^2}V$ and $y = -\dfrac{1}{H}U$ and after some computation and using the time rescaling given by $T = H\tau$ we obtain the following system

$$\bar{Z}_\nu : \begin{cases} \frac{dx}{dT} = -y - \frac{H^2}{4C^2(1-C)^3}xy \\ \frac{dy}{dT} = x - \frac{H}{C}xy - (4C+1)(1-C)y^2 - Hy^3 \end{cases}$$

Finally, with the help of a Mathematica routine[30] we compute the second Liapunov quantity[3]

$$\eta_2 = \frac{H}{8C}(-1 - 6C + 4C^2)$$

which in this case is negative for $0 < C < 1$, that is, a unique limit cycle exists surrounding the positive equilibrium point, hence $P_e = (u_e, v_e)$ is a first order weak (fine) focus. $\square$

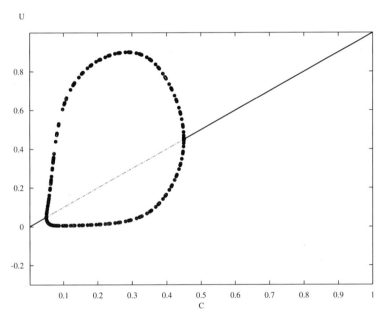

Figure 2. One parameter bifurcation diagram. The solid line indicates stable equilibrium points, the thin line indicates unstable equilibrium points, the dots indicate the maximum and minimum of the limit cycles.

Figures 1 and 2 show the stability results of Theorem 3.1 for the system (4). According to the sign of the derivative of the prey isocline function $G'(u)$. The slope of the function $G(u)$ is negative until it reaches a local minimum then turn positive changing the stability of the equilibrium point from stable to unstable, in which case is surrounded by a limit cycle, generating a Hopf bifurcation. Similar stability behavior occurs at the local maximum point of the prey isocline function. As the slope of the function $G(u)$ changes from positive to negative the stability of the equilibrium point changes from unstable surrounded by a limit cycle to stable, generating a second Hopf bifurcation point. Figure 3 shows a two parameter bifurcation diagram. For values of I (the scaled immigration) closed to zero there are two bifurcation points. As the value I increases, the prey isocline function stretches up and losses the bottom-valley and the hilltop points creating a graph with $G'(u) < 0$ for all values of $u > 0$ and a region where the equilibrium point is stable for all $0 < C < 1$.

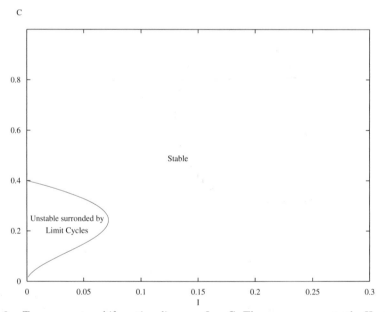

Figure 3.   Two parameters bifurcation diagram, I vs C. The curve represents the Hopf bifurcation points.

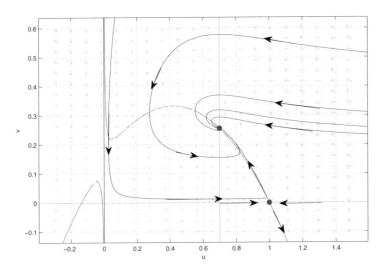

Figure 4. Global Stability of the equilibrium point $P_e(u_e, v_e)$ for $A = 0.1$, $B = 0.3$, $I = 0.14$, and $C = 0.7$

Next we have three results for the system (3) these results are similar to the later three results, therefore there is no need to present their demonstrations.

Lemma 3.3. *The set* $\Gamma = \left\{ (x, y) \in \left(\mathbb{R}_0^+ \right)^2 / 0 \leq x \leq K, y \geq 0 \right\}$ *is a invariant region.*

Lemma 3.4. *The equilibrium point* $(K, 0)$ *is :*

(a) a saddle point if $\frac{c}{p} < 1$

(b) an attractor node if $\frac{c}{p} > 1$ *which implies the nonexistence of an interior equilibrium point.*

(c) a non-hyperbolic attractor node if $\frac{c}{p} = 1$.

Theorem 3.2. *Assuming* $c < p$, *the singularity* (x_e, y_e) *is*

(a) an attractor equilibrium point if and only if $\alpha > \frac{(K(p-2c)-pa)c^2 K}{p(Kc^2 + p^2 a)}$.

(b) a repellor equilibrium point surrounded by an stable limit cycle if and only if, $\alpha < \frac{(K(p-2c)-pa)c^2 K}{p(Kc^2 + p^2 a)}$.

(c) a first order weak focus if and only if, $\alpha = \frac{(K(p-2c)-pa)c^2 K}{p(Kc^2 + p^2 a)}$.

236

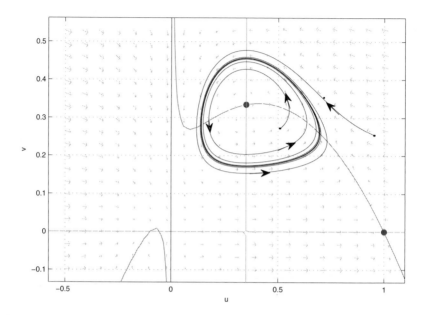

Figure 5. The equilibrium point $P_e(u_e, v_e)$ is a repellor surrounded by a limit cycle for $A = 0.05$, $B = 0.3$, $I = 0.1$, and $C = 0.35$.

Next, we present some results and comments regarding the system (7) or vector field Y_ν^E.

By defining the function $G(u) = \frac{(u-E)(1-u)(u+A)}{u}$ in a similar form as we did for the case $\alpha > 0$, we can redefined system (7) as follows;

$$\frac{du}{d\tau} = (G(u) - u)v$$
$$\frac{dv}{d\tau} = B(u - C)v$$

with Jacobian matrix similar to the one obtained in the case $\alpha > 0$

$$DY_\nu^E(u,v) = \begin{pmatrix} uG'(u) + G(u) - v & -u \\ Bv & B(u - C) \end{pmatrix} \qquad (9)$$

Lemma 3.5. (a) The set $\bar{\Gamma} = \{(u, v) \in \mathbb{R} \times \mathbb{R} : u \leq 1, v \geq 0\}$ is an invariant region.

(b) The solutions of system are bounded.

Proof. (a) 1. If $u = 1$, it has that

$$\frac{du}{d\tau} = -uv$$

$$\frac{dv}{d\tau} = B(1 - C)v$$

then the vector field Y_ν^E point out to inside of region Γ for any sign of $\frac{dv}{d\tau}$.

2. If $u = 0$, it has that

$$\frac{du}{d\tau} = -AE$$

$$\frac{dv}{d\tau} = -BCv$$

then, the trajectories of vector field Y_ν^E cross the v-axis towards the second quadrant.

3. Moreover, the u-axis is an invariant set.

(b) The proof of this part is similar the one for lemma 3.1 (b). $\qquad\square$

Lemma 3.6. *The equilibrium point $Q_A = (-A, 0)$ is an attractor for all parameter values.*

Proof. The proof is a direct analysis of the determinant and the trace of the jacobian matrix (9) evaluated at Q_A.

$$DY_\nu^E(-A, 0) = \begin{pmatrix} -(1 + A)(A + E) & A \\ 0 & B(-A - C) \end{pmatrix},$$

hence, we have that

$$\det(DY_\nu^E(-A, 0)) = (1 + A)(A + E)B(A + C) > 0$$

and

$$\text{tr}(DY_\nu^E(-A, 0)) = -(1 + A)(A + E) - B(A + C) < 0. \qquad\square$$

Remark 3.1. The equilibrium point $Q_A = (-A, 0)$ has no biological meaning but it can be a local attractor (locally asymptotically stable, see Figure 7), or a global attractor (globally asymptotically stable, see Figure 6). Therefore, it has a great influence on the behavior of the system taking account that the trajectories cross the v-axis towards the second quadrant.

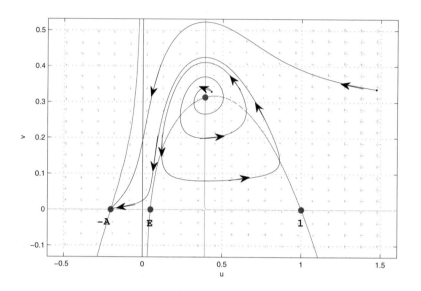

Figure 6. The equilibrium point $Q_A(-A, 0)$ is a global attractor and $Q_E(u_e, v_e)$ is repellor for $A = 0.2$, $B = 0.3$, $E = 0.0503$, and $C = 0.39$.

Lemma 3.7. *(a) The points $Q_1 = (1, 0)$ and $Q_E = (E, 0)$ are saddle points.*

(b) There exists a separatrix curve for trajectories in the phase plane determined by the stable manifolds of saddle point $(E, 0)$.

Proof. (a) The result is obtained analyzing the determinant and trace of jacobian matrix evaluated at the equilibrium points. The Jacobian matrix (9) evaluated at the critical points Q_1 and Q_E respectively is as follows;

$$DY_\nu^E(1, 0) = \begin{pmatrix} -(1+A)(1-E) & -1 \\ 0 & B(1-C) \end{pmatrix},$$

and

$$DY_\nu^E(E, 0) = \begin{pmatrix} (1-E)(A+E) & -E \\ 0 & -B(C-E) \end{pmatrix}$$

(b) Let denote by $W^s(P_E)$ the stable manifolds of saddle point $(E, 0)$, which determines a separatrix curve that divides the behavior of trajectories of system. All trajectories with initial conditions above this separatrix curve have the point $(-A, 0)$ as ω-limit. $\qquad\square$

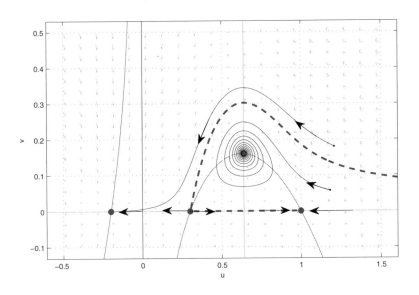

Figure 7.   The equilibrium points $Q_A(-A, 0)$ and $Q_E(u_e, v_e)$ are local attractors, and the dashed curve is the separatrix curve for $A = 0.2$, $B = 1.0$, $E = 0.3$, and $C = 0.64$.

Lemma 3.8. *For certain parameter values, the stable manifold of equilibrium point $(E, 0)$ and unstable manifold of equilibrium point $(1, 0)$ determine a heteroclinic curve joining both equilibrium points.*

Proof. Let $W^s(P_E)$, $W^u(P_1)$ the stable and unstable manifolds of $(E, 0)$ and $(1, 0)$, respectively[3,14]. It is clear that the α-limit of W^s and ω-limit of W^u are not at infinity on the direction of v-axis (since the solutions are bounded), hence no ω-limit of W^u is over u-axis. Then, for a fixed

240

$u = u^*$, there exist points $(u^*, v^s) \in W^s(P_E)$ and $(u^*, v^u) \in W^u(P_1)$, such that v^s and v^u dependent on the parameters, and letting $v^s = s(A, B, C, E)$ and $v^u = u(A, B, C, E)$, e can see that if $E < u^* << 1$ then $v^s < v^u$, and if $cE << u^* < 1$ then $v^s > v^u$. Now, since the vector field Y_ν^E is continuous with respect to parameter values, then the stable manifold $W^s(P_E)$ intersects the unstable manifold $W^u(P_1)$ which implies that there exists $(u^*, v^*) \in \bar{\Gamma}$, such that $v^{*s} = v^{*u}$ and the equation $s(A, B, C, E) = u(A, B, C, E)$ defines a surface at the parameters space for which the heteroclinic curve exists. □

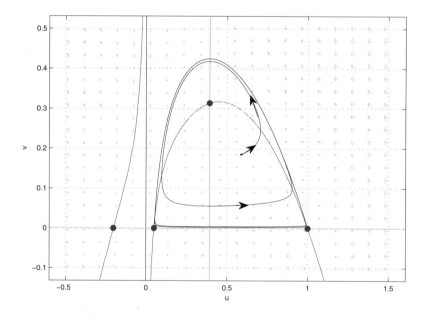

Figure 8. The Orbit approaching the heteroclinic curve for $A = 0.2$, $B = 0.3$, $E = 0.05031$, and $C = 0.3909$

Remark 3.2. The behavior of trajectories in system (7) at $\bar{\Omega}$, depend on their relative position with respect to; the separatrix curve, determined by $W^s(Q_E)$ the stable manifold of $(E, 0)$ and the separatrix curve determined by $W^u(Q_1)$ the unstable manifold of $(1, 0)$. When the trajectories are above $W^s(P_E)$ their ω-limit is the equilibrium point $(-A, 0)$, and if the trajectories are below $W^s(Q_E)$ then their ω-limit can be the positive equilibrium point $(C, v_e a)$, a limit cycle surrounding (C, v_e) or even the equilibrium point $(-A, 0)$.

Remark 3.3. In the case when the equation $s(A, B, C, E) = u(A, B, C, E)$ holds, a subregion the invariance $\check{\Gamma} \subset \bar{\Gamma}$ it is determined, where a limit cycle can exist. The diameter of this limit cycle increases when the parameters change until it collapses with the heteroclinic curve. Figure 8.

Theorem 3.3.

Let $(u^*, v^s) \in W^s(Q_E)$ and $(u^*, v^u) \in W^u(Q_1)$.

(a) If $v^s > v^u$, then the equilibrium point (C, v_e) is;

(i) a local attractor if $E < C^2 \frac{-1+2C+A}{C^2+A}$, (coexisting with the equilibrium point $(-A, 0)$, see figure 7).

(ii) a repellor if $E > C^2 \frac{-1+2C+A}{C^2+A}$, surrounded by a limit cycle (see figure 8).

(iii) a first order weak focus if $E = C^2 \frac{-1+2C+A}{C^2+A}$.

b) If $v^s < v^u$, then the equilibrium point (C, v_e) is a repellor without limit cycle and $(-A, 0)$ is globally asymptotically stable (see figure 6).

Proof. Since $\det(DY_\nu(C, v_e)) = BCG(C) = BCv_e > 0$, the nature of the point (C, v_e) depends on the behavior of the trace $\text{tr}(DY_\nu(C, v_e)) = BCG'(C) = \frac{T}{C}$ with

$$T = C^2 (1 - 2C - A) + E (C^2 + A),$$

(a) if $v^s > v^u$, we have a subregion $\hat{\Gamma} \subset \bar{\Gamma}$ determined by the stable manifold $W^s(P_E)$ and the unstable manifold $W^u(P_1)$, that is by the inequality $s(A, B, C, E) - u(A, B, C, E) > 0$ in the parameter space. Clearly,

(i) If $T < 0$, the equilibrium point (C, v_e) is a local attractor.

(ii) If $T > 0$, the equilibrium point (C, v_e) is a repellor, then there exists a limit cycle by Poincaré-Bendixon Theorem [2].

(iii) If $T = 0$ we have that $\text{tr}DY_\nu(C, v_e) = 0$, then the equilibrium point is a weak focus and $E = C^2 \frac{-1+2C+A}{C^2+A}$.

(b) if $v^s < v^u$ the heteroclinic curves determined by the intersection of the stable manifold $W^s(P_E)$ with the unstable manifold $W^u(P_1)$ (lemma

242

3.10) is broken, then $(-A, 0)$ is global attractor. □

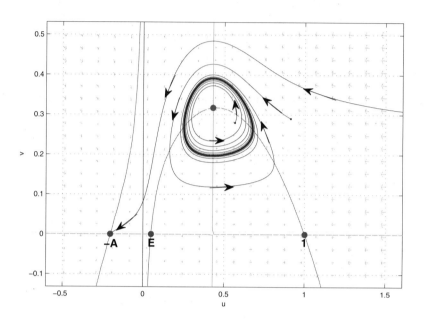

Figure 9.   The equilibrium point $Q_A(-A, 0)$ is a local attractor coexisting with a limit cycle for $A = 0.2$, $B = 0.3$, $E = 0.05031$, and $C = 0.43$.

Similarly as before here we reproduce four results for system (6).

Lemma 3.9. *(a) The set* $\Gamma = \{(x, y) \in \mathbb{R} \times \mathbb{R} / x \leq K, y \geq 0\}$ *is an invariant region.*

(b) The solutions are bounded.

Lemma 3.10. *The point* $(-a, 0)$ *is an attractor for all parameter values.*

Lemma 3.11. *(a) The points* $(K, 0)$ *and* $(\delta, 0)$ *are saddle points.*

(b) There exists a separatrix curve for trajectories in the phase plane determined by the stable manifolds of saddle point $(\delta, 0)$.

Lemma 3.12. *For certain parameter values, the stable manifold of equilibrium point $(\delta, 0)$ and unstable manifold of equilibrium point $(K, 0)$ determine a heteroclinic curve joining both equilibrium points.*

Theorem 3.4. *Let $(x^*, y^s) \in W^s(P_\delta)$ and $(x^*, y^u) \in W^u(P_K)$.*
 (a) Assuming that $y^s > y^u$, then the equilibrium point $(\frac{ac}{p-c}, y_e)$ is
 (i) a local attractor if $\delta < \frac{(K(-p+2c)+pa)c^2 K}{p(Kc^2+p^2a)}$, (coexisting with the equilibrium point $(-a, 0)$.
 (ii) a repellor if $\delta > \frac{(K(-p+2c)+pa)c^2 K}{p(Kc^2+p^2a)}$, surrounded by a limit cycle.
 (iii) a first order weak focus if $\delta = \frac{(K(-p+2c)+pa)c^2 K}{p(Kc^2+p^2a)}$.
 b) Supposing that $y^s < y^u$, then the equilibrium point $(\frac{ac}{p-c}, y_e)$ is a repellor without limit cycle and $(-a, 0)$ is globally asymptotically stable.

4. Discussion

We have analyzed the proposed model proving that it can be written as a Gause type model[6].

For the influx (immigration), when the entrance of immigrants is constant in each time unit, system ceases to be a Kolmogorov type system because the y-axis is not a invariant set, but we have shown that the dynamics is analogous to the well known Rosenzweig-MacArthur predator-prey model[22,28]. The proposed model can be considered as a modification of Rosenzweig-MacArthur model in which a new parameter in the prey growth rate is introduced, in order to express the entrance of prey to interaction. The systems derived from the model do not satisfy the conditions established in[15,31] for the uniqueness of limit cycle, however we have proved this property using Liapunov quantities[3].

Assuming the emigration phenomenon holds, we show that $(-A, 0)$ the equilibrium point at the second quadrant, has a great influence on behavior of the system, because for all parameter values, it is always a local attractor for a wide range of trajectories of model. Moreover, for a subset of parameter values this equilibrium point is globally asymptotically stable and is an attractor for all trajectories of the system, and the model has a similar properties as a model with strong Allee effect[10,11,13,19,29], in which the origin $(0, 0)$ had similar role as that of the point $(-A, 0)$. The high dependence of the trajectories respect to initial conditions require the existence of a threshold level of the population size. Furthermore, there is a coexistence of both biological phenomenon: the extinction of prey and predators population and the oscillations of both populations surrounding a unstable equilibrium point.

In this later case, we proved the existence of a separatrix curve determined by the stable manifold of the equilibrium point $(E, 0)$ that divides the behavior of trajectories of the system, which implies the high sensitivity of solutions of the system with respect to initial conditions. When there is emigration, it always exists the possibility of extinction of prey and predators, specially when the ratio prey/predator $\frac{x}{y}$ is small, that is, when the number (or density or biomass) of predator is high respect to number of prey.

We show that in the model there exists the bistability phenomenon[28], since it can coexist the attractor point $(-A, 0)$ either with a local attractor equilibrium point (C, v_e) (the positive equilibrium) or else, with a stable limit cycle. This model shows that both populations collapse to the origin over a large region of the phase plane, because the trajectories overlap or cross the y-axis for the existence of a separatrix for them, determined by the unstable manifold of point $(E, 0)$, at the first quadrant.

By mean of the diffeomorphisms φ, our result can be extended to systems (1) or (5), and we have that there exist a limit cycle coexisting with the separatrix curve of the saddle point $(\alpha, 0)$.

We note that the parameter B has no meaning in this model, but the parameters $C = \frac{\frac{c}{p}\frac{a}{K}}{1-\frac{c}{p}}$ and $E = \frac{\delta}{K}$ have a great influences on behavior of system.

In summary, the proposed model can be consider as a modification of Rosenzweig-MacArthur model when the immigration (influx) or emigration of prey population to constant rate is assumed, modifying r the biotic potential of prey, which has similar dynamic when immigration is assumed. If emigration is considered the behavior of system is similar to the Rosenzweig-MacArthur model with strong Allee effect on prey.

Acknowledgments

The authors wish to thank to the members of the Mathematical Ecology Group of Institute of Mathematics at the Pontificia Universidad Católica of Valparaíso, Chile, for their valuable comments and suggestions.

References

1. Andronov, A. A., Leontovich, E. A., Gordon I., and Maier, A. G., 1973. Qualitative theory of second-order dynamic systems, A Halsted Press Book, John Wiley & Sons, New York.

2. Arrowsmith, D. K. and Place, C. M., 1992. Dynamical System. Differential equations, maps and chaotic behaviour, Chapman and Hall.

3. Chicone, C., 1999. Ordinary differential equations with applications, Texts in Applied Mathematics 34, Springer.

4. Clark, C. W., 1990. Mathematical Bioeconomics: The optimal management of renewable resources, (2nd ed). John Wiley and Sons.

5. Courchamp F. , Clutton-Brock, T. and Grenfell, B., 1999. Inverse dependence and the Allee effect, Trends in Ecology and Evolution Vol. 14, N. 10, 405-410.

6. Freedman, H. I., 1980. Deterministic Mathematical Model in Population Ecology, Marcel Dekker, New York.

7. Gallego-Berrío, L. M., 2004. Consecuencias del efecto Allee en el modelo de depredación de May-Holling-Tanner, Maester thesis, Universidad del Quindío, Colombia.

8. W. M. Getz, 1996. A hypothesis regarding the abruptness of density dependence and the growth rate populations. Ecology, 77(7), 2014-2026.

9. González-Olivares E. and Ramos-Jiliberto, R., 2003. Dynamic consequences of prey refuges in a simple model system: more prey, fewer predators and enhanced stability, Ecological Modelling, Vol. 166, Issues 1-2, 135-146.

10. González-Olivares, E., and González-Yañez, B., 2005. Dynamics of Leslie Predator-Prey Model with Allee Effect on Prey, In R. Mondaini (Ed.) Proceedings of the Fourth Brazilian Symposium on Mathematical and Computational Biology, E-Papers Serviços Editoriais Ltda, Río de Janeiro, Vol 2.

11. González-Olivares, E., Meneses-Alcay, H., and González-Yañez, B., 2005. Metastable dynamics by considering strong and weak Allee effect on prey in Rosenzweig-MacArthur predator-prey model, submitted to Mathematical Biosciences.

12. González-Olivares, E.,González-Yañez, B., and R. Ramos-Jiliberto, 2006. Multiple attractors and uniqueness of limit cycles ina Gause predator-prey model with nomonotonic functional response and Allee effect, submitted.

13. González-Yañez, B. and González-Olivares E., 2004. Consequences of Allee effect on a Gause type predator-prey model with nonmonotonic functional response, In R. Mondaini (Ed.) Proceedings of the Third Brazilian Symposium on Mathematical and Computational Biology, E-Papers Serviços Editoriais Ltda, Río de Janeiro, Vol. 2, 358-373.

14. Guckenheimer, J., and Holmes, P., 1983. Nonlinear Oscillations, Dynamical Systems, and Bifurcations of Vector Fields. Springer-Verlag.

15. Hasík, K., 2000. Uniqueness of limit cycle in the predator-prey system with symmetric prey isocline. Mathematical Biociences 164 203-215.

16. Kent A., Doncaster C. P. and Sluckin T., 2003. Consequences for depredators of rescue and Allee effects on prey. Ecological Modelling 162, 233-245.

17. Kuznetsov, Y. A., 1995. Elements of Applied Bifurcation, AMS 112, Springer.

18. May, R. M. 1974, Stability and complexity in model ecosystems, Princeton University Press.

19. Meneses-Alcay, H. and González-Olivares, E., 2004. Consequences of the Allee effect on Rosenzweig-MacArthur predator-prey model, In R. Mondaini (Ed.) Proceedings of the Third Brazilian Symposium on Mathematical and

Computational Biology, E-Papers Serviços Editoriais Ltda, Río de Janeiro, Vol. 2, 264-277.

20. Meneses-Alcay, H. and González-Olivares, E., 2005. On the dynamics of a predator-prey model with threshold prey density for predator, In R. Mondaini (Ed.) Proceedings of the Fourth Brazilian Symposium on Mathematical and Computational Biology, E-Papers Serviços Editoriais Ltda, Río de Janeiro, Vol 2.

21. Minorsky, N., 1962, Nonlinear oscillations, Van Nostrand Company Inc.

22. Murdoch, W. W., Briggs, C. J. and Nisbet, R. M., 2003. Consumer-Resources Dynamics, Princeton University Press.

23. Rosenzweig, M. L., 1971, Paradox of enrichment: destabilization of exploitation ecosystem in ecological time, Science Vol. 171, 385-387.

24. Sotomayor, J. 1979. Lições de Equações Diferenciais Ordinárias. Projeto Euclides IMPA, CNPq.

25. Stephens, P. A. and Sutherland, W. J. , 1999. Consequences of the Allee effect for behaviour, ecology and conservation, Trends in Ecol. Evo., Vol. 14, N. 10, 401-405.

26. Stephens, P. A., Sutherland, W. J. and Freckleton, R. P. 1999, What is the Allee effect?, Oikos Vol. 87, 185-190.

27. Taylor, R. J. 1984. Predation. Chapman and Hall.

28. Turchin, P., 2003. Complex population dynamics. A theoretical/empirical synthesis, Monographs in Population Biology 35, Princeton University Press.

29. Wang, G., Liang, X-G. and Wang, F-Z., 1999. The competitive dynamics of populations subject to an Allee effect, Ecological Modelling 124, 183-192.

30. Wolfram Research, 1988, Mathematica: A System for Doing Mathematics by Computer, Champaing, IL.

31. Xiao, D., and Zhang, Z., 2003. On the uniquenes and nonexsitence of limit cycles for predator-prey systems, Nonlinearity 16, 1185-1201.

EPIDEMIC PREDICTIONS AND PREDICTABILITY IN COMPLEX ENVIRONMENTS

VITTORIA COLIZZA

Complex Networks Lagrange Laboratory (CNLL), Institute for Scientific Interchange Foundation (ISI), Turin, IT
E-mail: vcolizza@isi.it

ALAIN BARRAT

Complex Networks Lagrange Laboratory (CNLL), Institute for Scientific Interchange Foundation (ISI), Turin, IT, and
Unite Mixte de Recherche du CNRS UMR 8627, Batiment 210, Univ Paris-Sud, F-91405 Orsay, France

MARC BARTHÉLEMY

CEA-DIF Centre d'Etudes de Bruyeres-Le-Chatel, BP12, F- 91680, France

ALESSANDRO VESPIGNANI

Complex Networks Lagrange Laboratory (CNLL), Institute for Scientific Interchange Foundation (ISI), Turin, IT, and
School of Informatics, Indiana University, Bloomington, 47408 IN, USA

The spread of epidemics is inevitably entangled with human behavior, social contacts, and population flows among different geographical regions. The collection and analysis of datasets which trace the activities and interactions of individuals, social patterns, transportation infrastructures and travel fluxes, have unveiled the presence of connectivity patterns characterized by complex features encoded in large-scale heterogeneities and unbounded statistical fluctuations. These features dramatically affect the behavior of dynamical processes occurring on networks, and are responsible for the observed statistical properties of the processes' dynamics and evolution patterns. In the context of large-scale propagation of emerging infectious diseases, the air transportation network is known to play a major role in shrinking distances around the globe, by connecting far apart regions and allowing infectious travelers to potentially spread the disease to different geographic areas in a relatively short time. Here we will present a large-scale stochastic computational approach for the study of the global spread of emergent infectious diseases which explicitly incorporates real world transportation networks and census data. The

simulated spatio-temporal pattern of epidemic propagation is analyzed in relation
to the heterogeneous properties of the underlying complex architecture. Specific
quantitative indicators are introduced to evaluate the predictive capability of the
computational approach with respect to the intrinsic stochasticity of the disease
transmission and of human interactions and movements. The interplay of the complex properties of the transportation infrastructure with the disease dynamics leads
to the emergence of epidemic pathways as the most probable routes of propagation
of the disease, selected out of the huge number of possible paths the disease could
take by following airline connections. A case study for risk assessment analysis
and comparison with historical epidemics is analyzed.

1. Complexity and Epidemic Modeling

Epidemic forecast is crucially depending on our ability to model the spread
of epidemics in spatially extended systems and the movement of individuals at various levels, from the global scale of transportation flows to the
local scale of the activities and contacts of individuals. In this context,
modeling in mathematical and statistical epidemiology has evolved from
simple compartmental models into structured approaches in which the heterogeneities and details of the population and system under study are be-
coming increasingly important features[1]. In the case of spatially extended
systems, modeling approaches have been extended into schemes which explicitly include spatial structures and consist of multiple sub-populations
coupled by traveling fluxes, while the epidemic within the sub-population
is described according to approximations depending on the specific case
studied [2,3,4,5,6,7,8,9,10]. This patch or meta-population modeling framework
has then grown into a multi-scale framework in which the various possible
granularities of the system (country, inter-city, intra-city) are considered
through different approximations and coupled through interaction networks
describing the flows of people and/or animals [10,11,12,13,14,15,16,17,18,35]. At
the most detailed level, the introduction of agent based models (ABM) has
enabled to stretch even more the usual modeling perspective, by simulating
the propagation of an infectious disease individual by individual[19,20].

The above modeling approaches are based on actual and detailed data
on the activity of individuals, their interactions and movement, as well
as the spatial structure of the environment, transportation infrastructures,
traffic networks, and travel times. While for a long time this kind of data
was limited and scant, recent years have witnessed a tremendous progress
in data gathering thanks to the development of new informatics tools and
the increase in computational power. A huge amount of data, collected and
meticulously catalogued, has become finally available for scientific analysis

and study. Networks which trace the activities and interactions of individuals, social patterns, transportation fluxes and population movements on a local and global scale[19,21,22,23] have been analyzed and found to exhibit complex features encoded in large scale heterogeneity, self-organization and other properties typical of complex systems[24,25,26]. In addition, these features are shown to produce unexpected emergent behaviors that display a departure from the standard modeling perspective. Heterogeneous contact patterns induce the absence of any epidemic threshold below which no major outbreak occurs [27,28]. Also the time evolution of such processes is found to be strongly affected by the contact properties of the underlying network[29]. These considerations are particularly relevant given the ever increasing level of interconnectedness and globalization of our modern society along with a high level of diversity and heterogeneity, which induces a novel epidemiological context: the mathematical and computational modeling of disease spread needs to integrate such complex features.

2. Meta-population Models: Integrating Several Levels of Complexity

The link between infectious diseases and complex human movements and interactions leads us to investigate the role of complex connectivity patterns and traffic in meta-population models, where spatial structure is explicitly incorporated. Patch or meta-population modeling frameworks consider multiple sub-populations coupled by movements of individuals. These models are defined by the network describing the coupling among the populations along with the intensity of the coupling, which in general represents the rate of exchange of individuals between two populations. Meta-population models can be devised at various granularity levels (country, inter-city, intra-city) and the corresponding networks are therefore including very different systems and infrastructure. This implies scales ranging from the movement of people within locations of a city to the large flows of travelers among urban areas.

Let us consider as a prototypical case the world-wide spreading of epidemics through air travel. As a basic modeling strategy it is possible to use a meta-population approach[7,8,9] in which individuals are allowed to travel from one city to another by means of the airline transportation network while the disease within the city is described with opportune compartmental models or more detailed description of the disease dynamics. This amounts to write for each urban area the set of equations

$$\frac{\partial I_i}{\partial t} = K(S_i, I_i, R_i) + \Omega_i(\{I_j\}) \tag{1}$$

where the first term of the r.h.s. of the equation represents the variation of infected individuals due to the infection dynamics inside the city i (here for the sake of simplicity we consider a simple SIR model) and the second term corresponds to the net balance of infectious individuals traveling in and out of city i. This last term, the transport operator Ω_i, depends on the probability p_{ij} that an infected individual will go from city i to city j, and in the deterministic approach it can be simply written as

$$\Omega_i = \sum_{j \in V(i)} (p_{ji}I_j - p_{ij}I_i), \tag{2}$$

representing the total sum of infectious individuals arriving in city i from all neighboring cities j, subtracted of the amount of individuals traveling in the opposite directions, and acting as a coupling term among the evolution of the epidemics in the various urban areas.

This modeling program dates back to the work of Rvachev and Longini [11] and has been used along the years to simulate diseases such as pandemic influenza [30,31,32], HIV[33], and SARS[13]. While these earlier studies were considering a limited number of urban areas and travel connections, it has recently become possible to scale up this approach by including the full International Air Transport Association (IATA) [34] database. This has led to a modeling framework [16,17,18,35] considering up to 3100 airports with demographic data for the surrounding urban areas and 17182 connections among them, each representing the presence of a direct flight. This corresponds to more than 99% of the worldwide commercial traffic by plane. To each link connecting airports i and j is attached the weight w_{ij} given by the number of passengers traveling on that route in a given time (e.g. on a daily basis). The inclusion of such an extensive database is motivated by the various levels of complexity and heterogeneity present in the system composed by the worldwide air transportation network (WAN) and the associated urban areas. In particular it is possible to identify three relevant levels of strong fluctuations: the topology of the airport network, the distributions of the numbers of passengers and of the city populations (see refs.[16,17]). The model obtained by integrating all these data and the etiology of the disease within each city can be used to forecast the behavior of emerging diseases as well as to validate the approach. Fundamental to this approach is the consideration of stochasticity in the model, intrinsic both in the infection dynamics (chance related to the interaction events between individuals, to the transmission of the virus, etc.) and in the traveling behavior of individuals (for more details on the implementation, see refs.[16,17,18]). This allows for probabilistic predictions on the likelihood of local and global outbreaks, as well as quantitative predictions of their magnitude. Strikingly, this modeling approach appears to provide very good results in agreement with historical data[13,35], thus spurring the issue of identifying the fundamental limits in epidemic evolution predictability with computational modeling and their dependence on the underlying complex features of the system.

2.1. *Predictability*

A major question in the modeling of global epidemics consists in providing adequate information on the reliability of the obtained epidemic forecast; i.e. the epidemic predictability. The intrinsic stochasticity of the epidemic evolution, both in the infection dynamics and in the traveling behavior of individuals, will make each realization unique and reasonable forecast can be obtained only if all epidemic outbreak realizations starting with the same initial conditions and subject to different noise realizations are reasonably similar. A convenient quantity to monitor in this respect is the vector $\vec{\pi}(t)$ whose components are $\pi_j(t) = I_j(t)/\sum I_l$; i.e. the normalized probability that an infected individual is in city j. The similarity between two outbreak realizations (I and II) is quantitatively measured by the statistical similarity of the vectors $\vec{\pi}^I(t)$ and $\vec{\pi}^{II}(t)$. Such a measure of similarity $sim(\vec{\pi}^I(t), \vec{\pi}^{II}(t))$ is given by the standard Hellinger affinity $sim(\vec{\pi}^I(t), \vec{\pi}^{II}(t)) = \sum \sqrt{\pi_j^I \pi_j^{II}}$. Possible differences in the total (worldwide) epidemic prevalence $i = \sum I_j/P$ (where P is the worldwide population) are moreover measured by $sim(\vec{i}^I, \vec{i}^{II})$ where $\vec{i}^{I(II)} = (i^{I(II)}, 1 - i^{I(II)})$. The overlap function measuring the similarity between two different outbreak realizations is thus defined by:

$$\Theta(t) = sim(\vec{i}^I, \vec{i}^{II}) sim(\vec{\pi}^I, \vec{\pi}^{II}) \qquad (3)$$

The overlap is maximal ($\Theta(t) = 1$) when the very same cities have the very same number of infectious individuals in both realizations, and $\Theta(t) = 0$ if the two realizations do not have any common infected cities at time t. Clearly, a large overlap corresponds to a predictable evolution, providing a direct measure of the reliability of the epidemic forecast.

If we consider a model in which the cities are linked by a completely ho-

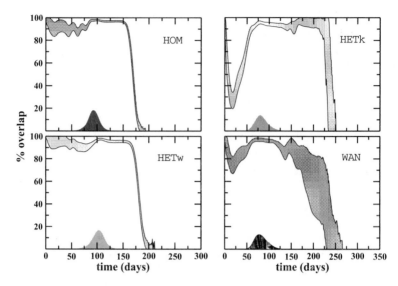

Figure 1. **Epidemics predictability.** Percentage of overlap as a function of time: the shaded area corresponds to the standard deviation obtained with $5\,10^3$ couples of different realizations. Topological heterogeneity plays a dominant role in reducing the overlap in the early stage of the epidemics. Large fluctuations at the end of the epidemics are observed when a heterogeneous topology is considered, due to the different lifetime of the epidemics in distinct realizations, induced by the large topological fluctuations of the network. We also report the prevalence profile as a function of time showing that the maximum predictability corresponds to the prevalence peak.

mogeneous transport network (HOM), where both number of connections of each airport and traffic flows on each connection are close to the average values measured in the WAN, we find a significant overlap ($\Theta(t) > 80\%$, see Fig. 1, the shaded area representing the standard deviation) even at the early stage of the epidemics - the most relevant phase for epidemic surveillance and the more prone to stochastic fluctuations. Similar results are obtained if the topologically homogeneous network is associated to heterogeneous traffic values (HETw) as the ones of the real air transportation network. The picture is different if we consider the real heterogeneous topology of the WAN (with both badly connected airports and hubs) associated to homogeneous travel fluxes equal to the average traffic value on each connection (HETk), since especially at the initial stage of the epidemics the predictability is much smaller. Finally, the values of the overlap for epidemics propagating on the real air-transportation network (WAN) show an intermediate situation. These results may be rationalized by considering the conflicting effects of the various levels of heterogeneity. On the one hand, the heterogeneity of the connectivity pattern (broad distribution of degrees), and in particular the existence of hubs, provides a multiplicity of equivalent channels for the travel of infected individuals, depressing the predictability of the evolution, as the comparison of HETk and HOM shows. On the other hand, the heterogeneity of traffic flows introduces dominant connections which select preferential pathways, increasing the epidemic predictability. The backbone of such dominant spreading channels thus defines specific "epidemic pathways" which are weakly affected by the stochastic noise. In the case of the worldwide airport network, the heterogeneity of the fluxes thus partially compensates for the decrease in predictability due to the topological heterogeneity.

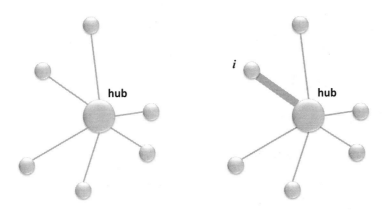

Figure 2. **Influence of heterogeneity on the predictability.** (A) Large degrees lower the predictability. Starting from the hub if all weights are equal, a disease can spread on all the nodes with equal probability. (B) For the same topology, weight heterogeneity selects a particular path (from the hub to the node i) and thus increases the predictability.

3. SARS: Risk Assessment Analysis and Forecast Reliability

The outbreak of severe acute respiratory syndrome (SARS) in 2002-2003 represented a serious public health threat for the International community. Its rapid spread to regions far apart from the initial outbreak created great concern for the potential ability of the virus to affect a large number of countries and required a coordinated effort aimed at its containment[36].

In order to study the epidemic pattern of SARS and how the complex structure of airline connections impacted the observed spread, we apply the stochastic meta-population epidemic model described before. The infection dynamics inside each urban area is based on a compartmentalization which considers the specific stages of SARS infection, including latency period, hospitalization, quarantine, etc. A schematic representation of the diagram flow is reported in Figure 3. The disease parameters are estimated from the Hong Kong outbreak in a way consistent with the global nature of the meta-population model by including the impact of infectious individuals traveling in and out of the city. Once the disease parameters are determined, no free adjustable parameters are left in the model.

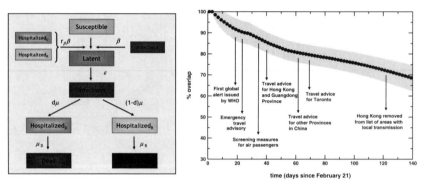

Figure 3. **Left: Flow diagram of the transmission model.** The population of each city is classified into seven different compartments, namely susceptible (S), latent (L), infectious (I), hospitalized who either recover (HR) or die (HD), dead (D) and recovered (R) individuals. Susceptible individuals exposed to SARS enter the latent class. Latents represent infected who are not yet contagious and are assumed to be asymptomatic. They become infectious after an average time ϵ^{-1} (mean latency period). The individual is classified as infectious during an average time equal to μ^{-1} from the onset of clinical symptoms to his admission to the hospital where he eventually dies or recovers. Patients admitted to the hospital are not allowed to travel. **Right: Overlap profile.** The value of the overlap is shown as a function of time, from the initial day of the simulations (February 21, 2003) to July 11, 2003. Details on relevant events occurring during SARS epidemics are shown for reference.

The temporal and geographic pattern of the disease can be analyzed, and the model's predictive power can be tested against available empirical data from the World Health Organization (WHO)[37]. Reference[35] reports on the introduction of specific indicators which consider the stochastic nature of the process in order to provide risk analysis scenarios and to compare simulation results with the empirically observed epidemic pattern. Here we would like to focus on the reliability of SARS epidemic forecasts and explore the role played by the airline transportation network in the observed worldwide propagation of the virus. The airline network structure explicitly incorporated into the model is composed by more than 17,000 different connections among 3,100 cities. Such a large number of connections produces a huge amount of possible different paths available for the infection to spread throughout the world. This in principle could easily result in a set of simulated epidemic outbreaks which are very different one from the other – though starting from the same initial conditions – thus leading to a poor predictive power for the computational model. The average overlap profile reported in Figure 3 together with the 95% confidence interval clearly shows a different trend. The overlap starts from a value equal to 1, since all stochastic realizations share the same initial conditions, and decreases monotonically with time. However, it assumes values larger than 0.7 in the time window investigated, pointing to the relatively strong computational reproducibility of the synthetic SARS outbreak. The simulated disease seems indeed to follow a very similar evolution at each realization of the process. As discussed in the previous section, the origin of such reproducibility lies in the emergence of epidemic pathways, i.e. preferential channels along which the epidemic will more likely spread[16,17]. In order to identify these pathways, we monitor the spreading path followed by the virus in 10^3 outbreaks starting from the same initial conditions. More precisely, starting from Hong Kong, we follow the propagation of the virus and identify for each infected country C_i the country C_j where the infection came from, thus defining a probability of origin of infection for each country. Results are reported in Figure 4 where the epidemic pathways are represented by arrows whose thickness accounts for the probability of infection. This information identifies for each country the possible origins of infection and provides a quantitative estimation of the probability of receiving the infection from each identified origin[35]. It is therefore information of crucial importance for the development and assessment of preparation plans of single countries. Travel advisories or limitations and medical screenings at the ports of entry – such as those put in place during SARS epidemic – might as well strongly benefit from the analysis and identification of such epidemic pathways.

256

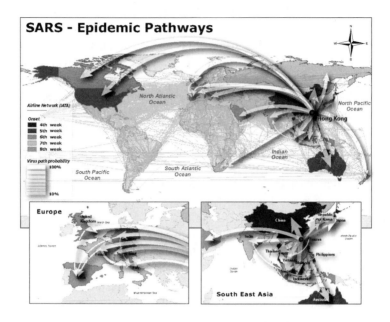

Figure 4. The SARS outbreak in 2003 has clearly demonstrated how the highly inter-
connected nature of our world can be a major disadvantage against the large-scale spread
of emerging infectious diseases. Simulation results describe a spatio-temporal evolution
of the disease in agreement with the historical data. Analysis on the robustness of the
model's forecasts leads to the emergence and identification of epidemic pathways as the
most probable routes of propagation of the disease. Only few preferential channels are
selected (arrows; width indicates the probability of propagation along that path) out
of the huge number of possible paths the infection could take by following the complex
nature of airline connections (source: IATA).

4. Conclusions

The presented computational approach shows that the integration of long
range mobility and demographic data provide epidemic models of a predictive power that can be consistently tested and theoretically motivated.
This computational strategy can be therefore considered as a general tool
in the analysis and forecast of emerging infectious diseases global spreading
and in the definition of containment policies aimed at reducing the effects
of potentially catastrophic outbreaks.

References

1. Anderson RM, May RM, *Infectious diseases in humans* (Oxford University Press, Oxford 1992).

2. H.W. Hethcote. An immunization model for a heterogeneous population. *Theor. Pop. Biol.* **14**, 338-349 (1978).

3. R.M Anderson, R.M May. Spatial, temporal and genetic heterogeneity in host populations and the design of immunization programs. *IMA J. Math. Appl. Med. Biol.* **1**, 233-266 (1984).

4. R.M May, R.M Anderson. Spatial heterogeneity and the design of immunization programs. *Math. Biosciences* **72**, 83-111 (1984).

5. B. M. Bolker, B. T. Grenfell. Chaos and biological complexity in measles dynamics. *Proc. R. Soc. Lond.* B **251**, 75-81 (1993).

6. B. M. Bolker, B. T. Grenfell. Space persistence and dynamics of measles epidemics. *Phil. Trans. R. Soc. Lond.* B **348**, 309-320 (1995).

7. A.L. Lloyd, R.M. May. Spatial heterogeneity in epidemic models. *J. Theor. Biol.* **179**, 1-11 (1996).

8. B. T. Grenfell, B. M. Bolker. Cities and villages: infection hierarchies in a measles meta-population. *Ecology Letters* **1**, 63-70 (1998).

9. M.J. Keeling, P. Rohani. Estimating spatial coupling in epidemiological systems: a mechanistic approach. *Ecology Letters* **5**, 20-29 (1995).

10. N.M. Ferguson, M.J. Keeling, W.J. Edmunds, R. Gani, B.T. Grenfell, R.M. Anderson, S. Leach. Planning for smallpox outbreaks. *Nature* **425**, 681-685 (2003).

11. L.A. Rvachev, I.M. Longini. *Mathematical Biosciences* **75**, 3-22 (1985).

12. M. J. Keeling, M.E.J. Woolhouse, D.J. Shaw, L. Matthews, M. Chase-Topping, D.T. Haydon, S.J. Cornell, J. Kappey, J. Wilesmith, B.T. Grenfell. Dynamics of the 2001 UK Foot and Mouth Epidemic: Stochastic Dispersal in a Heterogeneous Landscape. *Science* **294**, 813-817 (2001).

13. L. Hufnagel, D. Brockmann, T. Geisel. Forecast and control of epidemics in a globalized world. *Proc. Natl. Acad. Sci. (USA)* **101**, 15124-15129 (2004).

14. I.M. Longini, A. Nizam, S. Xu, K. Ungchusak, W. Hanshaoworakul, D.A.T. Cummings, M.E. Halloran. Containing pandemic infleunza at the source. *Science* **309** 1083 (2005).

15. N.M. Ferguson, D.A.T. Cummings, S. Cauchemez, C. Fraser, S. Riley, A. Meeyai, S. Iamsirithaworn, D.S. Burke. Strategies for containing an emerging inlfuenza pandemic in Southeast Asia. *Nature* **437**, 209 (2005).

16. V. Colizza, A. Barrat, M. Barthélemy, A. Vespignani. The role of the airline transportation network in the prediction and predictability of global epidemics. *Proc. Natl. Acad. Sci. USA* **103** 2015-2020 (2006).

17. V. Colizza, A. Barrat, M. Barthélemy, A. Vespignani.The modeling of global epidemics: stochastic dynamics and predictability. *Bull. Math. Biol.* **68** 1893-1921 (2006).

18. V. Colizza, A. Barrat, M. Barthélemy, A.-J. Valleron, A. Vespignani. Modeling the worldwide spread of pandemic influenza *PLoS Medicine* **4(1)** e13 (2007).

19. G. Chowell, J.M. Hyman, S. Eubank, C. Castillo-Chavez. Scaling laws for the movement of people between locations in a large city. *Phys. Rev. E* **68**, 066102 (2003).

20. S. Eubank, H. Guclu, V.S.A. Kumar, M.V. Marathe, A. Srinivasan, Z. Toroczkai, N. Wang. Modelling Disease Outbreaks in Realistic Urban Social Networks. *Nature* **429**, 180184 (2004).

21. A. Barrat, M. Barthélemy, R. Pastor-Satorras, A. Vespignani. The architecture of complex weighted networks. *Proc. Natl. Acad. Sci. (USA)* **101**, 3747-3752 (2004).

22. F. Liljeros, C.R. Edling, L.A.N. Amaral, H.E. Stanley, Y. Aberg. The web of human sexual contacts. *Nature* **411**, 907 (2001).

23. A. Schneeberger, C.H. Mercer, S.A. Gregson, N.M. Ferguson, C.A. Nyamukapa, R.M. Anderson, A.M. Johnson, G.P. Garnett. Scale-Free Networks and Sexually Transmitted Diseases: A Description of Observed Patterns of Sexual Contacts in Britain and Zimbabwe. *Sexually Transmitted Diseases* **31**, 380-387 (2004).

24. R. Albert, A.-L. Barabasi. Statistical mechanics of complex networks. *Rev. Mod. Phys.* **74**, 47-97 (2000).

25. S.N. Dorogovtsev, J.F.F. Mendes. *Evolution of Networks: From Biological Nets to the Internet and WWW* (Oxford Univ. Press, Oxford, 2003).

26. R. Pastor-Satorras, A. Vespignani. *Evolution and Structure of the Internet: A Statistical Physics Approach* (Cambridge Univ. Press, Cambridge, UK, 2003).

27. R Pastor-Satorras, A Vespignani. *Phys Rev Lett* **86**, 3200 (2001).

28. SL Lloyd, RM May. *Science* **292**, 1316 (2001).

29. M Barthélemy, A Barrat, R Pastor-Satorras, A Vespignani. *Phys Rev Lett* **92**, 178701 (2004).

30. I.M. Longini. A mathematical model for predicting the geographic spread of new infectious agents. *Mathematical Biosciences* **90**, 367-383 (1988).

31. R.F. Grais, J.H. Ellis, G.E. Glass. Assessing the impact of airline travel on the geographic spread of pandemic influenza. *Europ. J. Epidemiol.* **18**, 1065-1072 (1988).

32. R.F. Grais, J.H. Ellis, A. Kress, G.E. Glass. Modeling the Spread of Annual Influenza Epidemics in the US: The Potential Role of Air Travel. *Health Care Manag Sci* **7**, 127-134 (2004).

33. A. Flahault, A.-J. Valleron. A method for assessing the global spread of HIV-1 infection based on air-travel. *Math Pop Studies* **3**, 1-11 (1991).

34. International Air Transport Association. http://www.iata.org.

35. V. Colizza, A. Barrat, M. Barthélemy, A. Vespignani. Predictability and epidemic pathways in global outbreaks of infectious diseases: the SARS case study. *BMC Med* **5**, 34 (2007).

36. McLean AR, May RM, Pattison J, Weiss RA. *SARS. A case study in emerging infections.* (Oxford University Press; 2005).

37. World Health Organization. http://www.who.int/.

ASSESSING THE SPATIAL PROPAGATION OF WEST NILE VIRUS

NORBERTO A. MAIDANA, HYUN M. YANG

UNICAMP – IMECC / DMA
Caixa Postal 6065
CEP: 13083-859, Campinas, SP
E-mail: nmaidana@ime.unicamp.br

In this work we study a spatial model for the West Nile Virus (WNV) propagation across the USA from the east to the west. WNV is an arthropod-borne flavivirus that appeared at first time in New York city in the summer of 1999 and then spread prolifically within birds. Mammals, as human and horse, do not develop sufficiently high bloodstream titers to play a significant role in transmission, which is the reason to consider the mosquito-bird cycle. The proposed model aims to study this propagation in a system of partial differential reaction-diffusion equations considering the mosquito and the avian populations. The diffusion is allowed to both populations, being greater in avian than in the mosquito. When a threshold value R_0, depending on the model´s parameters, is greater than one, the disease remains endemic and could propagate to regions previously free of disease. The travelling wave solutions of the model are studied to determine the speed of the disease propagation. This wave speed is obtained as a function of the model´s parameters, for instance, vertical transmission rate and avian diffusion coefficient.

1. Introduction

West Nile Virus (WNV) is an arthropod-borne flavivirus. The primary vectors of WNV are *Culex spp* mosquitoes, although the virus has been isolated from at least 29 more species of ten genera, see Campbell *et al.*[9]. When an infected mosquito bites a bird, the virus is transmitted. A mosquito is infected when bites an infected bird. Also, the virus can be passed via vertical transmission, from a mosquito to its offspring.

One major feature of WNV spatial dissemination is the high velocity of geographic invasion and colonization. This is due to long distance flying of birds, and ubiquitous presence of mosquitoes. For instance, WNV was introduced in New York City in 1999, and then propagates across the USA. After five years, WNV was detected among birds in California, west side of USA. Some studies about the non spatial dynamic was developed by Kenkre *et al.*[11], wonham *et al.*[12], Cruz-Pacheco *et al.*[2] and Bowman *et al.*[13].

The models proposed consider different aspects of the WNV disease and determine threshold conditions to asses control strategies. Kenkre *et al.*[11] study the periodicity of the infection considering vertical transmission, mortality increase due to infection and time scale disparity. In the Wonham *et al.*[12] model is considered all the mosquito life cycle. Cruz-Pacheco *et al.*[2] analyze the mathematical model and use experimental data for several species of birds. In the Bowman *et. al*[13] model is considered the human population to asses preventive strategies.

A spatial model was study by Lewis *et al.*[3] considering for the non spatial dynamic the Wonham *et al.*[12] model. They study the WNV propagation using travelling wave solution for a simplified model which does not consider vertical transmission, WNV death rate and the avian recover subpopulation. Aiming to determine the biological invasion of WNV from east to west cost of USA, we develop a spatio-temporal model to study this propagation as a consequence of the zoonostic characteristic of WNV.

In the modeling for the spatial dynamics of WNV the diffusion is considered in avian and mosquito populations, taking into account the fact that the diffusion coefficient in the avian population is greater than the diffusion in the mosquitoes population. From the model we seek for the travelling waves connecting the two steady states, from which we determine the wave speed of propagation of the WNV disease. The depending of this wave speed on the vertical transmission and on the avian diffusion is obtained. Okubo[1] estimated the diffusion coefficient of birds situating between 0 and 14 km^2/day. Choosing a coefficient of avian diffusion equal to 6 km^2/day, and considering parameters regarded to two birds species, named Blue jay and Common grackle, we obtain for the velocity of the disease propagation approximately 3 km/day, which agrees with that observed from field data.

The paper is structured as follows. In section 2 the WNV spatial propagation model is presented, which is preceded by a brief description of the corresponding spatial homogeneous model. In section 3 the minimum speed of the travelling wave is determined, and conclusion is given in section 4.

2. Model for the West Nile Virus

Let us describe with some details the spatially homogeneous dynamics and the descriptions of the parameters of the model proposed in Cruz-Pacheco *et al.*[2]. From this model we derive the WNV geographic propagation model.

2.1. *Model for the spatially homogeneous WNV propagation dynamics*

The model proposed in Cruz-Pacheco *et al.*[2] includes cross-infection between the avian and the vector populations, which sizes are denoted by $N_a(t)$ and $N_v(t)$, respectively. The avian population was divided into susceptible, infective and recovered subpopulations, S_a, I_a and R_a, respectively, while for the vector population, the susceptible and infected subpopulations, S_v and I_v.

The mosquito population is taken constant, assuming that the birth and death rates are equal to μ_v. For the avian population, however, the total population size is allowed to vary, where Λ_a is a constant recruitment rate due to birth and migration, and death rate is μ_a. The differential equation for birds population is, then,

$$\frac{dN_a}{dt} = \Lambda_a - \mu_a N_a.$$

The biting rate b of mosquitoes is defined as the average number of bites per mosquito per day. β_a and β_v are the transmission probabilities from vector to birds and from birds to vector, respectively. Hence the infection rates per susceptible birds and susceptible vector are given by:

$$b\beta_a \frac{N_v}{N_a} \frac{I_v}{N_v} = b\frac{\beta_a}{N_a} I_v$$

and

$$b\beta_v \frac{I_a}{N_a}.$$

The birds are recovered at rate γ_a. The specific death rate associated with WNV in the avian population is α_a, with $\alpha_a \leq \gamma_a$. Another assumption is that mosquitoes can transmit WNV vertically. The fraction of progeny of infectious mosquitoes that is infectious is denoted by p, with $0 \leq p \leq 1$.

Based on the above parameters, the model is the following:

$$\frac{dS_a}{dt} = \Lambda_a - \frac{b\beta_a}{N_a}I_v S_a - \mu_a S_a \tag{1}$$

$$\frac{dI_a}{dt} = \frac{b\beta_a}{N_a}I_v S_a - (\gamma_a + \mu_a + \alpha_a)I_a \tag{2}$$

$$\frac{dR_a}{dt} = \gamma_a I_a - \mu_a R_a \tag{3}$$

$$\frac{dS_v}{dt} = \mu_v S_v + (1 - p)\mu_v I_v - \frac{b\beta_v}{N_a}I_a S_v - \mu_v S_v \tag{4}$$

$$\frac{dI_v}{dt} = p\mu_v I_v + \frac{b\beta_v}{N_a}I_a S_v - \mu_v I_v \tag{5}$$

$$\frac{dN_a}{dt} = \Lambda_a - \mu_a N_a - \alpha_a I_a. \tag{6}$$

The model has the disease free equilibrium and one endemic state, see Cruz-Pacheco et al.[2], which exists if:

$$R_0 = \frac{mb^2\beta_a\beta_v}{(1 - p)\mu_v(\gamma_a + \mu_a + \alpha_a)} > 1$$

In Table 1 we show the Basic Reproductive Number for three avian species, Blue jay, Common grackle and Fish crow.

Table 1. Basic Reproductive Number calculated from the epidemiological and demographic parameters

Common name	β_a	β_v	γ_a (day^{-1})	α_a (day^{-1})	μ_a (day^{-1})	μ_v (day^{-1})	$\sqrt{R_0}$
Blue jay	1.0	0.68	0.26	0.15	0.0002	0.06	5.89
Common grackle	1.0	0.68	0.33	0.07	0.0001	0.06	5.97
Fish crow	1.0	0.26	0.36	0.06	0.0002	0.06	3.60

2.2. Model for the spatial dynamics of WNV

WNV disease first appeared in North America in summer of 1999, with the simultaneous occurrence of an unusual number of deaths of exotic birds and crows in the New York City, see DeBiasi et al.[4]. We propose a model to study the propagation of WNV across the USA.

The diffusion among avians is denoted by D_a and D_v is regarded to the diffusion of mosquito population. We are not taking into account the long migratory movement of birds. The mosquitoes are considered as a sessible population, then $D_v << D_a$. For instance, the mean dispersal distance for *Aedes aegypty* was ranged from 28 to 199 meters, Harrington et al. [10]. From now on we consider the spatio-temporal dependence on the populations, e.g. $N_a(x,t)$ and $N_v(x,t)$, and their respective subpopulations. The model is the following:

$$\frac{\partial S_a}{\partial t} = D_a \frac{\partial^2 S_a}{\partial x^2} + \Lambda_a - \frac{b\beta_a}{N_a} I_v S_a - \mu_a S_a \tag{7}$$

$$\frac{\partial I_a}{\partial t} = D_a \frac{\partial^2 I_a}{\partial x^2} + \frac{b\beta_a}{N_a} I_v S_a - (\gamma_a + \mu_a + \alpha_a) I_a \tag{8}$$

$$\frac{\partial R_a}{\partial t} = D_a \frac{\partial^2 R_a}{\partial x^2} + \gamma_a I_a - \mu_a R_a \tag{9}$$

$$\frac{\partial S_v}{\partial t} = D_v \frac{\partial^2 S_v}{\partial x^2} + \mu_v S_v + (1-p)\mu_v I_v - \frac{b\beta_v}{N_a} I_a S_v - \mu_v S_v \tag{10}$$

$$\frac{\partial I_v}{\partial t} = D_v \frac{\partial^2 I_v}{\partial x^2} + p\mu_v I_v + \frac{b\beta_v}{N_a} I_a S_v - \mu_v I_v \tag{11}$$

$$\frac{\partial N_a}{\partial t} = D_a \frac{\partial^2 N_a}{\partial x^2} + \Lambda_a - \mu_a N_a - \alpha_a I_a. \tag{12}$$

Let us introduce the non dimensional parameters to the system (7) - (12). The time is scaled with respect to bm, where b is the biting rate of mosquitoes and $m = \frac{N_v}{\Lambda/\mu_a}$, the ratio between the vector population and the disease free equilibrium bird population. The spatial variable is scaled considering the bird's diffusion coefficient, according to $\sqrt{\frac{D_a}{bm}}$. Then non dimensional parameters are:

$$\tilde{S}_a = \frac{S_a}{\Lambda/\mu_a}, \ \tilde{I}_a = \frac{I_a}{\Lambda/\mu_a}, \ \tilde{R}_a = \frac{R_a}{\Lambda/\mu_a}, \ \tilde{N}_a = \frac{N_a}{\Lambda/\mu_a}, \ \tilde{S}_v = \frac{S_v}{N_v}, \ \tilde{I}_v = \frac{I_v}{N_v}$$

$$D = \frac{D_v}{D_a}, \ \tilde{\mu}_a = \frac{\mu_a}{bm}, \ \tilde{\gamma}_a = \frac{\gamma_a}{bm}, \ \tilde{\alpha}_a = \frac{\alpha}{bm} \ \tilde{\mu}_v = \frac{\mu_v}{bm}$$

$$\tilde{\beta}_a = \beta_a, \ \tilde{\beta}_v = \frac{\beta_v}{m}.$$

Therefore, omitting $\tilde{R}_a$ and $\tilde{S}_v$ (both are decoupled form the system), see Cruz-Pacheco et al.[2], the dimensionless model obtained is:

$$\frac{\partial \tilde{S}_a}{\partial t} = \frac{\partial^2 \tilde{S}_a}{\partial x^2} + \tilde{\mu}_a - \frac{\tilde{\beta}_a}{\tilde{N}_a} \tilde{I}_v \tilde{S}_a - \tilde{\mu}_a \tilde{S}_a \tag{13}$$

$$\frac{\partial \tilde{I}_a}{\partial t} = \frac{\partial^2 \tilde{I}_a}{\partial x^2} + \frac{\tilde{\beta}_a}{\tilde{N}_a} \tilde{I}_v \tilde{S}_a - (\tilde{\gamma}_a + \tilde{\mu}_a + \tilde{\alpha}_a) \tilde{I}_a \tag{14}$$

$$\frac{\partial \tilde{I}_v}{\partial t} = D \frac{\partial^2 \tilde{I}_v}{\partial x^2} + \frac{\tilde{\beta}_v}{\tilde{N}_a} \tilde{I}_a (1 - \tilde{I}_v) - (1 - p)\tilde{\mu}_v \tilde{I}_v \tag{15}$$

$$\frac{\partial \tilde{N}_a}{\partial t} = \frac{\partial^2 \tilde{N}_a}{\partial x^2} + \tilde{\mu}_a - \tilde{\mu}_a \tilde{N}_a - \tilde{\alpha}_a \tilde{I}_a. \tag{16}$$

The system of equations (13-16) has two steady states. The first is the disease free equilibrium point, given by:

$$P_0 = (1, 0, 0, 1).$$

The second is the endemic state:

$$P_1 = (S_a^*, I_a^*, I_v^*, N_a^*),$$

where S_a^*, I_v^* and N_a^* are given by:

$$S_a^* = \frac{\mu_a - (\gamma_a + \mu_a + \alpha_a)I_a^*}{\mu_a},$$

$$I_v^* = \frac{\mu_a \beta_v I_a^*}{(\beta_v \mu_a - \alpha_a(1 - p)\mu_v)I_a^* + (1 - p)\mu_v \mu_a}$$

and

$$N_a^* = \frac{\mu_a - \alpha_a I_a^*}{\mu_a},$$

where I_a^* is the positive root of the second degree polynomial

$$r(I_a) = EI_a^2 + FI_a + G,$$

with the coefficients

$$E = [\tilde{\beta}_v \tilde{\mu}_a - \tilde{\alpha}_a(1 - p)\tilde{\mu}_v] \frac{\tilde{\alpha}_a}{\tilde{\mu}_a}$$
$$F = 2\tilde{\alpha}_a(1 - p)\tilde{\mu}_v - \tilde{\beta}_v \tilde{\mu}_a - (1 - p)\tilde{\mu}_v(\tilde{\gamma}_a + \tilde{\mu}_a + \tilde{\alpha}_a)\tilde{R}_0$$
$$G = \tilde{\mu}_a(1 - p)\tilde{\mu}_v(\tilde{R}_0 - 1).$$

Notice that a positive solution always exists for $\tilde{R}_0 > 1$, where

$$\tilde{R}_0 = \frac{\tilde{\beta}_a \tilde{\beta}_v}{(1-p)\tilde{\mu}_v(\tilde{\gamma}_a + \tilde{\mu}_a + \tilde{\alpha}_a)}. \tag{17}$$

The original system has the equilibrium values given in Cruz-Pacheco *et al.*[2], which has positive solution when $R_0 > 1$.

In Figure 1 we show the variation of the basic reproductive number $\tilde{R}_0$ as a function of the vertical transmission. When p increases to 1, $\tilde{R}_0$ increases to infinity. The vertical transmission is an important fact on the spatially homogeneous situation.

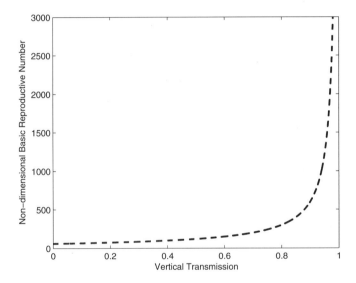

Figure 1. Graph of the variation of $\tilde{R}_0$ as a function of the vertical transmission p, for the parameters related to Blue jay given in Table 2. The Basic Reproductive Number increases to infinity with p. This reflects that the vertical transmission is an important factor in the homogeneous situation.

The following Theorem, equivalent to that in Cruz-Pacheco *et al.*[2], regarded to two equilibrium points, is established:

Theorem 2.1. *If $0 \le p < 1$, then the disease free equilibrium P_0 is unique and locally and asymptotically stable for $\tilde{R}_0 < 1$. When $\tilde{R}_0 > 1$, P_0 becomes unstable, and there appears a new endemic equilibrium P_1 which is locally and asymptotically stable. If $p = 1$, P_0 is always unstable, and P_1 is locally and asymptotically stable.*

3. Travelling Waves Solution

In this section we study the WNV geographic propagation, Murray *et al.*[6], determining the minimum wave speed connecting the disease free equilibrium point to the endemic state. The solution corresponding to minimum wave speed describes the observed dynamics of the system, see Sandstede[7], Volpert and Volpert[8].

The travelling waves solution, when exists, can be set in the usual form, see Murray [5]:

$$(s_a(x,t), i_a(x,t), i_v(x,t), n_a(x,t)) = (s_a(z), i_a(z), i_v(z), n_a(z)),$$

where $z = x + ct$. Considering that the diffusion of the avian population is greater than the mosquito population, we assume that $D = 0$. The corresponding first order ordinary differential equations with respect to variable z is:

$$\frac{ds_a}{dz} = u_1,$$

$$\frac{du_1}{dz} = cu_1 - \tilde{\mu}_a + \frac{\tilde{\beta}_a i_v}{n_a} s_a + \tilde{\mu}_a s_a,$$

$$\frac{di_a}{dz} = u_2,$$

$$\frac{du_2}{dz} = cu_2 - \frac{\tilde{\beta}_a i_v}{n_a} s_a + (\tilde{\gamma}_a + \tilde{\mu}_a + \tilde{\alpha}_a) i_a,$$

$$\frac{di_v}{dz} = (1/c)(\tilde{\beta}_v i_a \frac{(1 - i_v)}{n_a} - (1 - p)\tilde{\mu}_v i_v),$$

$$\frac{dn_a}{dz} = u_3,$$

$$\frac{du_3}{dz} = cu_3 - \tilde{\mu}_a + \tilde{\mu}_a \eta_a + \tilde{\alpha}_a i_a,$$

where the boundary conditions are:

$$\lim_{z \to -\infty} (s_a(z), u_1(z), i_a(z), u_2(z), i_v(z), n_a(z), u_3(z)) = (1, 0, 0, 0, 0, 1, 0)$$

and

$$\lim_{z \to \infty} (s_a(z), u_1(z), i_a(z), u_2(z), i_v(z), n_a(z), u_3(z)) = (S_a^*, 0, I_a^*, 0, I_v^*, N_a^*, 0).$$

The zeros in both equilibrium points deserve some words. The three zeros in the second equilibrium point correspond to derivatives of the subpopulations s_a, i_a and n_a. However, the first equilibrium point has two more zeros corresponding to infectious populations regarded to avians and mosquitoes, which must not be negative numbers. Due to this constraint, we impose to the linear system solutions that must not oscillate, i.e., the eigenvalues corresponding to this equilibrium point must assume real values.

The characteristic polynomial regarded to the linear system at the equilibrium point $(1, 0, 0, 0, 0, 1, 0)$ is $Q(\lambda) \times P(\lambda)$, where:

$$Q(\lambda) = (\lambda^2 - c\lambda - \tilde{\mu}_a)^2$$

and

$$P(\lambda) = \lambda^3 + A\lambda^2 + B\lambda + C,$$

where the coefficients are

$$A = c - \frac{\tilde{\mu}_v(1-p)}{c}$$
$$B = -c(\tilde{\alpha}_a + \tilde{\gamma}_a + \tilde{\mu}_a) - \tilde{\mu}_v c(1-p)$$
$$C = (1-p)\tilde{\mu}_v(\tilde{\gamma}_a + \tilde{\mu}_a + \tilde{\alpha}_a)\left(\tilde{R}_0 - 1\right),$$

with $\tilde{R}_0$ being given by (17). The polynomial $Q(\lambda)$ has always reals roots. Then the polynomial $P(\lambda)$ must carry on the conditions for the existence of the minimum speed. First, the existence of the endemic state implies that $\tilde{R}_0 > 1$, then we have that:

$$P(0) = (1-p)\tilde{\mu}_v(\tilde{\gamma}_a + \tilde{\mu}_a + \tilde{\alpha}_a)[\tilde{R}_0 - 1] > 0,$$

moreover, it is easy to verify that

$$\lim_{\lambda \to \pm\infty} P(\lambda) = \pm\infty, \quad \left.\frac{dP(\lambda)}{d\lambda}\right|_{\lambda=0} = -c(\tilde{\alpha}_a + \tilde{\gamma}_a + \tilde{\mu}_a) - \tilde{\mu}_v c(1-p) < 0,$$

which imply that $P(\lambda)$ always has one negative real root. Second, the remaining two roots can be either real positives or complex numbers. In order to obtain the minimum wave speed, we determine the condition that the imaginary part of the complex root must be zero (or, the roots must be real numbers). This condition is satisfied when the positive real roots are equal, from which we determine the wave speed, see Figure 2. The condition to obtain the double roots follows easy calculations: The polynomial evaluated at the unique local minimum, λ_+, is zero, that is, $P(\lambda_+) = 0$, where:

$$\lambda_+ = \frac{1}{3}\{-A + \sqrt{A^2 - 3B}\}.$$

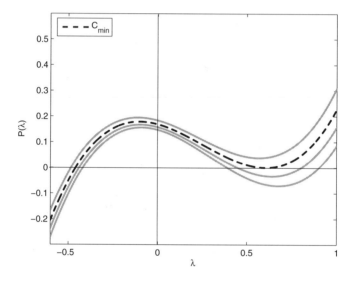

Figure 2.   Polynomial graphics for $c = 0.72$, $c_{min} = 0.78$, $c = 0.84$ and $c = 0.9$, taking into account the non dimensional parameters corresponding to those given in Table 2 for the Blue jay, with $p = 0.007$.

We calculate the non dimensional wave speed and the corresponding $\sqrt{\tilde{R}_0}$ for three species of birds, which are given in Table 2. We assume, as in Cruz-Pacheco et al.[2], the typical value of the biting rate, once every two days, $b = 0.5$ and the ratio $m = \frac{N_v}{\Lambda_a/\mu_a} = 5$. For Common grackle and Blue jay, $\sqrt{\tilde{R}_0}$ are different but the wave speeds are close. These species have the same importance in the spatial propagation, but they behave epidemiologically different. Figure 3 shows the wave speed as a function of the vertical transmission. For instance, letting $D_a = 6\ km^2/day$: (1) for $p = 0$, we have $V_{min} = 3.03$ km/day, and (2) for $p = 1$, we have $V_{min} = 3.09$ km/day.

The vertical transmission is not an important factor for the spatial dynamics, see Figure 3, due to the fact that mosquitoes movement is negligible compared with the avian movement, but it is important for the endemics level, as a local factor of the disease dissemination, see Figure 1. For Blue jay the wave speed increases from 0.784 to 0.798, when p increases form 0 to 1.

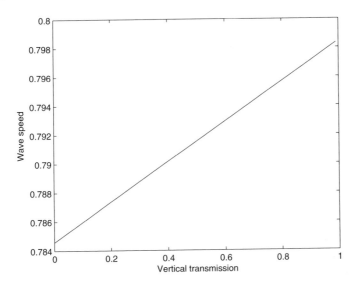

Figure 3. Graphic of the non dimensional speed wave as a function of p, the vertical transmission, for the Blue jays parameters listed in Table 2.

Table 2. Values of the non dimensional parameters used to calculate the non dimensional minimum wave speeds.

Common name	$\tilde{\beta}_a$	$\tilde{\beta}_v$	$\tilde{\gamma}_a$	$\tilde{\alpha}_a$	$\tilde{\mu}_a$	$\tilde{\mu}_v$	$\sqrt{\tilde{R}_0}$	c_{min}
Blue jay	1.0	0.136	0.104	0.06	0.00008	0.024	5.89	0.784
Common grackle	1.0	0.136	0.132	0.028	0.00004	0.024	5.97	0.789
Fish crow	1.0	0.052	0.144	0.024	0.00008	0.024	3.60	0.522

Figure 4 shows the wave speed as a function of the diffusion coefficient, for three birds species: Blue jay, Common grackle and Fish crow. The wave speeds for Blue jay and Common grackle are the same, although $\tilde{R}_0$ are different. The Fish crow has $\tilde{R}_0$ less than the other two species, and the wave speed is considerably lower. Okubo[1] estimates an interval for this diffusion between 0 and 14 km^2/day. Considering $p = 0.007$ and $D_a = 6$ km^2/day, we obtain 3.04 km/day as the velocity of the disease propagation, near to the 3 km/day observed from the field data, see maps in DeBiasi et $al.$[4] and the bounded above velocity estimated by Lewis et $al.$[3] assuming arbitrary values for some parameters.

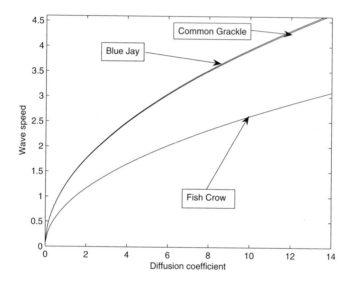

Figure 4.   Graphic of the dimensional speed wave as a function of the diffusion coefficient of the avian population D_a, considering $p = 0.007$.

From Figure 5 we can see the first peak of infection in the classes of infected mosquitoes and infected birds, for two values of the vertical transmission, $p = 0$ and $p = 0.8$. The wave speeds are close between them, but we arise an increasing in the proportion of the infected mosquitoes. This fact is due to the importance of p to the corresponding spatially homogeneous modeling.

In Figure 6 we show the numerical travelling waves solution for the first order system. We can observe the first peak of infection in the four classes.

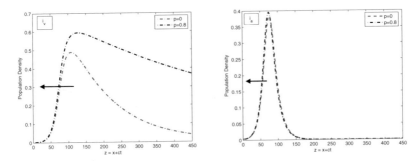

Figure 5. Graphics for the infected mosquitoes (left) and infected avian (right) population densities, for the parameters for the Blue jay listed in Table 2, considering $p = 0.8$ and absence of the vertical transmission ($p = 0$). The effect of vertical transmission is perceptible in the mosquitoes population, although the wave speed do not increase so much: $c_{min} = 0.785$ for $p = 0$ and $c_{min} = 0.796$ for $p = 0.8$.

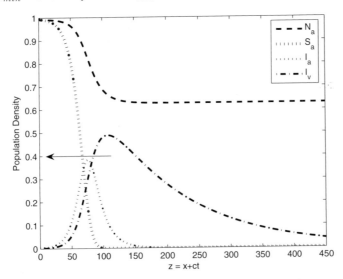

Figure 6. Travelling wave solution for WNV model, using the parameters related to the Blue jay bird listed in Table 2.

4. Conclusion

In this paper we develop and analyze a spatial propagation model in order to describe the spreading out of the WNV. For the spatially homogeneous dynamics we considered the non spatial model studied by Cruz-Pacheco *et al.* We determine, in non dimensional parameters, as the same way as in

Cruz-Pacheco et al.[2], the threshold value:

$$\tilde{R}_0 = \frac{\tilde{\beta}_a \tilde{\beta}_v}{(1-p)\tilde{\mu}_v(\tilde{\gamma}_a + \tilde{\mu}_a + \tilde{\alpha}_a)}.$$

When $\tilde{R}_0$ is greater than one, the endemic state of the disease exists. We study the conditions for the travelling waves solution connecting this endemic point with the disease free equilibrium point. An equation with respect to the minimum speed was determined as a function of the parameters of the model and the threshold $\tilde{R}_0$.

The depending of the wave speed on the vertical transmission was studied. As the mosquito movement is less than birds movement, we obtain that the vertical transmission is not an important factor in the spatial propagation, but it plays an important role as a local risk with respect to the incidence of disease.

Finally the wave speed was studied as a function of the avian diffusion. Choosing a value on the range estimated by Okubo[1], for the avian diffusion, we obtain the wave speed (3.03 km per day) which is very close to that observed from the field data. For instance, see maps in DeBiasi et al. [4], which is quite the same obtained by Lewis et al.[3] who studied a simplified model, which does not consider vertical transmission, WNV death rate and the avian recover subpopulation.

In future paper we will analyze the effects of other parameters than the vertical transmission and avian diffusion coefficient in order to determine the efficacy of control strategies, as well as the advection movements in birds and mosquitoes in the modeling.

References

1. Okubo, A. Diffusion-type models for avian range expansion. In: Henri Quellet, I. (Ed.), Acta XIX Congressus Internationalis Ornithologici. National Museum of Natural Sciences, University of Ottawa Press (1998) 1038-1049.
2. G. Cruz-Pacheco, L. Esteva, J.A. Montaño-Hirose and C. Vargas , Modelling the dynamics of the West Nile Virus, *Bulletin of Mathematical Biology*, **67** (2005) 1157-1172.
3. Lewis, M., Renclawowicz, J. and Van den Driessche, P. Travelling Waves and Spread rate for a West Nile Virus Model, *Bulletin of Mathematical Biology*, **68** pp. 3-23 (2006).
4. DeBiasi, R. L. and Tyler, K. West Nile Virus Meningoencephalitis. *Nature Clinical Pratice Neurology*, **Vol. 3 NO 5** pp. 264-275 (2006).
5. J.D. Murray, *Mathematical Biology*, Springer, Berlin, 2002.
6. J.D. Murray, F.R.S. Stanley and D.L. Brown, On the spatial spread of rabies among foxes, *Proc. R. Soc. Lond.*, **B229** (1986) 111-150.

7. B. Sandstede, Stability of traveling waves, In: B. Fiedler (Ed), *Handbook of Dynamical System II*, pp. 983-1059, Elsevier, Amsterdam, 2002.

8. A.I. Volpert and V.A. Volpert, *Traveling Waves Solutions of Parabolic System*, American Mathematical Society, Providence, RI, 1994.

9. L. G. Campbell, A. A., Martin, R. S., Lanciotti and D. J. Gubler, West Nile Virus, *The Lancet Infect Dis. 2* (2002) 519-529.

10. L. C. Harrington, T. W. Scott, K. Lerdthusnee, R. C. Coleman, A. Costero, G. G. Clarck, J. J. Jones, S. Kitthawee, P. K. Yapong, R. Sithiprasasna and J. D. Edman, Dispersal of the Dengue Vector *Aedes aegypti* Within and Between Rural Communities. *Am. J. Trop. Med. Hyg,* **72(2)** (2005) 209-220.

11. V. M. Kenkre, R. R. Parmenter, I. D. Peixoto and L. Sadasiv, A Theoretical Framework for the Analysis of the West Nile Virus Epidemic. *Computer and mathematics with applications* **42** (2005) 313-324.

12. M. J. Wonham, T. De-camino-Beck and M.A. Lewis, An epidemiological model for West Nile virus: invasion analysis and control applications. *Proc. R. Soc. Lond.* **B 271**, (2004) 501-507.

13. C. Bowman, A. B. Gumel, P. Van den Driessche, J. Wu and H. Zhu. A mathematical model for assessing control strategies against West Nile virus. *Bulletin of Mathematical Biology,* **67** pp. 1107-1133 (2005).

MANAGEMENT OF COMPLEX SYSTEMS: MODELING THE BIOLOGICAL PEST CONTROL

MARAT RAFIKOV

Department of Physics, Statistics and Mathematics,
Ijui University - UNIJUI, P.O. Box 560, 98700-00 Ijuí - RS, Brazil

JOSÉ MANOEL BALTHAZAR

Department of Statistics, Applied Mathematics and Computation,
Universidade Estadual Paulista - UNESP, P.O. Box 178, 13500-230,
Rio Claro, SP, Brazil

HUBERTUS F. VON BREMEN

Department of Mathematics and Statistics, California State Polytechnic
University, Pomona, USA

The aim of this paper is to study the cropping system as complex one, applying methods from theory of dynamic systems and from the control theory to the mathematical modeling of the biological pest control. The complex system can be described by different mathematical models. Based on three models of the pest control, the various scenarios have been simulated in order to obtain the pest control strategy only through natural enemies' introduction.

1. Introduction

Pests — arthropods, weeds, and pathogens — have been, are, and will continue to be major constraints to agricultural production in the world. Synthetic chemical pesticides were introduced in the 1940s and used widely on agricultural crops in the hope that they would control agricultural pests. It is now clear that their use has some unfortunate consequences. The trace residues of synthetic chemicals in food are undesirable and represent a significant food safety risk.

Biological control is the use of living organisms to suppress pest populations, making it less abundant and thus less damaging than they would otherwise be[14]. Pests are species that interfere with human activity or cause injury, loss, or irritation to a crop, stored product, animal, or people.

One of the main goals of the pest control is to maintain the density of the pest population in the equilibrium level below economic damages. Natural enemies play an important role in limiting potential pest populations. There are three categories of natural enemies of insect pests: predators, parasitoids and pathogens. Van den Bosch et al.[14] defined applied biological control as the manipulation of natural enemies by man to control pests, and, natural biological control as control that occurs without man's intervention.

From the ecological viewpoint, the specie is considered as a pest if its population density surpasses the economic injury level, i.e. the pest population density level at which an insect (or other organism) induced damage can no longer be tolerated and therefore the level at or before which it is desirable to initiate control activities. Thus, the premise of biological control is a reduction and establishment of the pest population density at equilibrium level lower the economic injury level.

There are many examples of success using of the biological control, such as the complex of imported parasites, which controls alfalfa weevil[11], or augmentative releases of natural enemies, which have been applied in greenhouses in Europe for control of many vegetables pests[9]. Unfortunately, there are also many cases where effective natural enemies simply haven't been found or haven't been successfully established in the target area. According to Thomas and Willis[11] less than 40% of introductions of biological agents against weeds and insects actually result in substantial control. In order for biological control to succeed, the dynamics of the pest and its enemy populations have to be understood.

The aim of this paper is to study the cropping system as complex one, applying methods from theory of dynamic systems and from the control theory to the mathematical modeling of the biological pest control. The complex system can be described by different mathematical models. Based on three models of the pest control, the various scenarios have been simulated in order to obtain the pest control strategy only through natural enemies' introduction.

2. Modeling of Complex Systems

New perspectives, concepts and tools developed within complex systems theory holds much promise for a deeper understanding and improved management of living systems[5]. An important aspect of the complex systems approach is the recognition that many different kinds of systems include self-regulation, feedback or adaptation in their dynamics and thus may have a common underlying structure despite their apparent differences. In addition to developing deeper understandings of specific systems, such interdisciplinary approaches should help elucidate the general structure and behavior of complex systems, and move us toward a deeper appreciation of the general nature of such systems.

An agricultural ecosystem consists of a dynamic web of relationships among crop plants or trees, herbivores, predators, disease organisms, weeds, etc. Organisms in a cropping system interact in many ways — through competition, molecular signaling, toxicity, host selection, predation, and antibiosis. These organisms constantly evolve and respond to each other, creating a diverse, complex, and ever-changing environment. In other words, a cropping system can be studied as complex one.

There are many definitions of complex system, but these definitions are not complete. Therefore many natural, artificial and abstract objects or networks can be considered to be complex systems. Complex systems are being found and studied at a huge range of time and space scales from those of a single cell to the entire globe and involve processes ranging from physical or chemical alone to the intersection of biophysics and socio-economics[5].

More formally, it is considered a phenomenon in the social, life, physical or decision sciences a complex system if it has a significant number of the following characteristics[12]: *agent-based* (the basic building blocks are the characteristics and activities of the individual agents in the environment under study); *heterogeneous* (these agents differ in important characteristics); *organization* (agents are organized into groups or hierarchies which are often rather structured, and these structures influence how the underlying system evolves over time); *dynamic* (these characteristics that change over time, as the agents adapt to their environment). Such systems are inherently *non-linear* and so may exhibit chaos and hysteretic or irreversible transitions between alternative states. In a complex system, the interaction between the parts or sub-systems allows the *emergence* of global behavior that would not be anticipated from the behavior of components in isolation. This emergent behavior depends upon the nature of the interactions as much as it does upon the character of the parts and changes when these interactions change. These changes are often the result of *feedback* that the agents receive as a result of their activities.

The characteristic non-linearity of complex systems means that mathematical and computer modeling and the concepts of dynamical systems theory play a major part in their study.

There are different approaches in regard the possibility of the modeling of the complex system. One of them argues that all complex systems may be united theoretically, because all may, in principle, be modeled with varying degrees of success by a certain kind of mathematics. It is therefore possible to state clearly what it is that these systems are supposed to have in common with each other, in relatively formal terms. Another approach suggests that complex systems cannot be fully represented by any description which is less complex than the system itself. This is not an argument against any attempt at representation since such attempts may be useful but does impose a limitation on the capacity of models, particularly in relation to prediction. The complex system can be described by different mathematical models. According Goldenfeld and Kadanoff[3], "modeling complex systems by tractable closure schemes or complicated free-field theories in disguise does not work. These may yield a successful description of the small-scale structure, but this description is likely to be irrelevant for the large-scale features. To get these gross features, one should most often use a more phenomenological and aggregated description, aimed specifically at the higher level". One of the famous exemples of simples model is logistic map which can model the complex dynamics of some real population system[7].

3. Mathematical Models of Biological Control in Population Systems

The dynamics of pest populations are complex, making the prediction of outbreaks difficult. Understanding the processes of these interactions can lead to a mathematical modeling playing a decisive role in controlling pest populations and contributing to the stability of natural systems[2]. Mathematical modeling and computer simulation provide ways to help unravel such complex systems and understand how pests interact with the environment and with other organisms. Modeling approaches are used to ensure efficient use of resources in integrated pest management programs, to predict the likely impact of new control measures, and to compare economic returns from different control strategies.

In general, biological control has been modeled as a two-species interaction[8]. In this case, the prey-predator or host parasitoid models ignore many important factors such as its age structure, interactions between another species of same ecosystem, interactions with environment, etc. Preliminary studies[1,4,15] have focused the importance of models that consider more than two species: in fact, for a continuous model, a large spectrum of complex behaviors may occur for systems with three or more species. In this section we consider the soybean cropping system. Supposing that there are soybean plants in abundance, we can model only relations between two soybean caterpillars (*Rachiplusia nu* and *Pseudoplusia includes*) and it's supposed parasitoid. It can be described, at least, by two different models. In the first, despite of the main characteristics of the biological cycle of two considered caterpillars are practically identical, two caterpillars are considered as different species, and the ecosystem can be described by three Lotka-Volterra equations[6,16]:

$$\frac{dx_1}{dt} = x_1(r_1 - \alpha_{11}x_1 - \alpha_{12}x_2 - \alpha_{13}x_3)$$

$$\frac{dx_2}{dt} = x_2(r_2 - \alpha_{21}x_1 - \alpha_{22}x_2 - \alpha_{23}x_3) \tag{1}$$

$$\frac{dx_3}{dt} = x_3(-r_3 + \alpha_{31}x_1 + \alpha_{32}x_2)$$

where x_1, x_2 and x_3 are first caterpillar, second caterpillar and parasitoid densities, respectively. The need to use the model (1) is motivated by the difference in the resistance of two considered caterpillars to diseases and by others environmental factors.

In the second model, taking into account that the main characteristics of the biological cycle of these pests are practically identical; two caterpillars are considered as same specie. In this case, the ecosystem can be described by two Lotka-Volterra equations[6,16]:

$$\frac{dy_1}{dt} = y_1(p_1 - \beta_{11}y_1 - \beta_{12}y_2)$$

$$\frac{dy_2}{dt} = y_2(-p_2 + \beta_{21}y_1) \tag{2}$$

where y_1 and y_2 are caterpillar and parasitoid densities, respectively.

The model (2) can be derived from (1), considering the following particular relations between the variables and parameters of two models:

$$y_1 = x_1 + x_2 \,; \ x_1 = x_2 \,; \ y_2 = x_3 \,; \ p_1 = r_1 + r_2 \,;$$
$$\beta_{11} = \alpha_{11} + \alpha_{12} + \alpha_{21} + \alpha_{22} \,; \ \beta_{12} = \alpha_{13} + \alpha_{12} + \alpha_{23} \,; \tag{3}$$
$$p_2 = r_3 \,; \ \beta_{21} = \alpha_{31} + \alpha_{32}$$

We remark that the relations (3) are not unique, and there are another one's that can reduce the model (1) to (2).

The numerical simulations based on the model (1) with the parameters

$$r_1 = 0.17\,,\ \alpha_{11} = 0.0002\,,\ \alpha_{12} = 0.0002\,,\ \alpha_{13} = 0.002\,,$$
$$r_2 = 0.17\,,\ \alpha_{21} = 0.0003\,,\ \alpha_{22} = 0.0002\,,\ \alpha_{23} = 0.0002\,, \qquad (4)$$
$$r_3 = 0.12\,,\ \alpha_{31} = 0.001\,,\ \alpha_{32} = 0.0001\,.$$

show the complex behavior of the caterpillar and parasitoid populations (Figures 1 and 2).

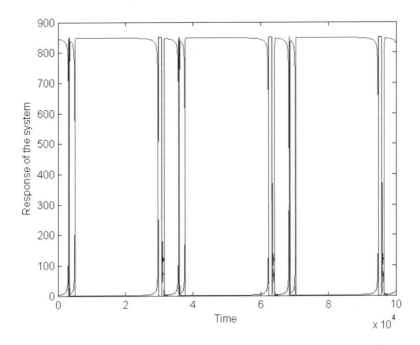

Figure 1.   System response simulations for a large value of time.

Figure 3 shows the strange attractor of the Lotka-Volterra system (1).

The Lyapunov Characteristic Exponents of the system (1) which were calculated, according to[13], are shown in Figure 4.

The Figure 4 suggests that for the time interval of integration the Lyapunov Characteristic Exponents values have converged. It also appears better in Figure 5 that the largest Lyapunov Characteristic Exponents is positive and the remaining two are negative that can characterize the chaotic motion of the system.

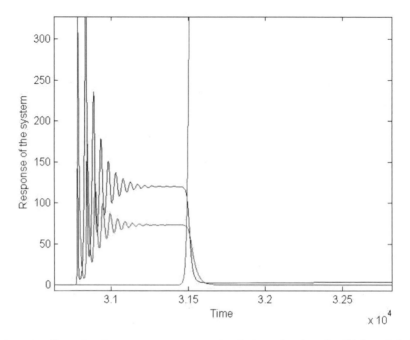

Figure 2. Snapshot from system response simulations showing the "high activity" period.

In Figure 6 the density variations of the caterpillar and parasitoid populations of the one host – one parasitoid model (1) are displayed for the parameters $p_1 = 0.34$, $p_2 = 0.12$, $\beta_{11} = 0.0009$, $\beta_{12} = 0.0024$, $\beta_{21} = 0.0011$.

Figure 7 shows the phase portrait of system the Lotka-Volterra (2).

The comparison of the behavior of the systems (1) and (2) shows that these models describe the same ecosystem in different ways. The model (2) shows the ecosystem tends a balanced behavior, but the model (1) attributes the chaotic behavior to the ecosystem. In this case, the detailed model (1) which has the rich dynamics, is seem more preferred, but only more detailed field experimental data can confirm or reject the considered model.

In next sections, we show that these models suggest the different pest control strategies.

Phase plot of the system

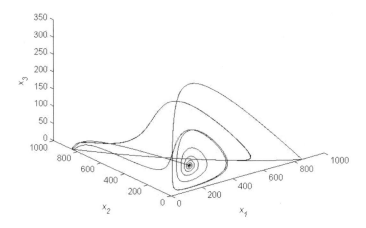

Figure 3. Phase portrait of system (1).

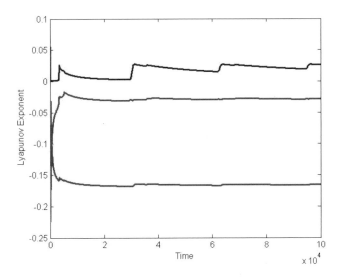

Figure 4. Lyapunov Characteristic exponents of the system (1).

4. Biological Pest Control Strategies

The objective of this section is to formulate the pest control strategy through natural enemies' introduction. This control moves the system to

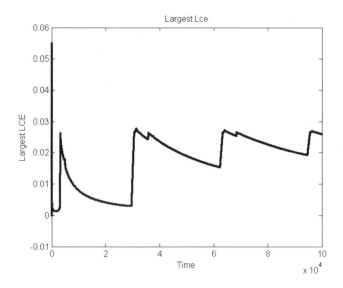

Figure 5.   Lyapunov Largest Characteristic exponents of the system (1).

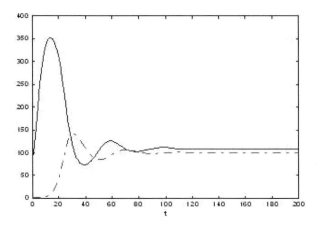

Figure 6.   Response simulations for the system (2).

the steady state in that the pest density is stabilized without causing economic damages, and that the natural enemies' population is stabilized in a level enough to control the pests.

We hope to illustrate the application of the proposed strategy to the

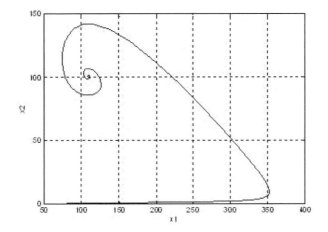

Figure 7. Response simulations for the system (2).

optimal pest control, considering an ecosystem which consist of two soybean caterpillars (*Rachiplusia nu* and *Pseudoplusia includes*) and its parasitoid that was modeled in Section 3. The biological cycles of the caterpillars are identical and the considered ecosystem can be modeled in different forms. We derive control strategies for three different models. The first involves finding a pest control strategy for one prey – one predator Lotka –Volterra model. In this case, two caterpillars are considered as same specie, and the ecosystem is described by two differential equations. In the second, the pest control strategy is formulated for two preys – one predator Lotka–Volterra model. In this case, two caterpillars are considered as different species, and the ecosystem is described by three differential equations. In the third, the pest control problem is resolved for two preys – two predators Lotka–Volterra model. In this case, the two caterpillars are considered as different species and one new parasitoid is introduced into system, and the ecosystem is modeled by four differential equations.

4.1. *Scenario 1: One prey – one predator Lotka –Volterra model*

In this case, the two caterpillars are considered as same specie, and the ecosystem is described by system (2). The Lotka-Volterra dynamic system

(2) with control has the following form:

$$\frac{dy_1}{dt} = y_1(p_1 - \beta_{11}y_1 - \beta_{12}y_2)$$

$$\frac{dy_2}{dt} = y_2(-p_2 + \beta_{21}y_1) + U \qquad (5)$$

where U is the control function.

There are two different manners to apply a biological control. The first of them is an inundative (or massive) control when the great number of enemy species is introduced in the ecosystem. The second is a seasonal introduction of a small population of natural enemies (inoculative control).

4.1.1. *Inundative biological control*

Mathematically, the inundative control can be interpreted by impulsive control function U that produces the discontinuous augmentation of the natural enemy population. The Figure 8 shows the inundative control applied in initial moment by introduction 500 parasitoids.

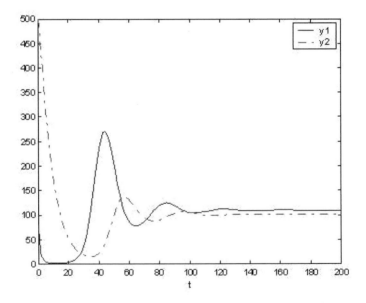

Figure 8. Inundative control application for the system (2).

It is clear from Figures 6 and 8, that the system (2) have the stable fix point equilibrium in $y_1 = 109.09$, $y_2 = 100.76$. The caterpillar density value $y_1 = 109.09$ is larger than the pest density threshold level $x_d = 20$ pests/m² recommended by EMBRAPA (Brazilian Agricultural Research Corporation) for the big soybean caterpillar with the length more than 1.5 cm. Densities equal and below the value $x_d = 20$ pests/m² do not cause economic damages the soybean crops. The Figure 8 shows that the inundative control, applied in initial moment by introduction 500 parasitoids, maintain the pest population below the value $x_d = 20$ pests/m² only 25 days. After this period, there is need to apply the control again.

4.1.2. Inoculative biological control

The goal of the pest control strategy maintains the pest population at desired level y_1^* below the value x_d. The control u^* that maintains the pest population at level y_1^* can be calculated from the following system

$$y_1^*(p_1 - \beta_{11}y_1^* - \beta_{12}y_2^*) = 0$$
$$y_2^*(-p_2 + \beta_{21}y_2^*) + u^* = 0 \qquad (6)$$

From the first equation of the system (6) we obtain the predator density value which is necessary to maintain the pest population at desired level y_1^*

$$y_2^* = (p_1 - \beta_{11}y_1^*)/\beta_{12} \qquad (7)$$

From the second equation of the system (6) we obtain the value of the constant control u^* that maintains the pest population at desired level y_1^*

$$u^* = y_2^*(p_2 - \beta_{21}y_2^*) \qquad (8)$$

The controlled system (5) with the constant control $U = u^*$ is not stable. It signifies that the control strategy has to consist of two parts

$$U = u^* + u \qquad (9)$$

where u^* is the constant control that maintains the pest population at desired level y_1^* and u is the control that stabilizes the pest population at level y_1^*.

The feedback control u can be calculated as[10]

$$u = -R^{-1}B^T P(y - y^*) \qquad (10)$$

where P the symmetric, positive definite matrix (for all $t \in [0, \infty)$, is the solution of the nonlinear, matrix algebraic Riccati equation

$$PA + A^T P - PBR^{-1}B^T P + Q = 0 \qquad (11)$$

Considering in our case

$$Q = \begin{bmatrix} 1 & 0 \\ 0 & 1 \end{bmatrix}, \ A = \begin{bmatrix} -0.0162 & -0.0432 \\ -0.1484 & 0.1002 \end{bmatrix}, \ B = \begin{bmatrix} 1 \\ 0 \end{bmatrix}, \ R = \begin{bmatrix} 1 \end{bmatrix} \qquad (12)$$

we will obtain

$$P = \begin{bmatrix} 16.685 & -0.8423 \\ -0.8423 & 1.1408 \end{bmatrix} \qquad (13)$$

from the solution of the Riccati equation (11).

Then, control function u has the following form

$$u = 0.8423y_1 - 1.1408y_2 \qquad (14)$$

The linear feedback control (15) is designed to stabilize the trajectory of Lotka-Volterra system (5) around of desired one, as shown in Figure 9.

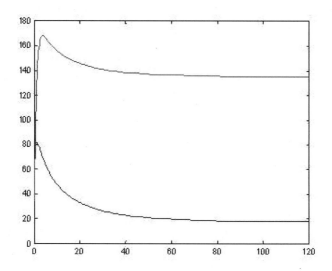

Figure 9.   Inoculative control application for the system (2).

The Figure 9 shows that inoculative control U can maintain the pest population below injury level during long time.

4.2. *Scenario 2: Two preys – one predator Lotka-Volterra model*

4.2.1. *Inundative biological control*

The simulations show that the introduction of a great amount of the parasitoid in the initial moment (inundative control) doesn't stabilize the system (1) (Figure 10). From Figure 10 we can see that the inundative control applied in initial moment by introduction 500 parasitoids can't control the pest population below of the economic injury level.

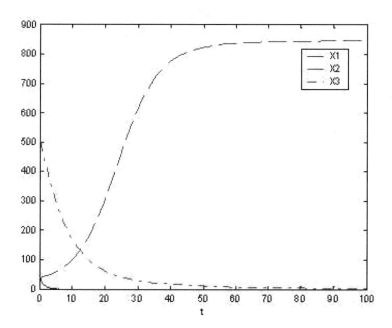

Figure 10. Density variations of the caterpillar and parasitoid populations due the model (1) with introduction 500 parasitoids in the initial moment.

4.2.2. *Inoculative biological control*

The Lotka-Volterra dynamic system (1) with control has the following form:

$$\frac{dx_1}{dt} = x_1(r_1 - \alpha_{11}x_1 - \alpha_{12}x_2 - \alpha_{13}x_3)$$

$$\frac{dx_2}{dt} = x_2(r_2 - \alpha_{21}x_1 - \alpha_{22}x_2 - \alpha_{23}x_3) \qquad (15)$$

$$\frac{dx_3}{dt} = x_3(-r_3 + \alpha_{31}x_1 + \alpha_{32}x_2) + U$$

The goal of the pest control strategy maintains the pest population at level

$$x_1^* + x_2^* = x_d \qquad (16)$$

by control u^*, where x_d is a pest population density below economic injury level (in soybeans can be considered $x_d = 18$).

The desired steady state can be calculated from the equation (16) which in this case has the following form

$$x_1^*(r_1 - \alpha_{11}x_1^* - \alpha_{12}x_2^* - \alpha_{13}x_3^*) = 0$$

$$x_2^*(r_2 - \alpha_{21}x_1^* - \alpha_{22}x_2^* - \alpha_{23}x_3^*) = 0 \qquad (17)$$

$$x_3^*(-r_3 + \alpha_{31}x_1^* + \alpha_{32}x_2^*) + u^* = 0$$

From the third equation of the system (17) we obtain the value of the control u^*

$$u^* = r_3 x_3^* - a_{31}x_1^* - a_{32}x_2^* \qquad (18)$$

Considering the parameter values of the ecosystem as (4), from two first equations of (17) we have:

$$x_1^* + x_2^* + 10x_3^* = 1000$$

$$1.5x_1^* + x_2^* + x_3^* = 1000 \qquad (19)$$

Eliminating x_3^* from (19) we have

$$x_1^* + 0.64286x_2^* = 642.857 \qquad (20)$$

Comparing (20) with

$$x_1^* + x_2^* \leq 18 \qquad (21)$$

we can conclude that it is impossible to control this system, satisfying (21). Biologically it is mean that there is no quantity of parasitoid can maintain the pest population density below economic injury level.

4.3. *Scenario 3: Two preys – two predators Lotka-Volterra model*

In this case, the two caterpillars are considered as different species and one new parasitoid is introduced into system, and the ecosystem is modeled by following four differential equations:

$$\frac{dx_1}{dt} = x_1(r_1 - \alpha_{11}x_1 - \alpha_{12}x_2 - \alpha_{13}x_3 - \alpha_{14}x_4)$$

$$\frac{dx_2}{dt} = x_2(r_2 - \alpha_{21}x_1 - \alpha_{22}x_2 - \alpha_{23}x_3 - \alpha_{24}x_4)$$

$$\frac{dx_3}{dt} = x_3(-r_3 + \alpha_{31}x_1 + \alpha_{32}x_2) \qquad (22)$$

$$\frac{dx_4}{dt} = x_4(-r_4 + \alpha_{41}x_1 + \alpha_{42}x_2)$$

The simulations show that the inundative control applied in initial moment by introduction 100 parasitoids can control the pest population of the system (22) during 100 days (Figure 10).

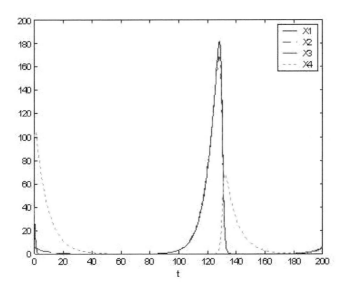

Figure 11. Density variations of the caterpillar and parasitoid populations due the model (22) with introduction 20 parasitoids in the initial moment.

The Figure 11 shows that the inundative control, applied in initial moment by introduction 100 parasitoids, maintain the pest population below the value $x_d = 20$ pests/m^2 , at least, 100 days. That is enough to guarantee a good crop. The application the inoculative control for the system (22) can be more expensive because it implies a creation of parasitoids in laboratory during the same 100 days period.

5. Concluding Remarks

In this paper, the cropping system was studies as complex one, applying methods from the theory of dynamic systems, control theory and mathematical modeling of the biological pest control. Based on three models of the pest control, three scenarios have been simulated in order to obtain the pest control strategy through natural enemies' introduction in ecosystem which consist of two soybean caterpillars (*Rachiplusia nu* and *Pseudoplusia includes*) and its parasitoid.

In the first scenario, based on one prey – one predator Lotka-Volterra model, two caterpillars were considered as same specie and the ecosystem was described by two differential equations. The numerical simulations showed that the inundative control, applied in initial moment by introduction 500 parasitoid/m^2, maintain the pest population below of the value $x_d = 20$ pests/m^2 only 25 days. After this period, there is need to apply the control again. In other hand, the numerical simulations showed that inoculative control strategy, based on control theory results, can maintain the pest population below injury level during long time.

In the second scenario, based on two preys – one predator Lotka-Volterra model, two caterpillars were considered as different species and the ecosystem was described by three differential equations. In this case, the analytical and numerical studies reveled that both control strategies, inundative and inoculative, could not control the pest population below of the economic injury level.

In the third scenario, the pest control problem was resolved for two preys – two predators Lotka-Volterra model. In this case, the two caterpillars were considered as different species and one new parasitoid was introduced into system, and the ecosystem was modeled by four differential equations. The numerical simulations showed that the inundative control, applied in initial moment by introduction 100 parasitoid/m^2, maintain the pest population below of the value $x_d = 20$ pests/m^2, at least, 100 days. That is enough to guarantee a good crop. The application the inoculative control for the system (22) can be more expensive because it implies a creation of parasitoids in laboratory during the same 100 days period.

Finally, we would like to remark that it is very important to choose the mathematical models corresponding to considered problems. These models can suggest the different pest control strategies.

Acknowledgments

The first and second authors thank CNPq (Brazilian National Science Foundation) and the second author thanks FAPESP (São Paulo State Science Foundation) for financial supports.

References

1. M.E. Gilpin, Spiral chaos in a predator-prey model, *American Naturalist*, **113** (1979) 306-308
2. B.S. Goh, *Management and analysis of biological populations*, Elsevier Scientific Publishing Company: Amsterdam, 1980
3. N. Goldenfeld, L. P. Kadanoff, Simple Lessons from Complexity, *Science*, **284**(87) (1999) 87-89
4. A. Hasting, T. Powell, Chaos in a three-species food chain, *Ecol.*, **72** (1991) 896-903
5. C.S. Holling, Understanding the Complexity of Economic, Ecological and Social Systems. *Ecosystems* 4: (2001) 390-405
6. A.J. Lotka, Fluctuations in the abundance of species considered mathematically (with comment by V. Volterra), *Nature*, **119** (1927) 12-13.
7. R.M. May, Simple mathematical models with very complicated dynamics, *Nature*, **261** (1976) 459-467
8. N.J. Mills, W.M. Getz, Modeling the biological control of insect pest: review of host-parasitoid models. *Ecol. Model.* **93**: (1996)121-143.
9. M.P. Parrella, K.M. Heinz, L. Nunney, Biological control through augmentative releases of natural enemies: a strategy whose time has come. *American Entomologist*, **38** (3) (1992) 172-179
10. M. Rafikov, J.M Balthazar, On control and synchronization in chaotic and hyperchaotic systems via linear feedback control. Communications on Nonlinear Science and Numerical Simulations, (in press) (2007) doi:10.1016/j.cnsns.2006.12.011
11. M.B. Thomas and A.J. Willis, Biocontrol- risky but necessary, *Trends in Ecology and Evolution*, **13** (1998) 325-329.
12. The Study of Complex Systems, http://cscs.umich.edu/old/complexity.html
13. F.E. Udwadia, , H.F. von Bremen, , An efficient and stable approach for computation of Lyapunov characteristic exponents of continuous dynamical systems, *Applied Mathematics and Computation*, **121** (2-3, p.219-259, June 15, 2001
14. R. Van den Bosh, P.S. Messenger, and A.P. Gutierrez, *An Introduction to Biological Control*, Plenum Press, New York, 1982.

15. R. Vance, Predation and resource partitioning in one predator-two prey model community, *American Naturalist*, **112** (1978) 797-813.
16. V. Volterra, Fluctuations in the abundance of the species considered mathematically, *Nature*, **118**, (1926) 558-560.

ON TURING-HOPF INSTABILITIES IN REACTION-DIFFUSION SYSTEMS

MARIANO RODRÍGUEZ RICARD

University of Havana, Cuba
E-mail: rricard@matcom.uh.cu

We examine the appearance of Turing instabilities of spatially homogeneous periodic solutions in reaction-diffusion equations when such periodic solutions are consequence of Hopf bifurcations. First, we asymptotically develop limit cycle solutions associated to the appearance of Hopf bifurcations in reaction systems. Particularly, we will show conditions to the appearance of multiple limit cycles after Hopf bifurcation. Then, we propose expansions to normal modes associated with Turing instabilities from spatially homogeneous periodic solutions associated to limit cycles which appear as a consequence of a Hopf bifurcation. Finally, we discuss examples of reaction-diffusion systems arising in biology and chemistry in which can be observed spatial and time-periodic patterning.

1. Introduction

The appearance of periodic patterns in reaction-diffusion systems is mainly a consequence of diffusive instabilities for non-trivial periodic solutions. Such periodic solutions are usually consequence of Hopf bifurcation (HB). Hopf Bifurcation Theorem[13] stands as a very important tool in mathematical modelling of biological and chemical systems. This theorem describes the conditions under which a limit cycle appears from the steady state as a consequence of a variation of some parameters in the system. On the other hand, the mathematical fundamentals of Turing instabilities (TI)[25] are recognized as basic principles in pattern formation and their implications in morphogenesis [3]. These instabilities describes an amazing behavior of the chemical reactants (morphogens) while the diffusion coefficients of the reactants in a reaction-diffusion system are different enough and the system behaves near the stable steady state. A great amount of papers concern about pattern formation in different situations. For simplicity, we only consider here two-component systems. More recently, but in many different scenarios, has been treated the question of periodic patterns as consequence of Turing instabilities for spatially homogeneous periodic solutions which appears due to Hopf bifurcations. Such instabilities are called *Turing-Hopf Instabilities* (THI) or Turing-Hopf bifurcations. Two parameters will be involved in the study of THI: the first is associated to the HB, and it can be taken as the trace τ_a of the jacobian matrix of the parameterized right-hand side evaluated at the steady state and the second is associated with the diffusion ratio d of the diffusion coefficients. Only if the pair (τ_a, d) belongs to a particular region in the corresponding parameter space will be possible the appearance of THI.

Due to many essential reasons, the usual analysis of these instabilities is done by focussing the attention in the region of the space of parameters in which bifurcations or instabilities coexist. In this paper we sketch an approach to diffusive instabilities in which the target is an asymptotic expansion to the normal modes at THI. After the identification of the region in the parameter space at which instabilities are associated, the knowledge of such normal modes may serve to get a better idea of the periodic patterns. Expansions are constructed on the basis of the expansion to the stable limit cycle solution due to a supercritical HB in the reaction system, which generate a spatially homogeneous periodic solution to the associated reaction-diffusion system defined in a bounded and fixed region assuming the non-flow condition across the boundary. As usual, averaging techniques are used to determine the expansions to the limit cycles solutions. We discuss here the appearance of THI in a couple of reaction-diffusion systems that have been studied extensively in literature from many different points of view: the Schnakenberg model and, a predator-prey system.

The plan of this paper is the following. First, we present an approach to the appearance of HB in reaction systems, obtaining asymptotic expansions to the corresponding periodic solutions to the selected examples. We also mention the example of a normal form in which multiple limit cycles appear after a HB. Later, we shall discuss conditions to the appearance of Turing instabilities from spatially homogeneous periodic solutions associated to HB in the examples.

2. Mathematical Formulation

We shall study the instabilities generated by a non-trivial spatially homogeneous periodic solution to the following system:

$$u_t = D_u \, \Delta u + f(u, v; a) \tag{1}$$
$$v_t = D_v \, \Delta v + g(u, v; a)$$

with Neumann boundary conditions

$$\frac{\partial u}{\partial n} = 0 \, , \ \frac{\partial v}{\partial n} = 0 \text{ on } \partial \Omega \tag{2}$$

in which such periodic solution appears due to HB in the reaction part of the system. Is included the fact that reactants diffuse randomly inside the spatial (fixed) bounded region Ω with regular boundary. Functions u, v represent the profiles of reactant concentrations under diffusion and τ_a be a scalar parameter in which are gathered all the parameters in the system, which is defined below. We shall consider τ_a as the *intrinsic* bifurcation parameter passing through the bifurcation value $\tau_a = 0$. For simplicity, we will assume that f and g are analytical in a neighborhood of the isolated steady state, and we will look for strong solutions to Eqs.1-2 with non-negative initial conditions. Here Δ is the laplacian operator. These periodic solutions are associated to the limit cycle which appears due to a supercritical HB while τ_a cross the bifurcation value $\tau_a = 0$. We shall take $d = D_v/D_u$ as the diffusion ratio and, assume that the sum $(D_u + D_v)$ is not small respect to the gauge parameter $\sqrt{\tau_a}$, i.e. $(D_u + D_v) = O(1)$ as $\sqrt{\tau_a} \to 0$.

2.1. *The Hopf bifurcation*

Let us consider a dynamical system on the plane

$$\begin{cases} \dot{u} = f(u, v; a) \\ \dot{v} = g(u, v; a) \end{cases} \tag{3}$$

in which f and g are assumed to be smooth enough to have Taylor expansions of higher order. Here the parameter a can be considered as multidimensional. Let the point

$$P_a = (u_0(a) \, ; v_0(a)) \tag{4}$$

represents the steady state for Eq.3 so,

$$\begin{cases} f(u_0(a), v_0(a) \, ; a) = 0 \\ g(u_0(a), v_0(a) \, ; a) = 0 \end{cases} \, .$$

Let

$$\dot{X} = J_a X \tag{5}$$

where $X = (U(t), V(t))^T$ be the linearization of Eq.3 at the steady state Eq.4, so $J_a = (j_{ij}^a)$ be the jacobian matrix of Eq.3, and let be

$\delta_a = \det(J_a) > 0$, $\tau_a =$ trace(J_a). So, in τ_a are assembled all the parameters in the system. We assume further that Eq.3 presents a limit cycle solution due to a HB[13]-3 respect to the parameter τ_a. Only *supercritical* bifurcation will be taken here in account, which means that the limit cycle appears for small positive values of τ_a. This is due to the fact that, at supercritical (respect to the parameter τ_a) HB, we are in presence of a non-catastrophic loss of stability. We take $j_{11}^a \cdot j_{22}^a < 0$ and, as the determinant is positive, we have $j_{12}^a \cdot j_{21}^a < 0$. We assume that the eigenvalues of J_a are complex conjugate numbers for all admissible values of a, so

$$\tau_a^2 - 4\delta_a < 0 . \tag{6}$$

It can be proved[15] that, if Eq.6 takes place, one can find an invertible analytical transform of coordinates

$$Y = \mathcal{H}(X) = \Gamma X + \mathcal{G}(X) \tag{7}$$

in which the matrix $\Gamma = (\gamma_{ij})$ is invertible and, the last term in Eq.7 contains a combination of higher-order monomials in the components of X. Our intention will be to consider the case where

$$Y = \begin{pmatrix} z \\ \dot{z} \end{pmatrix} \tag{8}$$

being $z(t)$ an unknown function. Imposing this condition, the matrix Γ will satisfy

$$\Gamma \, J_a \, \Gamma^{-1} = \begin{pmatrix} 0 & 1 \\ -\delta_a & \tau_a \end{pmatrix} .$$

In fact, there are infinite of such transforms. Assuming the existence of such a function z, the integration of the system Eq.3 can be reduced to the integration of a second order differential equation in the variable z:

$$\ddot{z} - \tau_a \dot{z} + \delta_a \, z = G\left(z, \dot{z}\right) \tag{9}$$

where

$$G\left(z, \dot{z}\right) = \Pi_2 \left\{ \Gamma \left(J_a \mathcal{K}(Y) + \Psi \left(\mathcal{H}^{-1}Y \right) \right) \right. \tag{10}$$

$$\left. + \langle \mathrm{grad}_X \, \mathcal{G} , \mathcal{F} \rangle \left(\mathcal{H}^{-1}Y \right) \right\}$$

being Π_2 the standard projector over the second component and, the inverse transform is written

$$X = \mathcal{H}^{-1}(Y) = \Gamma^{-1}Y + \mathcal{K}(Y) \tag{11}$$

where

$$K\left(Y\right) = \begin{pmatrix} \underline{\varphi}\left(z, \dot{z}\right) \\ \underline{\psi}\left(z, \dot{z}\right) \end{pmatrix} = \begin{pmatrix} \sum_{n=2}^{+\infty} \sum_{i+j=n} \underline{\varphi}_{ij} z^i \left(\dot{z}\right)^j \\ \sum_{n=2}^{+\infty} \sum_{i+j=n} \underline{\psi}_{ij} z^i \left(\dot{z}\right)^j \end{pmatrix}. \tag{12}$$

The next step is to do the change of variables and obtain the expansion of the right-hand side in

$$\dot{Y} = \left(\Gamma J_a \Gamma^{-1}\right) Y + \Gamma \left(J_a K\left(Y\right) + \Psi \left(\mathcal{H}^{-1}Y\right)\right) + \langle \mathrm{grad}_X \, \mathcal{G} \,, \mathcal{F} \rangle \left(\mathcal{H}^{-1}Y\right).$$

From Eq.9 we look for an oscillation with positive and small, but finite, amplitude ε. The small parameter ε will be chosen appropriately later. Making in Eq.9 the change of variables

$$z\left(t\right) = \varepsilon \varsigma\left(t\right) \tag{13}$$

follows the equation of a weakly nonlinear oscillator:

$$\ddot{\varsigma} - \tau_a \dot{\varsigma} + \delta_a \, \varsigma = \varepsilon \, G\left(\varsigma, \dot{\varsigma}; \varepsilon\right). \tag{14}$$

The non-trivial periodic solution to Eq.3 associated to the limit cycle will correspond to the non-trivial periodic solution to Eq.14. We shall use the Krylov-Bogoliubov-Mitropolski averaging method [19]-[27] in order to find an asymptotic expansion to this solution. From Eq.14 we take τ_a as the intrinsic bifurcation parameter, So, let us consider the new variables $r = r\left(t\right)$ and $\theta = \theta\left(t\right)$ defined as follows

$$\varsigma = r \cos\left(t + \theta\right) \tag{15}$$

$$\dot{\varsigma} = -r \sin\left(t + \theta\right) \tag{16}$$

then, the corresponding averaged equations are

$$\dot{r} = -\frac{1}{2\pi} \int_0^{2\pi} \sin\phi \, \{-\tau_a \, r \sin\phi + \varepsilon G\left(r \cos\phi, -r \sin\phi; \varepsilon\right)\} \, d\phi \tag{17}$$

$$\dot{\theta} = -\frac{1}{2\pi r} \int_0^{2\pi} \cos\phi \, \{-\tau_a \, r \sin\phi + \varepsilon G\left(r \cos\phi, -r \sin\phi; \varepsilon\right)\} \, d\phi \tag{18}$$

and finally,

$$\dot{r} = \frac{r}{2} \{\tau_a - p\left(r; \varepsilon\right)\} \tag{19}$$

$$\dot{\theta} = q\left(r; \varepsilon\right) \tag{20}$$

in which

$$p\left(r;\varepsilon\right) = \frac{\varepsilon}{\pi r} \int_0^{2\pi} \sin\phi \; G\left(r\cos\phi, -r\sin\phi;\varepsilon\right) \, d\phi \qquad (21)$$

$$q\left(r;\varepsilon\right) = -\frac{\varepsilon}{2\pi r} \int_0^{2\pi} \cos\phi \; G\left(r\cos\phi, -r\sin\phi;\varepsilon\right) \, d\phi \; . \qquad (22)$$

The existence of the limit cycle solution corresponds to the existence of a positive root r to Eq.19. Let us assume the existence of a positive root r to the equation

$$p\left(r;\varepsilon\right) - \tau_a = 0 \qquad (23)$$

The periodic solution to Eq.14 will be approximated via Eqs. 15 and 16 with the periodic solution to the averaged equations Eqs.17 and 18, respectively. Hence, we can follow a similar procedure to that in the study of the so-called Andronov-Hopf bifurcations[9] , in which the stability and bifurcation analysis is done studying the separate equation for the radius in polar coordinates. Particularly, we shall select the small parameter ε in connection with the bifurcation parameter τ_a. More precisely, we will take in the examples here

$$\varepsilon^2 = \tau_a \qquad (24)$$

and, the root to Eq.23 can be written as

$$r = r_0 + O\left(\tau_a\right) \; . \qquad (25)$$

From Eq.20 we will obtain the angular speed of the oscillation. Finally, going back in the substitutions given in Eq.13 and Eq.7 we shall obtain the uniform asymptotic expansion of the solution to Eq.9. We will denote the asymptotic expansion of the solution $\Theta\left(t\right) = \left(\overline{u}\left(t\right), \overline{v}\left(t\right)\right)$ to Eq.3 the orbit of which, is the limit cycle given by

$$\overline{u}\left(t\right) = u_0\left(a\right) + u_1\left(t\right)\left(\tau_a\right)^{\frac{1}{2}} + O\left(\tau_a\right) \qquad (26)$$

$$\overline{v}\left(t\right) = v_0\left(a\right) + v_1\left(t\right)\left(\tau_a\right)^{\frac{1}{2}} + O\left(\tau_a\right) \qquad (27)$$

where $\left(u_0\left(a\right); v_0\left(a\right)\right)$ are given in Eq.4, and

$$\begin{pmatrix} u_1\left(t\right) \\ v_1\left(t\right) \end{pmatrix} = r_0 \, \Gamma^{-1} \begin{pmatrix} \cos\left(\varpi \, t\right) \\ -\sin\left(\varpi \, t\right) \end{pmatrix} \qquad (28)$$

being

$$\varpi = 1 + q\left(r_0, \left(\tau_a\right)^{\frac{1}{2}}\right) \; .$$

2.2. *Turing instabilities for the stable steady state*

Turing showed that it was possible for diffusion to cause an instability under conditions in which the reaction kinetics admit a linearly stable spatially uniform steady state. Such instabilities lead to spatially varying profiles in reactant concentration. Again, the diffusion coefficients of reactants must be dissimilar for diffusive instability occur. Turing's standard procedure for the stability analysis is done for systems with two chemical reactants considering perturbation normal modes of the type

$$Z_k(x,t) = U_k(x) \exp(\sigma t) \; R_k \qquad (29)$$

as non-trivial solutions to the linearized equation

$$\frac{\partial Z}{\partial t} = D \, \Delta Z + J_a \, Z \qquad (30)$$

where $U_k(x)$ are eigenfunctions associated to the eigenvalue λ_k of the unbounded nonnegative linear operator $(-\Delta)$ with Neumann boundary conditions at $\partial \Omega$. We will call the λ_k as the *spatial eigenvalues*. Perturbations having the form in Eq.29 allows us to study the stability on bounded domains due to the fact that any perturbation Z can be expanded in terms of such basic functions

$$Z(x,t) = \sum_{k=1}^{\infty} U_k(x) \; \exp(\sigma_k \, t) \; R_k \qquad (31)$$

that is, in Fourier series. The vectors R_k are determined by the Fourier development of the initial condition. The temporal eigenvalues $\sigma = \sigma(\lambda_k)$ in Eq.29 are determined by the second degree equation

$$\det(J_a - \lambda_k \, D - \sigma I) = 0 \qquad (32)$$

and R is a non-zero eigenvector corresponding to σ. Here J_a, D and I are 2×2 matrixes which are the jacobian at the spatially homogeneous steady state solution to Eq.1, the diagonal matrix of diffusion coefficients and the identity matrix respectively. For instance, if we consider the parallelepiped $\Omega = [0, L_1] \times \cdots \times [0, L_n]$, $x = (x_1, \cdots, x_n)$, the eigenfunctions in Eq.31 can be taken as: $U_k(x) = \prod_{j=1}^{n} \cos k_j x_j$ where $k = (k_1, \cdots, k_n)$ is a multi-index and, feasible wavenumbers are those $k_j = \left(\frac{\pi}{L_j} p_j\right)$, such that $p_j \in \mathbb{N}$. The corresponding spatial eigenvalue to such eigenfunction will be $\lambda_k = |k|^2 = (k_1^2 + \cdots + k_n^2)$.

In this analysis, rather than study the general shape of the neutral stability curves on the plane of diffusion parameters will be most convenient to identify the value of the ratio d for which the steady state becomes locally unstable. The evolution to a spatially patterned state as d was varied, is the basic process generating spatial pattern in biology and chemistry[11]. The appearance of Turing instabilities is conditioned to the existence of roots σ to Eq.32 having positive real parts. In the stable steady state case we have $\tau_a < 0$, hence

$$\tau_T = \text{trace}\,(J_a - \lambda_k\,D) = \tau_a - \lambda_k\,(D_u + D_v) < 0 \tag{33}$$

for any eigenvalue λ_k. Then, we require

$$\delta_T = \det\,(J_a - \lambda_k\,D) < 0$$

as a condition for Turing instabilities. So δ_T is function of the wavenumber k. Whenever λ_k belongs to the open interval $\Lambda =]\lambda_-, \lambda_+[$ with extrema

$$\lambda_\pm = \frac{(D_v\,j_{11}^a + D_u\,j_{22}^a) \pm \sqrt{(D_v\,j_{11}^a + D_u\,j_{22}^a)^2 - 4D_uD_v\,\delta_a}}{2D_uD_v}, \tag{34}$$

we will get $\delta_T\,(\lambda_k) < 0$. The values $\lambda_\pm$ were determined by $\delta_T\,(\lambda_\pm) = 0$. So, in order that $\Lambda \neq \emptyset$ we take the condition

$$(D_v\,j_{11}^a + D_u\,j_{22}^a)^2 - 4D_uD_v\,\delta_a > 0 \tag{35}$$

and, to get $\Lambda \subset \mathbb{R}_+$ we put

$$D_v\,j_{11}^a + D_u\,j_{22}^a > 0 \tag{36}$$

determining together regions in the parameter space D_u, D_v in which instabilities could be expected. The following step is to consider the boundary curve of the region in the plane D_u, D_v, given by the equation

$$(D_v\,j_{11}^a + D_u\,j_{22}^a)^2 = 4D_uD_v\,\delta_a \ .$$

This curve is the boundary of the cone of negativeness of δ_T. Then, introducing the ratio $d = D_v/D_u$ we can obtain explicitly the *threshold diffusion ratios* substituting in the above equation. In the Fourier expansion of disturbances only the terms corresponding to eigenvalues $\lambda_k \in \Lambda$ can contribute to the appearance of Turing instabilities. More explicitly, only the referred terms of the Fourier expansion do not tend exponentially to zero as time goes to infinity. From this point follows the determination of the region of the parameter space within which diffusion driven instability turns possible. This region was called the *Turing space* [17].

2.3. *Asymptotic analysis for Turing-Hopf instabilities*

If we are doing the stability analysis for the stable limit cycle at supercritical HB in the reaction part of the system, we implicitly are taking unstable the steady state. In this situation nevertheless, there is a non-catastrophic loss of stability. In other words, for any initial condition close to the steady state the corresponding solution is bounded for all t and so, the instability of the steady state do not interfere with the formation of diffusive instabilities. In this Section we shall present an asymptotic approach to the analysis of diffusive instabilities for the stable limit cycle but taking in account the behavior near the unstable steady state. If the limit cycle appears as a consequence of a supercritical Hopf bifurcation, any orbit crossing a sufficiently small neighborhood of the steady state corresponds to a bounded solution. So, it is reasonable to study diffusive instabilities for an unstable steady state but in presence of a Hopf supercritical bifurcation, because a wavenumber-zero bifurcation is not a source of spatial instabilities.

Now we shall study small perturbations for the spatially homogeneous periodic solution

$$\Theta\left(t\right) = \left(\overline{u}\left(t\right), \overline{v}\left(t\right)\right) \tag{37}$$

to Eqs.1 and 2. As before, we are denoting by $\Theta\left(t\right)$ the corresponding solution to Eq.3 caused by a supercritical HB, the components of which are given in Eqs.26 and 27. Such perturbations should satisfy Eqs.1 and 2. Denoting the corresponding perturbations by capital letters we get,

$$u\left(t, x\right) = \overline{u}\left(t\right) + U\left(t, x\right)$$
$$v\left(t, x\right) = \overline{v}\left(t\right) + V\left(t, x\right)$$

and linearizing we get the linear system with periodic coefficients for the perturbations

$$\frac{\partial Z}{\partial t} = D\,\Delta Z + J_\Theta\left(t\right)\,Z \tag{38}$$

where $Z\left(x, t\right)$ is the column vector with components U and V. Substituting the development of Θ in Eqs.26 and 27 into the jacobian $J_\Theta\left(t\right)$ we get

$$J_\Theta\left(t\right) = J_a + \tau_a^{\frac{1}{2}}\,J_{\frac{1}{2}}\left(t\right) + O\left(\tau_a^{\frac{1}{N}}\right)$$

where $J_{\frac{1}{2}}\left(t\right)$ is a periodic matrix with coefficients being linear expression of $u_1\left(t\right)$ and $v_1\left(t\right)$ which are given in Eq.28.

Let us assume that the solutions to Eq.38 can be asymptotically developed in the small parameter as follows,

$$Z = Z_0\left(t, x\right) + \tau_a^{\frac{1}{2}}\ Z_1\left(t, x\right) + O\left(\tau_a\right)\ . \tag{39}$$

Then, substituting Eq.39 into the system Eq.38, we get a hierarchy of equations determining Z_j:

$$\frac{\partial Z_0}{\partial t} = D\ \Delta Z_0 + J_a Z_0 \tag{40}$$

which corresponds to $O\left(1\right)$ terms, and

$$\frac{\partial Z_1}{\partial t} = D\ \Delta Z_1 + J_a Z_1 + J_{\frac{1}{2}}\left(t\right)\ Z_0 \tag{41}$$

corresponding to the $O\left(\tau_a^{\frac{1}{2}}\right)$ terms. The functions Z_0, Z_1 are determined from Eq.40 and Eq.41 considering homogeneous Neumann boundary conditions in both cases. We expect the solution to Eq.41 to be in the form

$$Z_1\left(t, x\right) = \exp\left(\sigma t\right)\ U_k\left(x\right)\ R_k\left(t\right)\ . \tag{42}$$

and it can be proved[15] that the vector-valued function $R_k\left(t\right)$ uniformly tends to a $\left(2\pi/\varpi\right)$−periodic vector function. Eq.39 shows that the features of the function Z_0 will determine, up to the main term, the formation of Turing instabilities for the periodic solution due to Hopf bifurcation. At the beginning of the Hopf bifurcation the behavior of Z_1 is less transcendent to the whole Z, so the instabilities are basically managed by the "steady state" perturbation Z_0. From Eq.41 we may expect a gradual contribution of Z_1 due to secularity. The growth of Z_1 will depend on the effect of the forcing term in the right hand side of Eq.41.

3. Turing-Hopf Instabilities

In this Section we study the possible appearance of THI in reaction-diffusion systems following the procedure proposed above. The Schnakenberg model, a prey-predator system and a higher order system in normal form are considered.

3.1. The Schnakenberg model

A great attention has been given to patterning for the Schnakenberg's system because it has a simple structure but it is one of the few of the reaction-diffusion models in morphogenesis that exhibit patterns consistent with those in experiments, and has had strong influence on experimental design[3]-[17]. The study of conditions to the appearance of THI was treated, for instance, in[16]. Further, in[15] is shown that THI are possible for this system, and more, there are given values of the parameter d for which the diffusive instabilities start for values of the bifurcation parameter τ_a being greater than zero.

Let us consider the Schnakenberg's system

$$\begin{cases} \dot{u} = u^2 v - u + b \\ \dot{v} = a - u^2 v \end{cases} . \tag{43}$$

This system has a single stationary point

$$(u_0, v_0) = \left(a + b, \frac{a}{(a+b)^2} \right) . \tag{44}$$

Here the parameters a and b are both positive. and, the eigenvalues at this situation will be

$$\lambda_{1,2} = \pm (a + b) i .$$

It is usual to consider $b \ll a$ in order that the bifurcation parameter still remains $a_0 \approx 1$[3]. Note that for $a > a_0$ the steady state (u_0, v_0) is stable, and turns to be an unstable focus for $a < a_0$. The Jacobian matrix of the right hand part of Eq.43 at the steady state (u_0, v_0) is

$$J_0 = \begin{pmatrix} 2u_0 v_0 - 1 & u_0^2 \\ -2u_0 v_0 & -u_0^2 \end{pmatrix} . \tag{45}$$

The bifurcation occurs when takes place the relation $\tau_a = 0$. Considering the new variables (x, y) defined by the relations

$$x = u - u_0$$

$$y = v - v_0$$

we reflect perturbations near the stationary point, and get the following system

$$\begin{cases} \begin{pmatrix} \dot{x} \\ \dot{y} \end{pmatrix} = J_0 \begin{pmatrix} x \\ y \end{pmatrix} + \begin{pmatrix} v_0 x^2 + 2u_0 xy + x^2 y \\ -v_0 x^2 - 2u_0 xy - x^2 y \end{pmatrix} . \end{cases} \tag{46}$$

having the steady state at $(0,0)$. In[16] was shown that considering the change $z = x + y$ and $\dot{z} = -x$, Eq.46 can be reduced to the equation

$$\ddot{z} + (a + b)^2 z = \tau_a \dot{z} + \gamma \left(\dot{z} \right)^2 + 2(a + b) z \dot{z} - z \left(\dot{z} \right)^2 - \left(\dot{z} \right)^3$$

where

$$\tau_a = \frac{a - b}{a + b} - (a + b)^2$$

$$\gamma = 2(a + b) - \frac{a}{(a + b)^2}$$

and τ_a can be also considered as the bifurcation parameter.

Let us assume that the above equation represents an oscillation with small, but finite and positive, amplitude ε. Then, we do a change of variables $z(t) = \varepsilon\varsigma(t)$

$$\ddot{\varsigma} + (a + b)^2 \varsigma = \tau_a \dot{\varsigma} + G\left(\varsigma, \dot{\varsigma}\right)$$

where

$$G\left(\varsigma, \dot{\varsigma}\right) = \varepsilon\left(\gamma\left(\dot{\varsigma}\right)^2 + 2(a + b)\varsigma\dot{\varsigma} - \varepsilon\varsigma\left(\dot{\varsigma}\right)^2 - \varepsilon\left(\dot{\varsigma}\right)^3\right).$$

Finally,

$$\begin{cases} \dot{r} = \frac{r}{2}\left(\tau_a - \frac{3}{4}r^2\varepsilon^2\right) \\ \dot{\theta} = \frac{1}{8}r^2\varepsilon^2 \end{cases}. \tag{47}$$

From the first equation above it follows the existence of an orbitally asymptotically stable limit cycle if

$$r^2 = \frac{4\tau_a}{3\varepsilon^2} \tag{48}$$

$$\dot{\theta} = \frac{\tau_a}{6}.$$

We finally obtain the following uniform asymptotic expansion to the solution

$$z(t) = 2\sqrt{\frac{\tau_a}{3}} \cos\left(1 + \frac{\tau_a}{6}\right)t + O\left(\tau_a\right)$$

and we have,

$$\bar{u}(t) = u_0 + 2\sqrt{\frac{\tau_a}{3}} \sin\left(1 + \frac{\tau_a}{6}\right)t + O\left(\tau_a\right) \tag{49}$$

$$\bar{v}(t) = v_0 + 2\sqrt{\frac{\tau_a}{3}}\left(\cos\left(1 + \frac{\tau_a}{6}\right)t - \sin\left(1 + \frac{\tau_a}{6}\right)t\right) + O\left(\tau_a\right) \tag{50}$$

as the components of the periodic solution $\Theta(t) = (\overline{u}(t), \overline{v}(t))$ to Eq.43 leading to the limit cycle. Denoting the corresponding perturbation by capital letters we get,

$$u(t,x) \doteq \overline{u} + U(t,x)$$
$$v(t,x) = \overline{v} + V(t,x)$$

and linearizing we get the linear system with periodic coefficients for the perturbations

$$\frac{\partial U}{\partial t} = D_1 \Delta U + (2\overline{u}\,\overline{v} - 1)U + (\overline{u})^2 V \qquad (51)$$

$$\frac{\partial V}{\partial t} = D_2 \Delta V - (2\overline{u}\,\overline{v})U - (\overline{u})^2 V$$

where $\overline{u}(t)$ and $\overline{v}(t)$ denote the components of $\Theta(t)$ given in Eq.49 and Eq.50 respectively. We already know that the periodic solution $\Theta(t)$ to the reaction-diffusion system is orbitally asymptotically stable if the nonzero Flocquet coefficient of the linear system

$$\dot{Z} = J_\Theta(t)Z$$

where $Z = (U, V)^T$ and

$$J_\Theta(t) = \begin{pmatrix} (2\overline{u}\,\overline{v} - 1) & (\overline{u})^2 \\ -(2\overline{u}\,\overline{v}) & -(\overline{u})^2 \end{pmatrix}$$

is negative.

Remembering that $\tau_a > 0$ after bifurcation, let us assume that the solutions to Eq.51 can be asymptotically developed in the small parameter τ_a as follows,

$$U = U_0 + \tau_a^{\frac{1}{2}} U_1 + O(\tau_a) \qquad (52)$$

$$V = V_0 + \tau_a^{\frac{1}{2}} V_1 + O(\tau_a)$$

then, substituting Eqs.49-50-52 into the system Eq.51 we get the following hierarchy of equations:

$$\frac{\partial U_0}{\partial t} = D_1 \Delta U_0 + (2u_0 v_0 - 1)U_0 + (u_0)^2 V_0 \qquad (53)$$

$$\frac{\partial V_0}{\partial t} = D_2 \Delta V_0 - (2u_0 v_0)U_0 - (u_0)^2 V_0$$

which corresponds to terms $O(1)$, and

$$\frac{\partial U_1}{\partial t} = D_1 \Delta U_1 + (2u_0 v_0 - 1) U_1 \tag{54}$$
$$+ (u_0)^2 V_1 + 2 (u_0 v_1 + u_1 v_0) U_0 + 2u_0 u_1 V_0$$
$$\frac{\partial V_1}{\partial t} = D_2 \Delta V_1 - (2u_0 v_0) U_1$$
$$- (u_0)^2 V_1 - 2 (u_0 v_1 + u_1 v_0) U_0 - 2u_0 u_1 V_0$$

corresponding to the $O\left(\tau_a^{\frac{1}{2}}\right)$ terms. We remark the expressions of u_1 and v_1 participating in the coefficients of the above system, which follows from Eqs.49-50

$$u_1 = \frac{2}{3}\sqrt{3}\sin\left(1 + \frac{\tau_a}{6}\right)t$$
$$v_1 = \frac{2}{3}\sqrt{3}\left(\cos\left(1 + \frac{\tau_a}{6}\right)t - \sin\left(1 + \frac{\tau_a}{6}\right)t\right).$$

Hence, we may expect the appearance of spatio-temporal oscillations with frequency $\varpi = (1 + \tau_a/6)$.

3.2. *A predator-prey model with stable periodic solution*

A predator-prey system with logistic growth for prey and predators is

$$\begin{cases} \dot{u} = \lambda u - au^2 - bv\phi(u) \\ \dot{v} = \mu v - cv^2 + \delta v\phi(u) \end{cases}$$

$\phi(u) = u$ in the Lotka-Volterra case, and

$$\phi(u) = \frac{u}{1 + mu} \tag{55}$$

in the Holling's case. Here u and v are non-negative solutions. The parameters are: λ—the prey growth rate, μ—the predator growth rate, m-handling time of prey; a, c modulate the logistic effect and b, δ modulate the predation. Many interesting cases have this type of interaction. Let us consider the interaction in the following Holling's example in[9]

$$\begin{cases} \dot{u} = \rho u (1 - u) - cv\varphi(u) \\ \dot{v} = -\delta v + cv\varphi(u) \end{cases} \tag{56}$$

in which $\varphi(u) = u/(\alpha + u)$, all the coefficients are positive and, we assume further that $c > d$. Our approach to the qualitative analysis to this system will follow the procedure in Section 2.1. To do so, we will consider a polynomial system orbitally equivalent to the former Eq.56. More precisely, let us consider

$$\begin{cases} u' = \rho u \left(1 - u\right)\left(\alpha + u\right) - cuv \\ v' = -\alpha\delta v + \left(c - \delta\right) uv \end{cases} \tag{57}$$

which is orbitally equivalent to Eq.56 for $\alpha + u > 0$. All parameters are positive. To pass from the first system to the second, it is sufficient to consider a change of the time variable t for the new one t^1 given by the equation

$$dt = \left(\alpha + u\right) dt^1 . \tag{58}$$

The steady state to Eq.57 is

$$(u_0, v_0) = \left(\frac{\alpha\delta}{c - \delta}, \frac{\rho\alpha}{c - \delta} \left(1 - \frac{\alpha\delta}{c - \delta} \right) \right) \tag{59}$$

with jacobian matrix at this point given by

$$\widehat{J}_\alpha = \begin{pmatrix} \alpha\rho\delta \frac{(c+\delta)}{(c-\delta)^2} \left(\frac{c-\delta}{c+\delta} - \alpha \right) & -\alpha\frac{c\delta}{c-\delta} \\ \alpha\rho\frac{(c-\delta(1+\alpha))}{c-\delta} & 0 \end{pmatrix} . \tag{60}$$

So, the bifurcation parameter will be

$$\tau_\alpha = \frac{\alpha \rho \delta}{(c - \delta)^2} \left(c - \delta - \alpha\left(c + \delta\right)\right)$$

and, the bifurcation value $\tau_\alpha = 0$ corresponds to

$$\alpha_0 = \frac{c - \delta}{c + \delta} .$$

The determinant is

$$\delta_\alpha = \alpha \, c \, \delta \, v_0$$

and can be checked that $\delta_\alpha\left(\alpha_0\right) > 0$. So, in a neighborhood of the steady state Eq.6 holds, and can be written as

$$\left[\frac{\alpha \rho \delta}{(c - \delta)^2} \left(c - \delta - \alpha\left(c + \delta\right)\right) \right]^2 - 4\alpha^2\rho\left(c - \delta\left(1 + \alpha\right)\right) \frac{c\delta}{(c - \delta)^2} < 0 . \tag{61}$$

Eq.57 is rewritten after translating the steady state Eq.59 to the origin as

$$\begin{cases} U' = \tau_\alpha U - \alpha\frac{c\delta}{c-\delta}V + \rho\left(1 - \alpha - 3u_0\right) U^2 - \rho U^3 \\ V' = \left(c - \delta\right) v_0 U + \left(c - \delta\right) UV \end{cases} . \tag{62}$$

Let us consider the "near-Γ" transform of coordinates given by the equation

$$\begin{pmatrix} z \\ Z \end{pmatrix} = \begin{pmatrix} V + V^2 \\ \gamma U + 2UV\left[\gamma + (c - \delta)V\right] \end{pmatrix} \tag{63}$$

where

$$\gamma = (c - \delta)\, v_0 = \alpha\rho \frac{(c - \delta(1 + \alpha))}{c - \delta}. \tag{64}$$

Eq.63 gives an invertible variable transform between neighborhoods of the origin. Furthermore, by Eq.63 we may transform Eq.62 into a second order differential equation. Eq.63 is invertible because the matrix Γ in the linear part is non-singular. The second assertion follows from the fact that, the second component in Eq.63 is the time derivative of the first component respect to the vector field in Eq.62. So, instead of Z in Eq.63 we put z'. The connection between the direct and the inverse developments will be driven by the identity

$$Y = \mathcal{H} \circ \mathcal{H}^{-1}(Y).$$

More precisely, the undetermined coefficients in the inverse development were obtained from

$$\Gamma\left(\mathcal{K}(Y)\right) + \mathcal{G}\left(\mathcal{H}^{-1}(Y)\right) = 0.$$

Now, we obtain an approximate expansion of the inverse to the transform Eq.63 as

$$\begin{pmatrix} U \\ V \end{pmatrix} = \mathcal{H}^{-1}(Y) = \begin{pmatrix} \gamma^{-1}\left(z' - \left(2 + \frac{(c-\delta)}{\gamma}\right) z\, z' + \cdots\right) \\ z - z^2 + 2\,z^3 + \cdots \end{pmatrix}.$$

The next step is to do the change of variables in Eq.63 and, approximating the inverse transform with appropriate polynomials, we get,

$$z'' = \left(\gamma + 2V\left(\gamma + (c - \delta)V\right)\right)U' + 2U\left(\gamma + 2(c - \delta)V\right)V'$$

so, considering small but finite amplitudes for the oscillation $z = \varepsilon \varsigma$ follows

$$\ddot{\varsigma} - \tau_a \dot{\varsigma} + \delta_a\, \varsigma = \varepsilon\left(G_2\left(\varsigma, \dot{\varsigma}\right) + \varepsilon\, G_3\left(\varsigma, \dot{\varsigma}\right) + \cdots\right)$$

where

$$\begin{aligned}
G_3\left(\varsigma, \dot{\varsigma}\right) &= 2\alpha c\delta\left(-1 + \frac{\gamma}{c - \delta}\right)\varsigma^3 + \gamma^{-2}(c - \delta)\,\tau_a\,(c - \delta + 3\gamma)\,\varsigma^2\dot{\varsigma} \\
&\quad - \left(2\rho\,\gamma^{-2}\left((1 - \alpha)(c - \delta) - 3\alpha\delta\right) - 2\gamma^{-1}(c - \delta)\right) \\
&\quad + 2\rho\gamma^{-1}\left((1 - \alpha) - \frac{3\alpha\delta}{c - \delta} + 8\right)\varsigma\left(\dot{\varsigma}\right)^2 - \rho\gamma^{-2}\left(\dot{\varsigma}\right)^3.
\end{aligned}$$

The term $G_2\left(\varsigma, \dot{\varsigma}\right)$, after averaging, has null projection over the subspace of $L_2\left(0, 2\pi\right)$ spanned by $\{\sin\theta, \cos\theta\}$. From the above equation and Eqs.15-16, follows

$$p\left(r; \varepsilon\right) = \frac{1}{4}\gamma^{-2}\varepsilon^2 r^2\left(3\rho - (c - \delta)\tau_a\left(c - \delta + 3\gamma\right)\right) + O\left(\varepsilon^4\right)$$

$$q\left(r; \varepsilon\right) = O\left(\varepsilon^4\right)$$

then

$$\begin{cases} \dot{r} = \frac{r}{2}\left(\tau_a - \frac{3}{4}\gamma^{-2}r^2\varepsilon^2 A^2 + O\left(\varepsilon^4\right)\right) \\ \dot{\theta} = 0 + O\left(\varepsilon^4\right) \end{cases}$$

and, the inequality

$$A^2 = \rho - \frac{1}{3}(c - \delta)\tau_a\left(c - \delta + 3\gamma\right) > 0$$

is a sufficient condition to the existence of a limit cycle. This inequality can be rewritten as

$$\rho\left(M - N\rho\right) > 0$$

for appropriate M and N. We remark the fact that, for $\tau_a > 0$ sufficiently small, both M and N are positive. We conclude the existence of a limit cycle if

$$0 < \rho < \frac{M}{N}$$

being M and N positive. Taking $\tau_a = \varepsilon^2$, we conclude that the averaged limit cycle corresponds to the positive root to the equation

$$r = \frac{2\gamma}{3A}\sqrt{3} + O\left(\tau_a\right)$$

oscillating with the frequency $\varpi = 1$. The root appears when $\tau_a > 0$, so the HB is supercritical. Hence, the asymptotic expansion to the periodic solution leading to the limit cycle is

$$\bar{u}\left(t\right) = u_0 + 2\sqrt{\gamma^3\frac{\tau_\alpha}{3\rho}}\sin t + O\left(\tau_\alpha\right) \tag{65}$$

$$\bar{v}\left(t\right) = v_0 + 2\sqrt{\gamma^3\frac{\tau_\alpha}{3\rho}}\left(\cos t - \sin t\right) + O\left(\tau_\alpha\right) \tag{66}$$

where u_0 and v_0 are given in Eq.59.

Let us discuss now about the possible appearance of THI if in the system Eq.57 we introduce dissimilar diffusion coefficients

$$\begin{cases} u_t = D_u \, \Delta u + \rho u \, (1 - u) \, (\alpha + u) - cuv \\ \quad v_t = D_v \, \Delta v - \alpha \delta v + (c - \delta) \, uv \end{cases} \qquad (67)$$

corresponding to predators and preys. Considering $D_u = 1$, $D_v = d$, we obtain

$$\delta_a = -j_{12}^a \, j_{21}^a \text{ and } j_{11}^a = \tau_a$$

so the threshold for the diffusion ratio has a simple pole at $\alpha = \alpha_0$, and the THI are feasible only for $\tau_a > 0$ and highly dissimilar diffusion coefficients. A similar reasoning can be performed if in the system Eq.56 we introduce diffusion coefficients

$$\begin{cases} u_t = D_u \, \Delta u + \rho u \, (1 - u) - cv\varphi \, (u) \\ \quad v_t = D_v \, \Delta v - \delta v + cv\varphi \, (u) \end{cases} \qquad (68)$$

where $\varphi \, (u) = u / (\alpha + u)$. The point in Eq.59 is also a steady state for this system, but the jacobian slightly varies from the previous at the same point.

3.3. *Topological normal form for the Hopf Bifurcation*

Let us now consider the following example in which is represented a normal form of higher order. Higher order systems in biological or chemical modelling are rarely considered, because they are based mainly on the mass action law. Nevertheless, these are important because they may refer "generalized" Hopf bifurcations, in which several limit cycles coexist after Hopf bifurcation. This couple of examples corresponds, one to an strict Hopf bifurcation and the other to the so-called Bautin bifurcation[9] . Our intention here is only to show that multiple limit cycles can be appear after HB, and to analyze the possible THI generated by the limit cycles. These are examples of polynomial dynamical system with higher order interactions, one of which is featured by the appearance of a couple of limit cycles after the supercritical HB. The normal form in question is

$$\begin{pmatrix} \dot{x} \\ \dot{y} \end{pmatrix} = \begin{pmatrix} \beta & -1 \\ 1 & \beta \end{pmatrix} \begin{pmatrix} x \\ y \end{pmatrix} + \eta \, (x^2 + y^2) \begin{pmatrix} x \\ y \end{pmatrix} + \mu \, (x^2 + y^2)^2 \begin{pmatrix} x \\ y \end{pmatrix} \qquad (69)$$

in which we consider two possibilities. Let case (A) corresponds to $\eta = 1$, and let case (B) corresponds to $\eta = -1$. In each case we take $\mu = \pm 1$.

To analyze the system we first need to transform coordinates. This normal form can be easily rewritten in polar coordinates

$$x = r \ \cos\theta$$
$$y = r \ \sin\theta \ .$$

Taking dot products in both sides of Eq.69 by $(x, y)^T$ and later by $(-y, x)^T$ we obtain the exact equations

$$\dot{R} = f(R) = 2R\left(\beta + \eta R + \mu R^2\right)$$
$$\dot{\theta} = 1$$

being $R = r^2$, with analytic solutions. Limit cycles correspond to the roots of $f(R)$, which can be calculated explicitly. In case (A) we are in presence of a classical subcritical HB if $\mu = 1$, and two limit cycles exist after the subcritical HB if $\mu = -1$. In the last subcase only one of the two limit cycles is properly associated with the HB. The root of $f(R)$ which corresponds to the HB, i.e. that satisfies $R = 0$ if $\tau_\beta = 0$, can be estimated in each subcase by $R \simeq \tau_\beta/2$ for close to zero values of $\tau_\beta = 2\beta < 0$.More exactly, up to the main term in both branches of $R = R_A(\tau_\beta)$ and $R = R_B(\tau_\beta)$ we get

$$r = \frac{\sqrt{2}}{2} |\tau_\beta|^{\frac{1}{2}} + O(\tau_\beta)$$

as the estimate for the root of $f(R)$ associated with the HB. Finally, being subcritical the bifurcations in (A), we do not hope to determine THI. In case (B) we have the same situation, but with supercritical HB. So, for $\mu = 1$ and $\tau_\beta = 2\beta > 0$ two limit cycles will appear at supercritical HB. Then, being $\mu = 1$, we have two roots and, one root if $\mu = -1$. But, in both subcases, the root associated to the HB can be estimated as

$$r = \frac{\sqrt{2}}{2}\tau_\beta^{\frac{1}{2}} + O(\tau_\beta) \ .$$

These bifurcations are supercritical but, from the linear part in Eq.69, it can be checked that THI may exist only for higher values of d.

4. Conclusions

In this paper we have seen how to obtain asymptotic expansions to the periodic solutions associated to limit cycles at HB. We show how the limit cycles can be estimated the limit cycles in correspondence with the roots of appropriate equations for the radius. In dependence of the particularities of the system, it will be better to find these equations in standard form, as occurs in the Andronov-Hopf or Bautin bifurcation normal forms, or even for a very simple system as in the Schnakenberg reaction. We propose the use of the standard KBM-averaging method in the obtention of asymptotic expansions to the limit cycle solution. With the knowledge of this expansion, we propose the expansion of normal modes associated with Turing instabilities which include a time-periodic dependence. So, we may evaluate in a more precise way the influence of each parameter in the formation of THI.

References

1. D.L. Benson, P.K. Maini and J.A. Sherratt, (1998), Unravelling the Turing bifurcation using spatially varying diffusion coefficients, J. Math. Biol., 37, 381-417.
2. N.N. Bogoliubov, Y.A. Mitropolski (1961) Asymptotic Methods in the Theory of Nonlinear Oscillations, Gordon and Breach Sci. Pub., NewYork.
3. L. Edelstein-Keshet (1988) Mathematical Models in Biology, Birkhauser, NewYork.
4. M. Golubitsky, E. Knobloch, I. Stewart, (2000), Target Patterns and Spirals in Planar Reaction-Diffusion Systems, J. Nonlinear Sci., 10, 333-354.
5. D. Henry (1981) Geometric Theory of Semilinear Parabolic Equations, Springer-Verlag, New York.
6. J. Hofbauer, J.W.-H. So, (1994), Multiple Limit Cycles for Three Dimensional Lotka-Volterra Equations, Appl. Math. Lett., Vol.7, No.6, 65-70.
7. W. Just, M. Bose, S. Bose, H. Engel, E. Schöll, (2001), Spatiotemporal dynamics near a supercritical Turing-Hopf bifurcation in a two-dimensional reaction-difusion system, Phys.Rev. E, 64, (026219)1-12.
8. J. Kevorkian, J.D. Cole (1996) Multiple Scale and Singular Perturbation Methods, Springer-Verlag, NewYork.
9. Yu.A. Kuznetsov (1998) Elements of Applied Bifurcation Theory, Second Edition, Applied Mathematical Sciences, Vol.112, Springer Verlag, New York.
10. H. Leiva, (1996), Stability of a Periodic Solution for a System of Parabolic Equations, Applicable Analysis, 60, 277-300.
11. P.K. Maini, K.J. Painter and H.N.P. Chau, (1997), Spatial Pattern Formation in chemical and biological systems, J.Chem.Soc., Faraday Trans., 93(20), 3601-3610.
12. F. Marques, A. Yu. Gelfgat, J.M. Lopez, (2003), Tangent double Hopf bifurcation in a differentially rotating cylinder flow, Phys.Rev. Letters E, 68, (016310) 1-13.
13. J.E. Marsden, M. McCracken (1976) The Hopf Bifurcation and its Applications, Springer-Verlag, New York.
14. M. Meixner, A. De Wit, S. Bose, E. Schöll, (1997), Generic spatiotemporal dynamics near codimension-two Turing-Hopf bifurcations, Phys.Rev. E, 55, 6, 6690-6697.

15. M.Rodriguez Ricard, S. Mischler, (2007), Turing instabilities at Hopf bifurcation, in preparation.

16. M.R. Ricard, Y.H. Solano (2007) in BIOMAT 2006, R. Mondaini and R. Dilão (Eds.) Stability of Periodic Solutions to the Schnakenberg Model Under Diffusion, World Scientific, London, pp. 53-66.

17. J.D. Murray (1993) Mathematical Biology, Springer-Verlag, Berlin.

18. H. van der Ploeg, A. Doelman, (2005), Stability of Spatially Periodic Pulse Patterns in a Class of Singularly Perturbed Reaction-diffusion Equations, Indiana Univ. Math. J., Vol.54, No.5, 1219-1301.

19. J.A. Sanders, F.Verhulst (1985) Averaging Methods in Nonlinear Dynamical Systems, Applied mathematical sciences, vol.59, Springer-Verlag, NewYork.

20. B. Sandstede, (1997), Constructing dynamical systems having homoclinic bifurcation points of codimension two, J. Dyn. Diff. Eq. 9, 269-288.

21. B. Sandstede, A. Scheel, (2001), Essential instabilities of fronts: bifurcation and bifurcation failure, Dynamical Systems, Vol.16, No.1, 1-28.

22. R.A. Satnoianu, P.K. Maini and M. Menzinger, (2001), Parameter space analysis, pattern sensitivity and model comparison for Turing and stationary flow-distributed waves (FDS), Physica D, 160, 79-102.

23. J. Schnakenberg, (1979), Simple chemical reactions with limit cycle behaviour, J. Theor. Biol., 81, 389-400.

24. B. Schuman, J. Tóth, (2003), No limit cycle in two species second order kinetics, Bull. Sci. Mathématiques, 127, 3, 222-230.

25. A.M. Turing, (1952), The chemical basis for morphogenesis, Philos. Trans. R. Soc. London, B, 237, 37-72.

26. J.A. Vastano, J.E. Pearson, W. Horsthemke, H.L. Swinney, (1987), Chemical Pattern Formation with equal diffusion coefficients, Physics Letters A, Vol. 124, No.6-7, 320-324.

27. F. Verhulst (1990) Nonlinear Differential Equations and Dynamical Systems, Springer-Verlag, Berlin.

28. T. Wilhelm, R. Heinrich, (1996), Mathematical analysis of the smallest chemical reaction system with Hopf bifurcation, J. Math. Chem. 19, 2 , 1-14.

29. L.Yang, I.Berenstein, I.R. Epstein, (2005), Segmented Waves from a Spatiotemporal Transverse Wave Instability, Phys.Rev.Letters, 95,3, (038303)1-4.

GRASSHOPPER DENSITY POPULATION
CLASSIFICATION WITH NEURAL NETWORKS

ISAIAS CHAIREZ HERNÁNDEZ, J. NATIVIDAD GURROLA REYES,
CIPRIANO GARCIA GUTIERREZ

CIIDIR IPN Unidad Durango
Calle Sigma S/N Fracc. 20 de Noviembre II Durango, Dgo.
C.P. 34229 México
E-mails: `ichairez@hotmail.com`, `natigre1@hotmail.com`,
`garciaciprian@hotmail.com`

FRANCISCO ECHAVARRIA CHAIREZ

*Unidad Académica de Medicina Veterinaria y Zootecnia de la Universidad
Autónoma de Zacatecas*
Carretera Panamericana Zacatecas-Fresnillo km. 31.5 México
E-mail: `fechava@inifapzac.sagarpa.gob.mx`

Satellite images of the grassland area in Durango México were obtained of altitude,
slope, average annual temperature, annual precipitation, type of vegetation, type
of soil, normal vegetation index, percentage of herbaceous and percentage of bares
soil, in order to relate them with grasshopper density population (GDP) surveyed
in 35 sampling sites from June to November in 2003 in the study area. A stepwise
regression analysis was performed with the most abundant grasshopper species
Phoetaliotes nebrascensis (Thomas), *Melanoplus lakinus* (Scudder) and *Boopedon nubilum* (Say) with data extracted from the satellite images. Results showed
$R > 0.798$, $F(4, 27) > 9.86$ and $P > 0.000016$. The significant variables were normal vegetation index, type of vegetation, altitude and precipitation. GDP raster
maps were interpolated using the stepwise regression equations. Then, classification neural networks models were used in order to classify GDP maps. Analysis
of percentage of classification error showed that the adequate number of hidden
neurons was between six and twelve. Results of error classification were 21% for
P. nebrascensis, 5% for *M. lakinus* and 20% for *B. nubilum*. Neural networks are
practical tools to classify grasshopper population and it will help to take control
measurements in overpopulated areas.

1. Introduction

Grasshopper's populations in grassland areas change according to topo-
graphic, weather, soil and vegetation conditions and grasshopper densities

may be low or high, high densities could be so extreme that may cause damage to cropland and grassland. There have been a lot of studies trying to relate grasshopper populations and ecological factors. One of the earliest was the one carried out by Riley[15,16] where it was found out a relationship among heat, dry weather and number of grasshopper. Also, there were attempts to relate number of grasshoppers with fires, solar stains, earthquakes and volcanic activity, the correlations were not significant[4].

Many studies have been done to relate ecological factors with grasshopper population. For example, Edwards[7] found out a negative correlation between grasshopper density and precipitation, Samietz *et al.*[19] carried out a study relating *Melanoplus sanguinipes* density with altitude in laboratory in California, and they found out that mobility and bask are determined by parents altitude and Capinera and Sechrist[2] surveyed plant and grasshopper assembly and they realized that the total number of grasshoppers was related with high biomass content.

On the other hand, Skinner and Child[20] used stepwise multiple regressions in order to know which climate and soil variables were adequate to explain better the grasshopper population. They found that clay content, organic matter, salinity, three average month precipitations and four average months and precipitations were the variables that most influence the grasshopper population. Also, Capinera and Horton[1] carried out studies in Montana Wyoming Colorado and New Mexico and used correlation and regression analysis. Dependent variable was infestation levels and independent variables temperature, precipitation. Results showed that infestation levels were related wit three years average temperature of July and August and August precipitation in Montana. There were significant terms in the regression of three years average temperature of July and August in Colorado and three years average temperature of June and August in New Mexico.

There exist many applications of neural network in classification. For example the United States Army has used this tool to recognize faces and voices, in astronomy is used to recognize galaxies and currently in the Hospital of the University of Wisconsin, Madison, Dr. William H. Wolberg, used neural networks to recognize cancerous cell as benign or malign based on physical characteristics[21]. However, neural networks to classify GDP have been scarcely applied. The goal of this study is to classify GDP with neural networks in grassland area in Durango México.

2. Methodology

2.1. *Study area*

This study was carried out in grassland areas in Durango México in the Municipios (County); Panuco de Coronado, Gpe. Victoria, Nombre de Dios, Poanas, San Juan del Río, Rodeo, San Pedro del Gallo, San Luis del Cordero and Nazas. These municipios are located among latitude (23.916°, 25.983°) and longitude (-104.997°, -104.010°). There were established 35 sites (Figure 1) using the Berry *et al.* (2000) criteria. At each of these sites, twice a month a sampling survey was done from June to November 2003, this period of time was enough to detect the grasshopper life period.

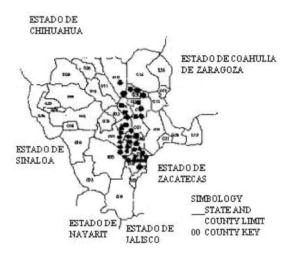

Figure 1.   Sampling sites Geographical distribution in grassland area in Durango México.

2.2. *Grasshopper survey*

At each site and sampling date 80 net sweeps at low step with a 50cm entomological net were done, the grasshoppers collected were kept in 250ml. plastic jar with 70% of alcohol and 30% of water. The Grasshoppers collected were identified using the keys of Capinera and Sechrist[2] and Richman *et al.*[14] and the more abundant species were sorted by instars using the size and wing shape criteria[5].

2.3. Raster and vector maps

Also, at each sampling site; altitude, latitude and longitude were de-termined using a GPS (GARMIN, Etrex personal navigator, 2000) and vector maps from the Internet site http://conabioweb.conabio.gob.mx were obtained, the maps were, type of soil (SEMARNAP, Subsecre-taria de Recursos Naturales 1998), type of vegetation (Comisión Na-cional para el conocimiento y Uso de la Biodiversidad CONABIO 1999), isotherms (García and CONABIO 1998), precipitation's isolines (García and CONABIO 1998). Maps of normal vegetation direct index, percentage of herbaceous and percentage of bare soil of North America were obtained from the internet site http://glcf.umiacs.umd.edu/. Finally the alti-tude map was gotten from the internet site http://www.ngdc.noaa.gov. Using the EXPAND and CONTRACT commands and heart projection transformations in IDRISI, it was able that the study area had UTM pro-jection with latitude (24.0478937, 26.0035961) and longitude (-104.9476515, -103.997292) with resolution of 300×150 and each pixel was of 0.006519 of latitude and 0.006335 of longitude.

2.4. Mathematical models

From the surveys of the 35 sampling sites, the number of grasshoppers col-lected of the three more abundant species *P. nebrascensis*, *M. lakinus* and *B. nubilum* in the instars 4th, 5th and adult were extracted, and in the same way data were extracted from maps of: altitude (m), slope (percentage), temperature (annual average °C), precipitation (total annual mm), type of vegetation (Forest = 6, Cropland = 5, Medium grassland = 4, Halo-phyte grassland = 3, Medium shrub = 2, Short shrub = 1 and Cities and body waters = 0) (Cotecoca, 1979), type of soil (Cities and water's bodies waters = 0, Rendzin = 1, Phaeozems = 2, Litosols = 3, Fluvisols = 4, Kas-tanozems = 5, Regosols = 6, Vertisols = 7, Planosols = 8, Xerosols = 9, Solonchaks = 10, Yermosols = 11, Chernozems = 12, Cambisols = 13 and Luvisols = 14) (FAO), vegetation index =)(close to infrared-red)/(close to infrared*red))*200 + 50, Percentage of herbaceous and percentage of bare soil.

A stepwise regression analysis was carried out, the enter criteria was $F = 1.0$ [10,6]. The dependents variables were grasshopper surveyed of the species *P. nebrascensis*, *M. lakinus* and *B. nubilum* for the fourth and fifth instars and adult stage and the independent variables were altitude (m), slope, temperature (°C), precipitation (mm), type of vegetation, type of soil, vegetation index, percentage of herbaceous and percentage of bare soil. GDP maps were generated using the equations obtained from stepwise regression.

In order to get a grasshopper density classification of the density maps based on the ecological factors, feed forward neural network with one hidden layer of N_h neurons or units was used (Equation 1, Figure 2).

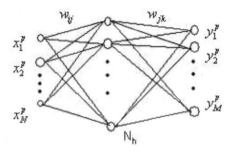

Figure 2. Neural network model.

$$y_k = f_k \left(\alpha_k + \sum_{j \to k} w_{jk} f_j \left(\alpha_j + \sum_{i \to j} w_{ij} x_i \right) \right) \qquad (1)$$

weights w_{ij} and w_{jk} were computed with the back propagation algorithm and the hidden layer transference function was logarithmic sigmoid (Equation 2) and the output layer transference function was lineal[17].

$$f(x) = \frac{e^x}{1 + e^x} \qquad (2)$$

The M output classes were coded as class one letting the first entry of the M vector being one and the others being zero, the second class letting the second entry of the M vector being one an the others zero and so on. The N inputs were normalized and the Neural network Toolbox in MATLA R12 were used to calculate the weights. The specie $B.$ $nubilum$ had nine classes, $M.$ $lakinus$ five and $P.$ $nebrascensis$ 10 classes the first class was from 0 to 9 second from 10 to 19 and so on.

3. Results

Three species were the most abundant $P.$ $nebrascensis$ (23%)[11], $M.$ $lakinus$ (20%)[12] and $B.$ $nubilum$ (5%)[13]. The specie $M.$ $lakinus$ held the normality assumption without bizarre data (Kolmogorv Smirnov, $d = 0.1579$, $p > 0.2$), and the regression results showed significant coefficients for normal vegetation index, type of vegetation, altitude and precipitation (Table 1). A natural logarithm transformation was applied to the specie *B. nubilum* since this did not hold the normality assumption. The significant coefficients were normal vegetation index, type of vegetation, altitude, precipitation and bare soil (Table 2). Finally, a natural logarithm transformation was applied to the species *P. nebrascensis* and the significant coefficients were normal vegetation index, altitude, type of vegetation, precipitation, percentage of herbaceous and type of soil (Table 3).

Table 1. Stepwise regression results for the specie *M. lakinus* without, $R = 0.82$, $F(4, 27) = 14.13$ and $P < 0.000002$.

	B	Standard error of B	t	P
Intersection	-40.3304	9.096674	-4.43353	0.000139
Nvdi	0.2371	0.040510	5.85234	0.000003
Altitude	0.0174	0.005900	2.94685	0.006543
Vegetation	-3.3716	1.304673	-2.58426	0.015488
Precipitation	-0.0188	0.018204	-1.03455	0.310058

Table 2. Stepwise regression results for the logarithm of the specie *B. nubilum*, $R = 0.798$, $F(5, 28) = 9.86$ and $P < 0.000016$.

	B	Standard error of B	t	P
Intersection	-4.62178	1.914443	-2.41416	0.022557
Nvdi	0.04107	0.007231	5.68025	0.000004
vegetation	-0.77764	0.212759	-3.65501	0.001051
Altitude	0.00289	0.001025	2.81848	0.008756
Precipitation	-0.00581	0.003110	-1.86844	0.072197
Bare soil	-0.02369	0.018877	-1.25487	0.219899

Note: Nvdi = normal vegetation direct index, Vegetation = type of vegetation.

Classification results of the specie *B. nubilum* shows that the 12 neurons model gave the minimal percentage error 20.51 and that the best classification classes were one, two, three and nine (Figure 3).

Classification results of the specie *M. lakinus* shows that the eight neu-

Table 3. Stepwise regression results for the logarithm of the specie $P.$ *nebrascensis*, $R = 0.88$, $F(6, 27) = 16.06$ and $P < 0.000001$.

	B	Standard error of B	t	P
Intersection	−9.32540	1.676630	−5.56199	0.000007
Nvdi	0.04012	0.005984	6.70482	0.000000
Altitude	0.00370	0.000875	4.22589	0.000243
Vegetation	−0.48604	0.179330	−2.71034	0.011537
Precipitation	−0.00509	0.002663	−1.91015	0.066790
Herbaceous	0.02583	0.017163	1.50479	0.143988
Soil	0.04490	0.039492	1.13693	0.265559

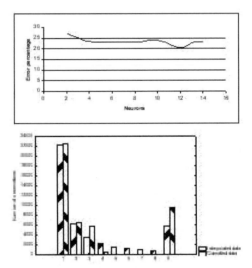

Figure 3. Error percentage vs. number of neurons in the *B. nubilum* neural network model classification. Histogram of real (left) and classified data (right) classes of the specie *B. nubilum* with 12 neurons, 1000 iterations, 0.0296 of mean square error and 20.51 of error Percentage.

rons model gave the minimal percentage error 5.32 and the six neuron model gave 5.4. Also, shows that the four classes are well classified (Figure 4).

Classification results of the specie *P. nebrascensis* shows that the six neurons model gave the minimal percentage error 18.75 and shows that the best classification classes were one, two, three and nine (Figure 5).

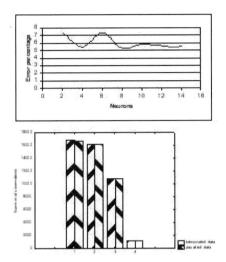

Figure 4. Error percentage vs. number of neurons in the *M. lakinus* neural network model classification. Histogram of real (left) and classified data (right) classes of the specie *M. lakinus* eight neurons, 1000 iterations, 0.02177 mean square error and 5.32 of error Percentage.

4. Discussion

From the stepwise regression results, it can be conclude that GDP variations are because variations of normal vegetation index, altitude, type of vegetation, precipitation, bare soil, type of soil and percentage of herbaceous and the three grasshopper species are affected by normal vegetation index, altitude, type of vegetation and precipitation, where normal index vegetation and altitude have positive index and type of vegetation and precipitation negative index.

The above means that normal vegetation index increases as GDP increases, Capinera and Sechrist[2] found that the total number of GDP were related with high biomass grassland, which is congruent with the results found in this study.

Precipitation relation with GDP is negative, that means that low precipitation values imply high GDP. Also Edwards[7] found a negative correlation and Rodell[18] testify that "Precipitation is one the most important factors that modify the grasshopper population dynamic where this factor is related with the embryo development and quality an quantity of plants".

Vegetation is related negatively with GDP. According to Vegetation

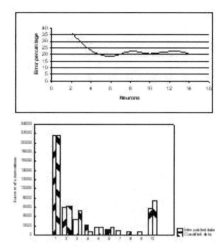

Figure 5. Error percentage vs. number of neurons in the *P. nebrascensis* neural network model classification. Histogram of real (left) and classified data classes (right) of the specie *P. nebrascensis* , six neurons, 1000 Iterations, 0.02382 mean square error 1000 and 18.75 error percentage (right).

category: Forest = 6, Cropland = 5, Medium grassland = 4, Halophyte grassland = 3, Medium shrub = 2, Short shrub = 1 and Cities and body waters = 0, high GDP are related to Medium grassland where around 28% grasshopper population are located in this type of vegetation, 22% in short shrub and 15% in cropland. Capinera and Sechrist[2] found out that the number of fitophyle of the *Gomphocerinae* and *Catantopinae* subfamilies are highly related with grass and shrubs, which is congruent with the results in this study, where *B. nubilum* belongs to the *Gomphocerinae* subfamily and *M. lakinus* and *P. nebrascensis* belong to the genus *Melanoplus* and this belongs the *Catantopinae* subfamily.

Two hidden layer model with logarithmic sigmoid transference functions, with different number of neurons in each layer were analyzed. However, the one hidden layer model showed less error percentage of classification.

The three species classification results of the specie showed error percentages bigger than 20.51 and lower than 5.4. A range from two to 20 of neurons in the hidden layer was tested. However, more than 12 neurons neither less than six neurons did not decrease the error percentage. Back propagation showed quick convergence after 100 iterations and the mean quadratic error did not decrease significatively after that.

From Figures 3, 4 and 5 can be conclude that the specie with less classification error was *M. lakinus*, since it has just four classes and *B. nubilum* got the biggest classification error since this class has nine classes, it can be noticed that the best classified classes in *B. nubilum*, were one, two, three and nine so this situation suggest to regrouping classes four, five, six, seven, eight and nine.

On the other hand, *P. nebrascencis* has ten classes, however showed lower error than *B. nubilum* which has nine classes, this because a bigger number of classes one, two, three, five and ten were correctly classified and four, seven, eight and nine had poor or null classification. Thus, the above suggest regrouping the classes four, seven, eight and nine with ten in order to get better classification. The above suggest that an appropriate classification can be done with a few numbers of neurons and classes.

References

1. Capinera, J. L., and D. R. Horton. 1989. Geographic Variation in effects of weather on grasshopper infestation. Entomological Society of America 18(1), 8-14.
2. Capinera, J. L., and T. S. Sechrist. 1982. Grasshoppers (Acrididade) of Colorado, Identification, Biology and management. Colorado State University, Exp. Stat. Fort Collins. Bull. No. 584S.
3. Comisión Técnico Consultiva para la Determinación Regional de los Coeficientes de Agostadero. 1979. Secretaria de Agricultura y Recursos Hidráulicos, Subsecretaría e Ganadería , Durango, México.
4. Cridle, N. 1932. The correlation of sunspot periodicity with grasshopper fluctuations in Manitoba. Can. Field-Nat 46:195-199.
5. Cushing, W. 2000. Hopper Helper. VI.7. *In* United States Department of Agriculture and Animal and Plant Health Inspection Service (eds.,) Grasshopper Integrated Pest Management User Handbook. Tech. Bull. No. 1809. pp: 3-4.
6. Derksen, S., and H. J. Keselman. 1992. Backward, forward and stepwise automated subset selection algorithms: Frequency of obtaining authentic and noise variables. British Journal of Mathematical and Statistical Psychology 45: 265-282.
7. Edwards, R. L. 1960. Relationship between grasshopper abundance and weather conditions in Saskatchewan, 1930-1956. The Canadían Entomology. XCII: 619-624.
8. IDRISI, 1999. Clark Labs, The Idrisi Project, 950 Main Street, Worcester MA. 01610-1477 USA.
9. MATLAB 6.0.0.88 Release 12. 2000. The language of Thechnical Computing Copyrigth: 1984-2000, The Math Works, inc.

10. Miller, A. J. 1990. Subset Selection in Regression, Chapman & Hall, New York.

11. Pfadt, R. E. 1994. Largeheaded Grasshopper. Phoetaliotes nebrascensis (Thomas). In Field Guide to Common Western Grasshoppers, Third Edition. Wyoming Agriculture Experimental Station. Bulletin 912. 3 p.

12. Pfadt, R. E. 1997a. Ebony Grasshopper, Boopedon nubilum (Say). In Field Guide to Common Western Grasshoppers, Third Edition. Wyoming Agriculture xperimental Station. Bulletin 912. 3 p.

13. Pfadt, R. E. 1997b. Lakin Grasshopper. Melanoplus lakinus (Scudder). In Field Guide to Common Western Grasshoppers, Third Edition. Wyoming Agriculture Experimental Station. Bulletin 912. 3 p.

14. Richman, D. B., C. A. Sutherland and D. J. Ferguson. 1993. A Manual of the Grasshoppers of New Mexico; Orthoptera: Acrididae and Romaleidae. Cooperative Extension Service. New Mexico State University. Handbook No. 7.

15. Riley, C. V. 1877. The locust plague in the United States: Being more particularly a treatise on the Rocky Mountain locust or so called grasshopper, as it occurs east of the Rocky Mountain, with practical recommendations for its destruction. Rand, McNally & Co., Chicago.

16. Riley, C. V. 1891. Destructive locusts. USDA Division of Entomol. Bull 25.

17. Ripley, B. D. 1996. Pattern Recognition and Neural Networks. United Kingdom. Cambridge University Press.

18. Rodell, C. F. 1977. A Grasshopper model for grassland ecosystem. Ecology 58. pp:227-245.

19. Samietz, J., M. A. Salser, and H. Dingle. 2005. Altitudinal variation in behavioral thermoregulation: local adaptation vs. plasticity in California. Journal of Evolutionary Biology 18(4): 1087 p.

20. Skinner, K.M. y R.D. Child. 2000. Multivariate Analysis of the Factors influencing changes in Colorado grasshopper abundance. Journal of Othoptera Research. 9:103-110.

21. Wolberg, W. H., K.P. Bennett and O.L. Mangasarian. 1992. Breast Cancer Diagnosis and Prognostic Determination from Cell Analysis. Manuscript, 1992, Departments of Surgery and Human Oncology and Computer Sciences, University of Wisconsin, Madison, WI 53706.

A STORY OF GROWING CONFUSION: GENES AND THEIR REGULATION

SONJA J. PROHASKA

*Biomedical Informatics, Arizona State University, PO-Box 878809, Tempe, AZ
85287, USA, and
Center for Evolutionary Functional Genomics, The Biodesign Institute, Arizona
State University, PO-Box 875301, AZ 85287, USA
E-mail: sonja.prohaska@asu.edu*

PETER F. STADLER

*Bioinformatics Group, Department of Computer Science, and Interdisciplinary
Center of Bioinformatics, University of Leipzig
Härtelstrasse 16-18, D-04109 Leipzig, Germany, and
RNomics Group, Fraunhofer Institut für Zelltherapie und Immunologie — IZI
Deutscher Platz 5e, D-04103 Leipzig, Germany, and
Institute for Theoretical Chemistry, University of Vienna, Währingerstrasse 17,
A-1090 Wien, Austria, and
Santa Fe Institute, 1399 Hyde Park Rd., Santa Fe, NM 87501, USA
E-mail: studla@bioinf.uni-leipzig.de*

High-throughput experiments have produced convicing evidence for an extensive
contribution of diverse classes of RNAs in the expression of genetic information.
Instead of a simple arrangement of mostly protein-coding genes, the human transcriptome features a complex arrangement of overlapping transcripts, many of
which do not code for proteins at all, while others "sample" exons from several
different "genes". The complexity of the transcriptome and the prevalence of non-
coding transcripts forces us to reconsider both the concept of the "gene" itself and
our understanding of the mechanisms that regulate "gene expression".

1. Introduction

The prevailing picture of genome organization is a rather simple linear arrangement of separated individual genes which are predominantly protein-
coding. Albeit a few exceptional cases of overlapping or interleaving genes
have long been known, this paradigm has been dominating research in
molecular genetics so far. The ENCODE Pilot Project[86], the mouse cDNA
project FANTOM[60,45], and a series of other large scale transcriptome

studies[34,65] have amassed a robust body of data and profoundly change the picture in a way that might be difficult to accept: The "obscure exceptions" are in fact the rule! The mammalian transcriptome is characterized by an complex mosaic of overlapping, bi-directional transcripts and a plethora of non-protein coding transcripts arising from the same locus.

This newly discovered complexity is not unique to mammals. Similar high-throughput studies in invertebrate animals[47,28] demonstrate the generality of the mammalian genome organization among higher eukaryotes. Even the yeast *S. cerevisiae*, whose genome has been treated as conceptionally well understood, catches up with the more complex picture and surprises us with a much richer repertoire of transcripts than previously thought [27,14,57].

Both experimental and computational evidence suggest that many of the newly discovered transcripts and processing products are functional, although at present we have very little knowledge of the potentially different types of molecular mechanisms and function integrating the individual transcripts. However, both theoretical considerations and a set of examples[98] hint at a role in a variety of different regulation mechanisms.

The unexpected diversity of the transcriptome and the unexpected complexity of genomic organization also forces us to re-think our understanding of genes and their regulation. We not only have most likely overlooked many of the players by focusing on three dominating and well-separated layers of regulation (transcriptional regulation, signaling networks, and microRNA-based gene silencing), but we also grossly underestimate the complexity of the regulatory circuitry by assuming a hierarchical structure of well-separated layers. Indeed, several systems demonstrate an intimate interplay of distinct regulatory mechanisms. The interleaved usage of protein- and ncRNA-based mechanisms in apoptosis control [11,29] shows that transcriptional networks alone are not sufficient to understand the regulatory processes despite their great success in the area of early development[3,50].

This contribution is organized as follows. In the next section we briefly argue that most of the observed transcription has biological function, and more strictly, at least a large fraction of the non-coding transcripts themselves are functional. We then consider the structure of mammalian transcriptome, concluding that the very notion of the "gene" itself becomes problematic, begging the question what is regulated in "gene regulation". We then refine our claim about transcript function, arguing that many of the "novel" transcripts have functions as RNAs, and that these functions are most likely regulatory. Given that gene regulation occurs in space and time to pattern a three-dimensional organism, we will briefly discuss how spatial regulation is achieved and integrated into the whole picture.

2. Function of Noise?

The dramatic expansion of transcript repertoire has been argued to be at least largely functional[52,55,95,7]. However, noise is unavoidable in any physical system. Therefore, the novel non-coding transcripts could be interpreted as *"transcriptional noise"* resulting from accidental initiation of transcription or the accumulation of malformed RNAs as a by-product of transcription of the truly functional mRNAs. This explanation is highly unlikely for a variety of reasons.

The amount of observed non-coding transcripts, at least half of the transcriptome, would imply a very unfavorable signal to noise ratio. The effort for the cell to sort the wheat from the chaff on transcript level should be a sufficient basis for a selection to reduce over time. The structure and organization of the majority of the non-coding transcripts is also striking: they are well-defined and do not seem to differ from the previously known ones in terms of their processing (e.g. capping, splicing, and polyadenylation)[86] or in terms of regulated expression[66]. Their transcription start sites, as identified by CAGE/ditags, are very similar to known start sites judged by the presence of proximal regulatory elements, as identified via ChIP-on-Chip experiments[86]. Consequently, many of the novel transcripts show differential expression[8,66] in differentiation and disease[51]. Also phylogenetic considerations support their retention: in many cases there is at least syntenic conservation (i.e., transcripts arising from corresponding positions relative to homologous neighboring features, even though their sequences might not be well-conserved)[35,70,46]; in many cases there is also extensive sequence conservation. In fact, many of the ultraconserved sequence elements in mammals give rise to ncRNAs of hitherto unknown function[6].

Given a set of transcripts that show a well-defined genomic location and pattern of expression but a lack of functional annotation, one might be tempted to handle the "act of transcription" as the transcripts' function while the RNA molecules themselves are meaningless. It is well known that *transcriptional interference*[79,53] may play an important role when transcripts from the same locus compete for transcription. It has been demonstrated, for instance, that perturbation of an anti-sense RNA can alter the expression of the sense messenger RNAs[36]. Furthermore, it is yet undecided whether chromatin has to be in open conformation to allow for transcription or whether the act of transcription opens the closed chromatin conformation[23]. As in the case of transcriptional interference, the regulation of chromatin structure may explain non-coding transcription without a function for the transcript itself. There are, however, several lines of evidence suggesting strongly that the transcripts are not irrelevant. In addition to transcriptional interference, a recent review[39] discusses four distinct mechanisms for anti-sense function: RNA masking (in which the anti-sense transcripts regulates alternative splicing), RNAi (in which RNA double strand formation triggers RNA degradation), RNA editing of double stranded regions, and the involvement in chromatin remodeling.

Most of the transcripts show evidence of post-processing. Typical non-coding RNAs appear to be spliced, polyadenlyated, and exported just like regular mRNAs. Furthermore, small RNAs ≤ 200nt in length[35] have been shown to be processed from longer precursors. It is unclear why the cell would carry the burden of processing when it could make use of several efficient degradation mechanisms[32] to get ride of non-functional RNAs. While this is certainly the fate of *some* "cryptic" pol-II transcripts[97], it does not seem to apply to the many stable mRNA-like ncRNAs.

At least a fraction of the long non-coding primary transcripts are precursors of small functional ncRNAs. Starting with microRNAs in the year 2000, several newly discovered classes of small non-coding RNAs, among them piRNAs[40,1], 21-U RNAs[72], and the promoter and terminator associated paRNAs and taRNAs[35] add to the diversity of ncRNAs. For each of these classes there is at least circumstantial evidence that they are functional. Classical genetic studies have stumbled across only a small number of small ncRNAs, among them the microRNAs *lin-4*, *let-7*, or *iab-4*, for several reasons: they are hard to find in mutagenesis experiments because of their size, many functions (e.g. in antiviral defense or specific stress response) do not produce a discernible phenotype under most lab conditions, and ncRNAs are often redundant because they appear as multicopy genes.

A sizable fraction of both large and small ncRNAs shows weak but significant sequence conservation. Bioinformatics-based surveys of genomic DNA show that there is a large number of loci which show unexpected conservation of RNA secondary structure[91,92,63,89,93]. Stabilizing selection of RNA structure strongly suggests specific function(s) at the RNA level.

Summarizing these argument, we conclude that the view of predominantly non-functional transcripts is not a parsimonious explanation of the available data. Of course, we cannot (yet) ascribe a *specific* function to the majority of transcripts, a fact that does not imply that they are not functional. Conversely, we do not claim that all transcripts or transcription is biologically functional — we only argue that non-coding transcripts are functional often enough that we cannot treat them as odd special cases.

3. What Are Genes Anyway?

The data briefly reviewed in the previous section draw a picture of the mammalian transcriptome that forces us to rethink the notion of well-separated individual genes. The human genome no longer presents itself as linear arrangement of mostly protein coding genes (together with a few exceptional RNAs), littered by more the 90% of useless junk DNA that make up huge "gene deserts" that nicely separate functional regions. Instead, at least the non-repetitive fraction of the genome is transcribed almost in its entirety; regions with high density of protein coding genes form complex clusters of multiple overlapping transcripts in both sense and anti-sense direction, Fig. 1. These primary transcripts are then processed into a plethora of different mature products of diverse functions and localizations. Only a small fraction of the transcripts gives rise to "classical" protein-coding mRNAs.

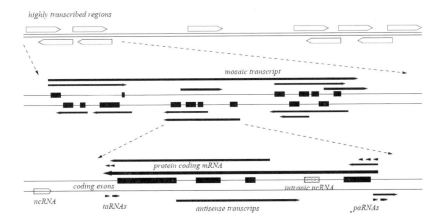

Figure 1.   Sketch of the post-ENCODE view of a mammalian transcriptome (adapted from[35]). Highly transcribed regions consist of a complex mosaic overlapping transcripts (arrows) in both reading-directions. These transcripts link together the locations of several protein coding genes (coding exon indicated by black rectangles). Conversely, multiple transcription products, many of which are non-coding, are processed from the same locus as a protein coding mRNA.

With overlapping transcripts emerging as the rule rather than the exception, and with functional non-protein-coding RNAs arising as a major class of transcripts we are forced to rethink the concept of the *gene* itself. This implication has of course been realized in the field[22,24], but so far no satisfactory solution has arisen that really encompasses the non-protein-coding transcription products.

Conceptually, one would like to use the term *gene* for a unit in both functional and structural terms. We argue that function is primarily associated with the transcripts and their processing products rather than with the DNA locus, hence it becomes futile to assign a single function or a single functional product to a DNA locus. In human, for instance, both the non-coding co-activator SRA, whose function has been well studied experimentally, and a protein SRAP, which is conserved at least throughout chordates[42], are produced as overlapping sense-transcripts from the same locus. A gene concept advocated by Gerstein *et al.*[22] is based on the assumption that the "important" function is encoded in the protein-coding exon(s), if present at all. Abandoning this "proteinocentric" point of view, we argue that functional units on the level of the transcriptome or proteome and structural units, most naturally intervals on the DNA, are typically in a many-to-many relation. This implies, for instance, that one and the same mutation may have several distinct functional consequences. A recent example of this type[31] thus might as well be a common phenomenon.

Another open question is whether "nearby" regulatory elements (such as promoters, cis-regulatory regions, or elements responsible for positional effects) should be considered part of a "gene". While appealing from a functional point of view (since a gene product's spatial and temporal expression pattern plays a role in its biological function), such a definition makes it nearly impossible in practise to annotate a "complete gene" on the genomic DNA. Furthermore, thinking of a "gene" as single DNA interval might not satisfy our needs to annotate the template of a functional unit together with its dispersed regulatory context.

4. Regulation by RNA

Although the overwhelming majority of both the experimentally discovered non-protein-coding transcripts and the computationally predicted structured RNA elements have resisted functional annotation attempts, it appears safe to assume that a large fraction of them has some *regulatory* function:

(1) Catalytic functions are expected to be rare. While some of the "classical" ncRNAs have catalytic functions (the ribosome after all is essentially a ribozyme, and snRNAs are actively involved in splicing), such an active role appears to be very rare outside the world of viruses and viroids. A genome-wide survey resulted in a single example, namely a HDV-like sequence in the human CPEB3 gene[73]. Furthermore, proteins are in general more efficient and versatile catalysts than natural and engineered ribozymes (although this limitation might be overcome in theory)[5], and non-coding transcripts in particular are often evolutionarily very young[99]. It would thus be very surprising if a large set of recently-evolved ribozymes were discovered.

(2) RNA is an ideal molecule for regulatory functions from a theoretical perspective. It is capable of "digital" information processing by specifically interacting with other nucleic acids under the rules of base pairing[51]. On the other hand, it can form elaborate three-dimensional structures which can be recognized by proteins with high specificity. Indeed, several known ncRNAs, including RNAse P RNA, microRNAs, and snoRNAs, do not exert their function in isolation but as part of specific ribonucloprotein complexes. The role of the RNA component is to provide specificity to a target, while the generic function (e.g. creating a base modification as in the case of snoRNAs) is performed by the protein component. In summary, ncRNAs could serve as specific though exchangeable links between nucleic acids and proteins.

(3) Recent attempts to identify classes of RNAs that are seemingly unrelated in sequence but share common structures have not only recovered the expected groups of tRNAs, box H/ACA snoRNAs, and microRNAs, but also provided strong evidence for the existence of previously undescribed classes of structured RNAs[94,71]. The structural similarities within these groups are at least indicative of a common function.

(4) A computational analysis of the binding energies of predicted structured RNA candidates[90] with mRNAs that are not transcribed from the same locus shows that these candidate ncRNAs are enriched in sequences that can bind particularly tightly to near-complementary mRNA targets[85], Fig. 2. Although these data provide only statistical evidence at present, they give a strong indication that direct RNA-RNA interaction is not a rare phenomenon.

(5) Synthetic "modifier RNAs" have been used as experimental techniques for changing the gene expression patterns independent of the RNAi pathway[10,54,62]. One possible mechanism is the modification of (m)RNA

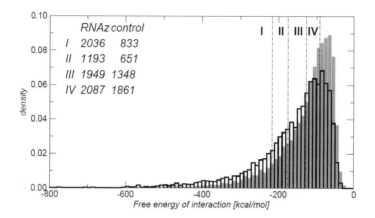

Figure 2. Distribution of interaction free energies of predicted ncRNAs with their best mRNA targets. The density of the interaction free energy distribution of predicted interactions of RNAz hits and mRNA is shown in black, the corresponding distribution of randomized control data-set in which the sequences of the RNAz-predictions were shuffled is shown in grep. Dotted lines indicate the 0.05, 0.10, 0.25, 0.50 quantiles of the randomized distribution. The insets lists the absolute number of true and randomized interaction in the four energy classes defined by the four quantiles. Figure adapted from[85].

structure as a consequence of the duplex formation between modifier and its target (m)RNA, which in turn can dramatically affect the binding affinity of the (m)RNA and a protein[25].

In summary, the available evidence points towards the existence of a large and very diverse pool of functional RNAs whose roles are predominantly regulatory. On the other hand, there is no evidence at present that these functions are performed by means of a uniform mechanism.

5. Elusive Networks: Many More Layers

Ten years ago, gene regulation was essentially understood as the result of two well-separated layers: (1) Signaling pathways would transmit the signal to the nucleus via direct protein-protein interactions and protein modifications (e.g. phosphorylation); (2) Transcription factors would then bind to sequence-specific DNA motifs and recruit the transcription complex to start sites, causing the specific transcription of target mRNA(s). This dichotomy is highlighted by a large number of publications exclusively dealing with *transcription networks*[30,83] and dedicated data bases such as TRANSPATH[75] for cell signaling. Taken together, these two subsystems form the *gene reg-*

ulatory networks which resulted in a major advance in our understanding of developmental regulation[15].

While our knowledge on signaling and transcription factor binding rapidly grew over time, the general concept stayed the same until two more "layers" of ubiquitous importance were discovered: epigenetic regulation and post-transcriptional regulation. Again, these additional types of gene regulation were perceived as separate layers, coming with their own set of regulatory molecules and mechanisms.

While regulatory RNAs became known first for their post-transcriptional effects, different classes of ncRNAs are instrumental at most or all layers of the regulatory network[12,64]. Beyond their effect on translation as microRNAs or siRNAs, they could serve as "transport shuttles" for bound proteins, activate and inhibit transcription[43,26], bind to transcription factors[9], act as transcriptional enhancers[18], take part in structuring the nucleus[67], trigger epigenetic activation and inactivation of chromosomal regions (e.g. Xist[59]), DNA elimination during development of the somatic macronucleus in tetrahymena[58], and may even serve as "backup copies" that transmit information to the next generation independent of genomic DNA[44]. While some of these mechanism are observed in diverse eukaryotes or even throughout all kingdoms of life, we have to-date only seen a few individual examples of others, such as the TRE RNAs[77,74] which, tethered to their site of origin, function by anchoring histone-modifying enzymes to their target, or the evf-2 RNA[18], which acts as a transcriptional enhancer. Furthermore, several surprising effects have been observed *in vitro* in reaction to artificial small RNAs, such as "RNAa", i.e., transcriptional *activation* by means of siRNAs[43] and the up-regulation of protein production in response to small RNAs that modify mRNA structure and thereby alter mRNA-protein binding properties[62,54,25].

The structure, content, and diversity of the transcriptome of higher eukaryotes thus forces us to consider a mechanistically much more diverse system in which a substantial amount of the information is stored and processed by the interaction of many different regulatory mechanisms. The direct interaction of regulators from different mechanistic "levels" enables the synchronization of their joint action. For instance, the integration of small ncRNAs and transcription factors in feed-forward loops, which appears to be a common phenomenon, can tighten regulation by combination of both, (slow) transcriptional and (fast) post-transcriptional regulation[80]. The large number and the diversity of distinct functional units encoded by the same DNA locus further suggest dependencies which most likely facilitate additional regulatory mechanisms.

The assumption that a sizable fraction of players and mechanism in cellular regulation so far has escaped discovery and the expectation that these additional molecules strongly interact with known parts of the regulatory network have far-reaching consequences for our ability to reconstruct regulatory networks from data. In fact, existing methods[48] invariably assume that the vertices of a reasonably autark sub-system are known and complete, so that only the interaction rule need to be inferred. The changing view of the transcriptome, however, implies several important research questions, with potentially dramatic consequences for systems biology: What is the relation between the true and reconstructed networks when vertices are missing in the reconstruction? In other words, to what extent do our current network models represent true interaction and physical regulation? How can we infer missing players in network reconstruction, e.g. from discrepancies between data and incomplete networks? Can such studies actually aid the targeted search for missing components?

6. Space Matters

Molecular interactions partly represented by regulatory networks actually occur in space and require directly interacting partners to physically meet. Integration of the molecular players and their regulation into space and time eventually governs the formation of a three-dimensional organism.

Regulatory molecules are typically not homogeneously distributed in a living cell. In fact, generation of asymmetry where only symmetry has been before is the key to differentiation and an initial divergence in spatial distribution of expression products. It is relevant to all levels of organization from large multicellular organisms to small subcellular compartments. The origin of asymmetry might be found in the dynamic embedding of one-dimensional (e.g. linearly ordered genes on the genome) and two-dimensional (e.g. surfaces of membranes) objects in a three-dimensional space and their physically determined relations. During sporulation of *Bacillus subtilis*, it has been traced back to chromosomal spatial asymmetry and the relative position of two loci (oriC and spoIIAB) on the genome[17] and concentration differences of surface and volume associated factors (spoIIE and spoIIAA)[2]. However, the picture is still incomplete.

Viewing asymmetry as a sorting event, maintenance of an established pattern requires additional efforts as sequestration and/or localization and turns them into important regulatory mechanisms. Incongruity of mRNA and protein distribution have revealed the relevance of translational inhibition (e.g. bcd inhibits cad translation) and cytoskeleton-dependent mRNA localization (e.g. Staufen localizes bcd mRNA) in *Drosophila* embryogenesis[16]. Translational inhibition by miRNAs, mRNA localization[61] and regulated local translation have been demonstrated to be of considerable importance in the differentiation of dendrites[49,13]. More attention has been payed to subcellular localization of proteins as reflected by the LOCATE Database[19]. The mechanism of intra- and inter-cellular trafficking by protein targeting via signal peptides, direct protein synthesis into the target compartment and usage of the secretory pathway have been described decades ago.

Once a set of regulators is asymmetrically distributed, sometimes in form of a gradient, combinatorial patters of their spatial expression can define new compartments or segments and cause the expression of segment-specific factors[96,4]. These relative positions of spatial expression patterns are biologically meaningful, distinguish biological structures and reveal rules of hierarchical (protein) network organization[37,78]. Combinatorial molecular phenotypes have been successfully mapped with a technique termed multi-epitope-ligand cartography (MELC or MELK)[78]. Alterations of expression patterns relative to each other have been shown to characterize diseases (e.g. inflammatory skin disease[78]) and are probably instrumental in major evolutionary innovations (e.g. fin-limb transition[56]) indicating clearly that spatial (and temporal) organization of expression products crucially contributes to function. One implication is that subfunctionalization of duplicated genes may also occur in space (e.g. colinear expression of Hox gene paralogs along the anterior-posterior axis).

All possible layers of gene regulation are potentially involved in creating spatial organization. Among the best-know examples are transcriptional enhancers which bind a number of regulators cooperatively[41]. The resulting expression pattern is the evaluation of the expression map of these regulators. A gene can be under the control of several enhancers, each of which represents a sub-pattern of the complete spatial and temporal expression pattern[69,81,82,21] as in the example of eve stripe formation, Fig. 3. Evaluation of the expression map might allow different sets of regulators to cause the same spatial distribution of the regulated gene. Post-transcriptional control as well es a mixture of both can give rise to specific expression patterns[29] but as transcription only occurs in the nucleus, post-transcriptional regulation enables a fine-grained, sub-cellular localization. It has been proposed that splice variants with alternative

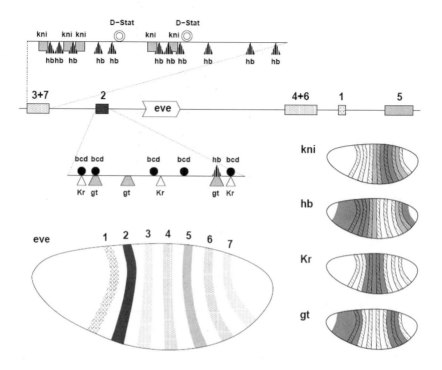

Figure 3.   Regulation of the seven even-skipped stripes in early Drosophila development. Five regulatory regions located upstream and downstream of the eve gene, stripe 3+7 enhancer, stripe 2 enhancer, stripe 4+6 enhancer, stripe 1 enhancer and stripe 5 enhancer, each regulate a sub-pattern of the eve expression pattern in the cellularizing syncytial blastoderm stage of the embryo.   These enhancers are composed of binding sites for transcriptional activators (above the line) and repressors (below the line).   While the activators (e.g. bcd, cad and D-Stat) are rather ubiquitously distributed, the spatially located repressors (e.g. kni, hb, Kr and gt – expression patterns on the right side) define the anterior and posterior borders of eve stripes[81,82,21].   The current picture offers no direct control for the observed even width and spacing of eve stripes.

UTRs probably combine the coding sequence with different elements for post-transcriptional regulation and, consequently, localization[84].   Epigenetic regulation predominantly connects to localized expression on a multicellular level.   Fully differentiated cells will only utilize a fraction of all genes for the rest of their lifetime.   To repress de-differentiation (and the risk of cancer) early developmental genes and pluripotency-associated genes can be permanently silenced by DNA methylation[68].

Even though much exemplary knowledge about spatial patterns of expression has be gathered over past years, large-scale data sets and comprehensive statistical analysis are not yet available. High-throughput *in-situ* hybridization methods now allow the collection of spatial information (e.g. gene expression patters during the early stages of *Drosophila* development (e.g. BDGP[88]) and search for similar patterns on the basis of image comparison (e.g. FlyExpress[38]).

Differences in expression of transcripts, RNAs and proteins are due to environmental factors, position effects and the genetic program of the cell. The differences in expression of regulators themselves regulate the spatial distribution of others together with ATP dependent localization mechanisms.

7. Concluding Remarks

In the few years since the completion of the human genome sequence[87], a diverse set of high-throughput experiments has amassed data that profoundly change our understanding of molecular biology. Not only are there an order of magnitude fewer protein coding genes than predicted a decade ago, but the transcriptional output of eukaryotic genome is unexpectedly diverse and covers an unexpectedly large fraction of the genomic DNA. The data confront us with a surprisingly diverse collection of processed RNAs with — as we argue here — diverse, predominantly regulatory, functions which they perform by means of many different molecular mechanisms.

The sheer extent of this regulatory "layer" of ncRNAs implies that ncRNA-based regulation mechanisms should be seen as paradigms of gene regulation rather than as the odd exceptions to a "proteinocentric" picture of molecular biology. Indeed, RNA-related regulation does not appear to be well-separated from the essentially protein-based transcriptional and signaling networks but to incorporate them into a tightly woven web of complex interactions which cannot be understood comprehensively by artificially focusing on particular molecules or interaction mechanisms. We argue that this complexity needs to be integrated into our current modeling approaches[76] for gene regulatory networks.

While the long-known discrepancy between mRNA and protein expression levels has been largely ignored in the interpretation of gene expression experiments, the large number of transcripts and processing products arising from the same genomic locus emphasizes that such a simplification is even conceptually untenable. From the influence of nuclear organization on transcriptional regulation to the spatial expression patterns of RNAs and proteins on a multicellular level, spatial organization will gain importance as data are accumulating. The formation of spatially differentiated expression patterns is ultimately the cause of cell differentiation, tissue formation, and hence the fundamental molecular process in development.

With pattern formation being inherently a dynamical phenomenon, it becomes imperative to consider in more details dynamical aspects of regulation. For instance, it is well know that transcriptional regulation is much slower than post-transcriptional gene silencing by means of microRNAs. We cannot not expect that detailed kinetic data will be available for large fractions of the human transcriptome in the near future (although this is feasible for organism such as yeast because of the much small genome size and simpler transcriptome organization). A much more detailed mechanistic understanding of the interactions and at least an approximately complete picture of the types players and their interaction rules therefore is an indispensable prerequisite for modeling in Systems Biology. In the same context, the consequences of fluctuations in transcript numbers, and in particular of the variation across individuals, can only be understood based on a comprehensive map of the regulatory networks. One would expect that the need for developmental robustness in the presence of substantial noise at the molecular level has shaped the interplay of regulation mechanisms of evolutionary time-scales[33,20].

The changing paradigms in gene regulation imply the need for a corresponding experimental and theoretical research agenda, with topics ranging from large scale tracing of processing pathways, high-throughput functional assays for non-coding transcripts, to computational approaches for unbiased transcript detection, network reconstruction from incomplete data, and comprehensive modelling of spatio-temporal expression patterns.

Acknowledgments

PFS acknowledges the hospitality of the EFG @ ASU in September 2007, where part of this work was conceived.

References

1. A. Aravin, D. Gaidatzis, S. Pfeffer, M. Lagos-Quintana, P. Landgraf, N. Iovino, P. Morris, M. J. Brownstein, S. Kuramochi-Miyagawa, T. Nakano, M. Chien, J. J. Russo, J. Ju, R. Sheridan, C. Sander, M. Zavolan, and T. Tuschl. A novel class of small RNAs bind to MILI protein in mouse testes. *Nature*, 442:203–207, 2006.
2. I. Barák and A. J. Wilkinson. Where asymmetry in gene expression originates. *Mol Microbiol*, 57(3):611–620, Aug 2005.

3. S. Ben-Tabou de Leon and E. H. Davidson. Gene regulation: gene control network in development. *Annu Rev Biophys Biomol Struct.*, 36:191–212, 2007.

4. S. Bondos. Variations on a theme: Hox and Wnt combinatorial regulation during animal development. *Sci STKE*, 355:pe38, 2006.

5. R. R. Breaker, G. M. Emilsson, D. Lazarev, S. Nakamura, I. J. Puskarz, A. Roth, and N. Sudarsan. A common speed limit for RNA-cleaving ribozymes and deoxyribozymes. *RNA*, 9:949–957, 2003.

6. G. A. Calin, C. G. Liu, M. Ferracin, T. Hyslop, R. Spizzo, C. Sevignani, M. Fabbri, A. Cimmino, E. J. Lee, S. E. Wojcik, M. Shimizu, E. Tili, S. Rossi, C. Taccioli, F. Pichiorri, X. Liu, S. Zupo, V. Herlea, L. Gramantieri, G. Lanza, H. Alder, L. Rassenti, S. Volinia, T. D. Schmittgen, T. J. Kipps, M. Negrini, and C. M. Croce. Ultraconserved regions encoding ncRNAs are altered in human leukemias and carcinomas. *Cancer Cell*, 12:215–229, 2007.

7. P. Carninci. Constructing the landscape of the mammalian transcriptome. *J. Exp. Biol.*, 210:1497–1506, 2007.

8. P. Carninci, T. Kasukawa, S. Katayama, J. Gough, M. Frith, N. Maeda, R. Oyama, T. Ravasi, B. Lenhard, C. Wells, and *et al.*; FANTOM Consortium; RIKEN Genome Exploration Research Group and Genome Science Group (Genome Network Project Core Group). The transcriptional landscape of the mammalian genome. *Science*, 309:1559–1563, 2005.

9. L. A. Cassiday and L. J. Maher. Having it both ways: transcription factors that bind dna and rna. *Nucleic Acids Res*, 30(19):4118–4126, Oct 2002.

10. J. L. Childs, M. D. Disney, and D. H. Turner. Oligonucleotide directed misfolding of RNA inhibits *Candida albicans* group I intron splicing. *Proc. Natl. Acad. Sci. USA*, 99:11091–11096, 2002.

11. H. A. Coller, J. J. Forman, and A. Legesse-Miller. "Myc'ed messages": *myc* induces transcription of E2F1 while inhibiting its translation via a microRNA polycistron. *PLoS Genet.*, 3:e146, 2007.

12. F. F. Costa. Non-coding RNAs: new players in eukaryotic biology. *Gene*, 357(2):83–94, 2005.

13. R. Dahm, M. Kiebler, and P. Macchi. RNA localisation in the nervous system. *Semin Cell Dev Biol*, 18(2):216–223, Apr 2007.

14. L. David, W. Huber, M. Granovskaia, J. Toedling, C. J. Palm, L. Bofkin, T. Jones, R. W. Davis, and L. M. Steinmetz. A high-resolution map of transcription in the yeast genome. *Proc. Natl. Acad. Sci. USA*, 103:5320–5325, 2006.

15. E. Davidson. *The Regulatory Genome.* Elsevier, Amsterdam, 2006.

16. W. Driever and C. Nüsslein-Volhard. A gradient of bicoid protein in drosophila embryos. *Cell*, 54(1):83–93, Jul 1988.

17. J. Dworkin and R. Losick. Differential gene expression governed by chromosomal spatial asymmetry. *Cell*, 107(3):339–346, Nov 2001.

18. J. Feng, C. Bi, B. S. Clark, R. Mady, P. Shah, and J. D. Kohtz. The Evf-2 noncoding RNA is transcribed from the Dlx-5/6 ultra conserved region and functions as a Dlx-2 transcriptional coactivator. *Genes Dev*, 20:1470–1484, 2006.

19. J. L. Fink, R. N. Aturaliya, M. J. Davis, F. Zhang, K. Hanson, M. S. Teas-

dale, C. Kai, J. Kawai, P. Carninci, Y. Hayashizaki, and R. D. Teasdale. Locate: a mouse protein subcellular localization database. *Nucleic Acids Res*, 34(Database issue):213–217, 2006.

20. M. A. Fuentes and D. C. Krakauer. The evolution of developmental patterning under genetic duplication constraints. *J R Soc Interface*, Jun 2007. 10.1098/rsif.2007.1074.

21. M. Fujioka, Y. Emi-Sarker, G. L. Yusibova, T. Goto, and J. B. Jaynes. Analysis of an even-skipped rescue transgene reveals both composite and discrete neuronal and early blastoderm enhancers, and multi-stripe positioning by gap gene repressor gradients. *Development*, 126(11):2527–2538, Jun 1999.

22. M. B. Gerstein, C. Bruce, J. S. Rozowsky, D. Zheng, J. Du, J. O. Korbel, O. Emanuelsson, Z. D. Zhang, S. Weissman, and M. Snyder. What is a gene, post-ENCODE? history and updated definition. *Genome Res.*, 17:669–681, 2007.

23. N. Gilbert and B. Ramsahoye. The relationship between chromatin structure and transcriptional activity in mammalian genomes. *Brief Funct Genomic Proteomic*, 4(2):129–142, Jul 2005.

24. T. R. Gingeras. Origin of phenotypes: Genes and transcripts. *Genome Res.*, 17:682–690, 2007.

25. J. Hackermüller, N.-C. Meisner, M. Auer, M. Jaritz, and P. F. Stadler. The effect of RNA secondary structures on RNA-ligand binding and the modifier RNA mechanism: A quantitative model. *Gene*, 345:3–12, 2005.

26. J. Han, D. Kim, and K. V. Morris. Promoter-associated RNA is required for RNA-directed transcriptional gene silencing in human cells. *Proc. Natl. Acad. Sci. USA*, 104:12422–12427, 2007.

27. M. Havilio, E. Y. Levanon, G. Lerman, M. Kupiec, and E. Eisenberg. Evidence for abundant transcription of non-coding regions in the sac charomyces cerevisiae genome. *BMC Genomics*, 6:93, 2005.

28. H. He, J. Wang, T. Liu, X. S. Liu, T. Li, Y. Wang, Z. Qian, H. Zheng, X. Zhu, T. Wu, B. Shi, W. Deng, W. Zhou, G. Skogerbø, and R. Chen. Mapping the *C. elegans* noncoding transcriptome with a whole-genome tiling microarray. *Genome Res.*, 2007. DOI: 10.1101/gr.6611807.

29. O. Hobert. Common logic of transcription factor and microRNA action. *Trends Biochem Sci*, 29(9):462–468, Sep 2004.

30. J. Ihmels, S. Bergmann, M. Gerami-Nejad, I. Yanai, M. McClellan, J. Berman, and N. Barkai. Rewiring of the yeast transcriptional network through the evolution of motif usage. *Science*, 309:938–940, 2005.

31. Y. Ikeda, R. S. Daughters, and L. P. Ranum. Bidirectional expression of the SCA8 expansion mutation: One mutation, two genes. *Cerebellum*, 2007. doi: 10.1080/14734220701413781.

32. O. Isken and L. E. Maquat. Quality control of eukaryotic mRNA: safeguarding cells from abnormal mRNA function. *Genes Dev.*, 21:1833–1856, 2007.

33. L. J. Johnson and J. F. Brookfield. Evolution of spatial expression pattern. *Evol Dev*, 5(6):593–599, Nov-Dec 2003.

34. D. Kampa, J. Cheng, P. Kapranov, M. Yamanaka, S. Brubaker, S. Cawley, J. Drenkow, A. Piccolboni, S. Bekiranov, G. Helt, H. Tammana, and T. R.

Gingeras. Novel RNAs identified from an in-depth analysis of the transcriptome of human chromosomes 21 and 22. *Genome Res.*, 14:331–342, 2004.

35. P. Kapranov, J. Cheng, S. Dike, D. Nix, R. Duttagupta, A. T. Willingham, P. F. Stadler, J. Hertel, J. Hackermüller, I. L. Hofacker, I. Bell, E. Cheung, J. Drenkow, E. Dumais, S. Patel, G. Helt, G. Madhavan, A. Piccolboni, V. Sementchenko, H. Tammana, and T. R. Gingeras. RNA maps reveal new RNA classes and a possible function for pervasive transcription. *Science*, 316:1484–1488, 2007.

36. S. Katayama, Y. Tomaru, T. Kasukawa, K. Waki, M. Nakanishi, M. Nakamura, H. Nishida, C. C. Yap, M. Suzuki, J. Kawai, H. Suzuki, P. Carninci, Y. Hayashizaki, C. Wells, M. Frith, T. Ravasi, K. C. Pang, J. Hallinan, J. Mattick, D. A. Hume, L. Lipovich, S. Batalov, P. G. Engström, Y. Mizuno, M. A. Faghihi, A. Sandelin, A. M. Chalk, S. Mottagui-Tabar, Z. Liang, B. Lenhard, C. Wahlestedt, RIKEN Genome Exploration Research Group, Genome Science Group (Genome Network Project Core Group), and FANTOM Consortium. Antisense transcription in the mammalian transcriptome. *Science*, 309:1564–1556, 2005.

37. D. Kosman, C. M. Mizutani, D. Lemons, W. G. Cox, W. McGinnis, and E. Bier. Multiplex detection of RNA expression in drosophila embryos. *Science*, 305(5685):846–846, 2004.

38. S. Kumar, K. Jayaraman, S. Panchanathan, R. Gurunathan, A. Marti-Subirana, and S. J. Newfeld. Best: a novel computational approach for comparing gene expression patterns from early stages of drosophila melanogaster development. *Genetics*, 162(4):2037–2047, Dec 2002.

39. M. Lapidot and Y. Pilpel. Genome-wide natural anti-sense transcription: coupling its regulation to its different regulatory mechanisms. *EMBO Rep.*, 7:1216–1222, 2006.

40. N. C. Lau, A. G. Seto, J. Kim, S. Kuramochi-Miyagawa, T. Nakano, D. P. Bartel, and R. E. Kingston. Characterization of the piRNA complex from rat testes. *Science*, 313:363–367, 2006.

41. D. Lebrecht, M. Foehr, E. Smith, F. J. Lopes, C. E. Vanario-Alonso, J. Reinitz, D. S. Burz, and S. D. Hanes. Bicoid cooperative dna binding is critical for embryonic patterning in drosophila. *Proc Natl Acad Sci U S A*, 102(37):13176–13181, Sep 2005.

42. E. Leygue. Steroid receptor RNA activator (SRA1): unusual bifaceted gene products with suspected relevance to breast cancer. *Nucl. Recep. Signaling*, 5:e006, 2007.

43. L. C. Li, S. T. Okino, H. Zhao, D. Pookot, R. F. Place, S. Urakami, H. Enokida, and R. Dahiya. Small dsRNAs induce transcriptional activation in human cells. *Proc. Natl. Acad. Sci. USA*, 103:17337–17342, 2006.

44. S. J. Lolle, J. L. Victor, J. M. Young, and R. E. Pruitt. Genome-wide non-mendelian inheritance of extra-genomic information in arabidopsis. *Nature*, 434(7032):505–509, Mar 2005.

45. N. Maeda, T. Kasukawa, R. Oyama, J. Gough, M. Frith, P. G. Engström, B. Lenhard, R. N. Aturaliya, S. Batalov, K. W. Beisel, C. J. Bult, C. F. Fletcher, A. R. Forrest, M. Furuno, D. Hill, M. Itoh, M. Kanamori-Katayama,

S. Katayama, M. Katoh, T. Kawashima, J. Quackenbush, T. Ravasi, B. Z. Ring, K. Shibata, K. Sugiura, Y. Takenaka, R. D. Teasdale, C. A. Wells, Y. Zhu, C. Kai, J. Kawai, D. A. Hume, P. Carninci, and Y. Hayashizaki. Transcript annotation in FANTOM3: Mouse gene catalog based on physical cdnas. *PLoS Genetics*, 2:e62, 2006. doi:10.1371/journal.pgen.0020062.

46. G. Mainguy, J. Koster, J. Woltering, H. Jansen, and D. A. Extensive polycistronism and antisense transcription in the mammalian Hox clusters. *PLoS ONE*, 2:e356, 2007.

47. J. R. Manak, S. Dike, V. Sementchenko, P. Kapranov, F. Biemar, J. Long, J. Cheng, I. Bell, S. Ghosh, A. Piccolboni, and T. R. Gingeras. Biological function of unannotated transcription during the early development of Drosophila melanogaster. *Nat Genet*, 38:1151–1158, 2006.

48. F. Markowetz and R. Spang. Inferring cellular networks — a review. *BMC Bioinformatics*, 8(Suppl 6):S5, 2007.

49. K. C. Martin and R. S. Zukin. RNA trafficking and local protein synthesis in dendrites: an overview. *J Neurosci*, 26(27):7131–7134, 2006.

50. S. C. Materna and E. H. Davidson. Logic of gene regulatory networks. *Curr Opin Biotechnol.*, 18:351–354, 2007.

51. J. S. Mattick. A new paradigm for developmental biology. *J. Exp. Biol.*, 210:1526–1547, 2007.

52. J. S. Mattick and I. V. Makunin. Non-coding RNA. *Hum Mol Genet.*, 15:R17–29, 2006.

53. A. Mazo, J. W. Hodgson, S. Petruk, Y. Sedkov, and H. W. Brock. Transcriptional interference: an unexpected layer of complexity in gene regulation. *J Cell Sci.*, 120:2755–2761, 2007.

54. N.-C. Meisner, J. Hackermüller, V. Uhl, A. Aszódi, M. Jaritz, and M. Auer. mRNA openers and closers: A methodology to modulate AU-rich element controlled mRNA stability by a molecular switch in mRNA conformation. *Chembiochem.*, 5:1432–1447, 2004.

55. L. M. Mendes Soares and J. Valcárcel. The expanding transcriptome: the genome as the 'Book of Sand'. *EMBO J.*, 25:923–931, 2006.

56. B. D. Metscher, K. Takahashi, K. Crow, C. Amemiya, D. F. Nonaka, and G. P. Wagner. Expression of hoxa-11 and hoxa-13 in the pectoral fin of a basal ray-finned fish, polyodon spathula: implications for the origin of tetrapod limbs. *Evol Dev*, 7(3):186–195, May-Jun 2005.

57. F. Miura, N. Kawaguchi, J. Sese, A. Toyoda, M. Hattori, S. Morishita, and T. Ito. A large-scale full-length cDNA analysis to explore the budding yeast transcriptome. *Proc. Natl. Acad. Sci. USA*, 103:17846–17851, 2006.

58. K. Mochizuki and M. A. Gorovsky. Small RNAs in genome rearrangement in tetrahymena. *Curr Opin Genet Dev.*, 14:181–187, 2004.

59. K. Ng, D. Pullirsch, M. Leeb, and A. Wutz. *Xist* and the order of silencing. *EMBO Rep.*, 8:34–39, 2007.

60. Y. Okazaki, M. Furuno, T. Kasukawa, J. Adachi, H. Bono, S. Kondo, I. Nikaido, N. Osato, R. Saito, H. Suzuki, and *et al*. Analysis of the mouse transcriptome based on functional annotation of 60,770 full-length cDNAs. *Nature*, 420:563–573, 2002.

61. I. M. Palacios and D. St Johnston. Getting the message across: the intracellular localization of mRNAs in higher eukaryotes. *Annu Rev Cell Dev Biol*, 17:569–614, 2001.

62. M. Paulus, M. Haslbeck, and M. Watzele. RNA stem-loop enhanced expression of previously non-expressible genes. *Nucl. Acids Res.*, 32:9/e78, 2004. doi 10.1093/nar/gnh076.

63. J. S. Pedersen, G. Bejerano, A. Siepel, K. Rosenbloom, K. Lindblad-Toh, E. S. Lander, J. Kent, W. Miller, and D. Haussler. Classification of conserved RNA secondary structures in the human genome. *PLoS Comput. Biol.*, 2:e33, 2006.

64. K. V. Prasanth and D. L. Spector. Eukaryotic regulatory RNAs: an answer to the 'genome complexity' conundrum. *Genes Dev.*, 21:11–42, 2007.

65. T. Ravasi, H. Suzuki, K. C. Pang, S. Katayama, M. Furuno, R. Okunishi, S. Fukuda, K. Ru, M. C. Frith, M. M. Gongora, S. M. Grimmond, D. A. Hume, Y. Hayashizaki, and J. S. Mattick. Experimental validation of the regulated expression of large numbers of non-coding RNAs from the mouse genome. *Genome Res.*, 16:11–19, 2006.

66. T. Ravasi, H. Suzuki, K. C. Pang, S. Katayama, M. Furuno, R. Okunishi, S. Fukuda, K. Ru, M. C. Frith, M. M. Gongora, S. M. Grimmond, D. A. Hume, Y. Hayashizaki, and J. S. Mattick. Experimental validation of the regulated expression of large numbers of non-coding RNAs from the mouse genome. *Genome Res.*, 16:11–19, 2006.

67. S. V. Razin, A. Rynditch, V. Borunova, E. Ioudinkova, V. Smalko, and K. Scherrer. The 33 kb transcript of the chicken alpha-globin gene domain is part of the nuclear matrix. *J Cell Biochem*, 92:445–457, 2004.

68. W. Reik. Stability and flexibility of epigenetic gene regulation in mammalian development. *Nature*, 447(7143):425–432, May 2007.

69. G. Riddihough and D. Ish-Horowicz. Individual stripe regulatory elements in the drosophila hairy promoter respond to maternal, gap, and pair-rule genes. *Genes Dev*, 5(5):840–854, 1991.

70. J. L. Rinn, M. Kertesz, J. K. Wang, S. L. Squazzo, X. Xu, S. A. Brugmann, L. H. Goodnough, J. A. Helms, P. J. Farnham, E. Segal, and H. Y. Chang. Functional demarcation of active and silent chromatin domains in human HOX loci by noncoding RNAs. *Cell*, 29:1311–1323, 2007.

71. D. R. Rose, J. Hackermüller, S. Washietl, S. Findeiß, K. Reiche, J. Hertel, P. F. Stadler, and S. J. Prohaska. Computational RNomics of drosophilids. *BMC Genomics*, 2007. accepted.

72. J. G. Ruby, C. Jan, C. Player, M. J. Axtell, W. Lee, C. Nusbaum, H. Ge, and D. P. Bartel. Large-scale sequencing reveals 21U-RNAs and additional microRNAs and endogenous siRNAs in *C. elegans*. *Cell*, 127:1193–1207, 2006.

73. K. Salehi-Ashtiani, A. Lupták, A. Litovchick, and J. W. Szostak. A genomewide search for ribozymes reveals an HDV-like sequence in the human CPEB3 gene. *Science*, 313:1788–1792, 2006.

74. T. Sanchez-Elsner, D. Gou, E. Kremmer, and F. Sauer. Noncoding RNAs of trithorax response elements recruit Drosophila Ash1 to Ultrabithorax. *Science*, 311:1118–1123, 2006.

75. F. Schacherer, C. Choi, U. Götze, M. Krull, S. Pistor, and E. Wingender. The TRANSPATH signal transduction database: a knowledge base on signal transduction networks. *Bioinformatics*, 17:1053–1057, 2001.

76. T. Schlitt and A. Brazma. Current approaches to gene regulatory network modelling. *BMC Bioinformatics*, 8 (Suppl 6):S9, 2007.

77. S. Schmitt and R. Paro. RNA at the steering wheel. *Genome Biol*, 7(5):218–218, 2006.

78. W. Schubert, B. Bonnekoh, A. J. Pommer, L. Philipsen, R. Böckelmann, Y. Malykh, H. Gollnick, M. Friedenberger, M. Bode, and A. W. Dress. Analyzing proteome topology and function by automated multidimensional fluorescence microscopy. *Nat Biotechnol*, 24(10):1270–1278, Oct 2006.

79. K. E. Shearwin, B. P. Callen, and J. B. Egan. Transcriptional interference—a crash course. *Trends Genet.*, 21:339–345, 2005.

80. Y. Shimoni, G. Friedlander, G. Hetzroni, G. Niv, S. Altuvia, O. Biham, and H. Margalit. Regulation of gene expression by small non-coding RNAs: a quantitative view. *Mol. Syst. Biol.*, 3:138, 2007.

81. S. Small, A. Blair, and M. Levine. Regulation of even-skipped stripe 2 in the drosophila embryo. *EMBO J*, 11(11):4047–4057, Nov 1992.

82. S. Small, A. Blair, and M. Levine. Regulation of two pair-rule stripes by a single enhancer in the drosophila embryo. *Dev Biol*, 175(2):314–324, May 1996.

83. S. Soneji, S. Huang, M. Loose, I. J. Donaldson, R. Patient, B. Gottgens, T. Enver, and G. May. Inference, validation, and dynamic modeling of transcription networks in multipotent hematopoietic cells. *Ann N Y Acad Sci.*, 1106:30–40, 2007.

84. A. Stark, J. Brennecke, N. Bushati, R. B. Russell, and S. M. Cohen. Animal microRNAs confer robustness to gene expression and have a significant impact on 3'UTR evolution. *Cell*, 123(6):1133–1146, Dec 2005.

85. The Athanasius F. Bompfünewerer RNA Consortium:, R. Backofen, C. Flamm, C. Fried, G. Fritzsch, J. Hackermüller, J. Hertel, I. L. Hofacker, K. Missal, S. J. Mosig, Axel Prohaska, D. Rose, P. F. Stadler, A. Tanzer, S. Washietl, and W. Sebastian. RNAs everywhere: Genome-wide annotation of structured RNAs. *J. Exp. Zool. B: Mol. Dev. Evol.*, 308B:1–25, 2007.

86. The ENCODE Project Consortium. Identification and analysis of functional elements in 1% of the human genome by the ENCODE pilot project. *Nature*, 447:799–816, 2007.

87. The Human Genome Sequencing Consortium. Finishing the euchromatic sequence of the human genome. *Nature*, 431:931–945, 2004.

88. P. Tomancak, A. Beaton, R. Weiszmann, E. Kwan, S. Shu, S. E. Lewis, S. Richards, M. Ashburner, V. Hartenstein, S. E. Celniker, and G. M. Rubin. Systematic determination of patterns of gene expression during drosophila embryogenesis. *Genome Biol*, 3(12), 2002.

89. E. Torarinsson, M. Sawera, J. Havgaard, M. Fredholm, and J. Gorodkin. Thousands of corresponding human an mouse genomic regions unalignable in primary sequece contain common RNA structure. *Genome Res.*, 16:885–889, 2006.

90. S. Washietl, I. L. Hofacker, M. Lukasser, A. Hüttenhofer, and P. F. Stadler. Mapping of conserved RNA secondary structures predicts thousands of functional non-coding RNAs in the human genome. *Nature Biotech.*, 23:1383–1390, 2005.

91. S. Washietl, I. L. Hofacker, and P. F. Stadler. Fast and reliable prediction of noncoding RNAs. *Proc. Natl. Acad. Sci. USA*, 102:2454–2459, 2005.

92. S. Washietl, J. S. Pedersen, J. O. Korbel, A. Gruber, J. Hackermüller, J. Hertel, M. Lindemeyer, K. Reiche, C. Stocsits, A. Tanzer, C. Ucla, C. Wyss, S. E. Antonarakis, F. Denoeud, J. Lagarde, J. Drenkow, P. Kapranov, T. R. Gingeras, R. Guigó, M. Snyder, M. B. Gerstein, A. Reymond, I. L. Hofacker, and P. F. Stadler. Structured RNAs in the ENCODE selected regions of the human genome. *Gen. Res.*, 17:852–864, 2007.

93. C. Weile, P. P. Gardner, M. M. Hedegaard, and J. Vinther. Use of tiling array data and RNA secondary structure predictions to identify noncoding RNA genes. *BMC Genomics*, 8:244, 2007.

94. S. Will, K. Missal, I. L. Hofacker, P. F. Stadler, and R. Backofen. Inferring non-coding RNA families and classes by means of genome-scale structure-based clustering. *PLoS Comp. Biol.*, 3:e65, 2007.

95. A. T. Willingham and T. R. Gingeras. TUF love for "junk" DNA. *Cell*, 125:1215–1220, 2006.

96. X. Wu, R. Vakani, and S. Small. Two distinct mechanisms for differential positioning of gene expression borders involving the drosophila gap protein giant. *Development*, 125(19):3765–3774, 1998.

97. F. Wyers, M. Rougemaille, G. Badis, J. C. Rousselle, M. E. Dufour, J. Boulay, B. Régnault, F. Devaux, A. Namane, B. Séraphin, D. Libri, and J. A. Cryptic pol II transcripts are degraded by a nuclear quality control pathway involving a new poly(A) polymerase. *Cell*, 121:725–737, 2005.

98. O. Yazgan and J. E. Krebs. Noncoding but nonexpendable: transcriptional regulation by large noncoding RNA in eukaryotes. *Biochem Cell Biol.*, 85:484–496, 2007.

99. Z. Zhang, A. W. Pang, and M. Gerstein. Comparative analysis of genome tiling array data reveals many novel primate-specific functional RNAs in human. *BMC Evol Biol.*, 7 (Suppl 1):S14, 2007.

GEODESIC CURVES FOR BIOMOLECULAR STRUCTURE MODELLING

RUBEM P. MONDAINI, ROBERTO. A. C. PRATA

Federal University of Rio de Janeiro, CT/COPPE
Ilha do Fundão, 21.945-972, P.O. Box 68511, Rio de Janeiro, Brazil

The present literature has several contributions to the understanding of biomolecular structure by adopting helix curves through the atom sites as a paradigm. In the present contribution we introduce the weaker paradigm of geodesic curves of the manifold along ordered sets of points. This alternative allows us to use many interesting curves with some basic properties of helices in the modelling of protein and DNA structures.

1. Introduction

This work follows the guidelines of searching for alternative forms of representing the tertiary form of proteins and the structure of double helix of DNA molecule.

We start from a set of points in space corresponding to the sites of carbon and nitrogen atoms. The experimental methods will search always for sets of ordered points. This was essencially the guideline of Crick and Watson[1] in their study of DNA molecule or Pauling[2] with α-helices of proteins. A fundamental step in this kind of modelling is to identify a regularity of the set by a curve or surface. These ideas come from Schroedinger's approach to the origin of life[3] and his fundamental question about the internal manifold of biomolecules. The literature has many contributions to this research program in which the emphasis is put on helices for modelling special sequences of atomic sites. This is the aim of any scientific modeller after analyzing data on coordinates of the sites. We shall propose here, instead of assuming the existence of helices a priori, to look for curves characterized by the property that points evenly spaced along them are also evenly spaced according the Euclidean metric. This class of curves will contain helices as representatives of a sub-class, but we intend to know if the sub-class of helices is unique.

2. Smooth Curve Through an Ordered Set of Points

Let us consider a smooth curve which goes through all the points of an ordered set in $\mathbb{R}^3$ such that the points result to be evenly spaced according the metric along the curve. The curve will be characterized as usual by this metric, by its curvature and torsion. We notice that the converse problem of marking points evenly spaced along a given curve sounds obviously trivial but this does not correspond to the spirit which inspires the modelling of biomolecular structures.

Let $S_{j+1} - S_j$ the measure of the arc length of a continuous and differentiable curve between consecutive points. We have

$$S_{j+2} - S_{j+1} = S_{j+1} - S_j, \quad \forall j. \tag{1}$$

This difference equation can be solved by

$$S_j = j\sigma, \quad \forall j. \tag{2}$$

where σ is the arc length between consecutive points. See fig. 1

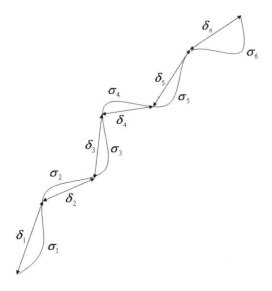

Figure 1. An ordered set of points on a smooth curve. The restrictions imposed by eqs. (2) and (12) will correspond to $\sigma_1 = \sigma_2 = \ldots = \sigma$ and $\delta_1 = \delta_2 = \ldots = \delta$.

Let $\vec{r}(\phi)$ be the position vector of any point of a $\mathbb{R}^3$ smooth curve parameterized by ϕ. The metric along this curve is given by

$$g = \frac{\partial \vec{r}}{\partial \phi} \cdot \frac{\partial \vec{r}}{\partial \phi} \tag{3}$$

and the arc length between consecutive points by

$$S_{j+1} - S_j = \int_{\phi_j}^{\phi_{j+1}} g^{1/2} \, d\phi. \tag{4}$$

We can always consider evenly spaced values of the parameter

$$\phi_{j+2} - \phi_{j+1} = \phi_{j+1} - \phi_j, \quad \forall j \tag{5}$$

or

$$\phi_j = j\varphi, \quad \forall j. \tag{6}$$

From eq. (4), a necessary and sufficient condition for the validity of eqs. (1) and (5) or (2) and (6) is

$$g = a^2 \tag{7}$$

where a is a constant.

Any coordinate can be taken as a parameter for describing the curve. The jth value of this coordinate, say the z coordinate will satisfy to

$$z_j = jh, \quad \forall j. \tag{8}$$

If we impose now the same behaviour to the other coordinates, $y_j = jy$; $x_j = jx$, $\forall j$, we shall have,

$$\vec{r}_{j+2} - \vec{r}_{j+1} = \vec{r}_{j+1} - \vec{r}_j \tag{9}$$

or

$$\vec{r}_j = j\vec{\rho}, \quad \forall j. \tag{10}$$

where $\vec{\rho} = (\rho_x, \rho_y, \rho_z)$ is a constant vector.

This corresponds to the equation of a straight line in $\mathbb{R}^3$

$$\frac{x - x_j}{\rho_x} = \frac{y - y_j}{\rho_y} = \frac{z - z_j}{\rho_z}. \tag{11}$$

However, there are solutions to eq. (7) which satisfy a weaker condition than eq. (9). This will be written as

$$\|\vec{r}_{j+2} - \vec{r}_{j+1}\| = \|\vec{r}_{j+1} - \vec{r}_j\| = \delta, \quad \forall j. \tag{12}$$

where $\| \cdot \|$ stands for Euclidean norm.

In order to obtain an Ansatz for the generic vector position of these curves, we shall make a restriction of the form of eq. (8) on two arbitrary coordinates. As an example, we introduce a system of cylindrical coordinates. The jth position vector can be written as

$$\vec{r}_j = \left(\sqrt{x_j^2 + y_j^2} \cos \arctan \frac{y_j}{x_j}, \sqrt{x_j^2 + y_j^2} \sin \arctan \frac{y_j}{x_j}, z_j \right) . \tag{13}$$

The two assumed restrictions will be characterized by

$$x_j = r_j(\varphi) \cos(j\varphi) ; \quad y_j = r_j(\varphi) \sin(j\varphi) ; \quad z_j = z(j\varphi_= jh(\phi), \quad \forall j . \tag{14}$$

If in addition, we assume cylindrical symmetry, or

$$r_j(\varphi) = r(j\varphi) = R(\phi) , \tag{15}$$

we then get a curve such that evenly spaced points along it are also evenly spaced according the Euclidean metric, namely,

$$\vec{r}_j = (R(\phi) \cos(j\varphi), R(\phi) \sin(j\varphi), jh(\varphi)) \tag{16}$$

and from eqs. (6), we can write for the generic position vector of the curve

$$\vec{r} = (R(\phi) \cos \phi, R(\phi) \sin \phi, z(\phi)) . \tag{17}$$

After using esq. (17), (7) and (3), we get

$$R'^2 + R^2 + z'^2 = a^2, \quad (') = \frac{d}{d\phi} . \tag{18}$$

Some remarks are now in order. Firstly, the assumption of cylindrical symmetry made above does not imply that we consider the configuration of points to stand on the surface of a cylinder. There is a sub-sequence of points to be modelled by a curve for each ϕ-value. We can even divide the point set configuration into subsets each one of them corresponding to a ϕ-value. Another remark is that we are not considering the functions r and z as homogeneous functions of degrees 0 and 1, respectively, as the development above may suggest. In this case we would have $\phi \frac{\partial r}{\partial \phi} = 0$, $\phi \frac{\partial z}{\partial \phi} = z$ or $r = r_0$, $z = \alpha \phi$ with r_0 and α constants, corresponding to a helix. Finally, we should note that eq. (18) is the first integral for geodesic equations of $\mathbb{R}^3$ with a paramenter ϕ, the polar angle of cylindrical coordinates. If we restrict the search of curves to those on the surface of a cylinder $R = R_0$, we get $z = \alpha \phi = \sqrt{a^2 - R_0^2}$ from eq. (18) and the helix would be the unique solution, or

$$\vec{r} = (R_0 \cos \phi, R_0 \sin \phi, \alpha \phi) . \tag{19}$$

Other helix configurations can be obtained from eq. (18) if we make $z = \alpha\phi$. A look-up technique will show that

$$R(\phi) = R_0 \cos(\phi - \phi_0) \tag{20}$$

where R_0, ϕ_0 are constants, is also a solution. Actually it corresponds to a helix with a half radius of the helix given by eq. (19) and a shifted axis as an elementary substitution into eq. (17) will show up:

$$\vec{r} = \left(\frac{1}{2} R_0 \cos(2\phi - \phi_0), \frac{1}{2} R_0 \sin(2\phi - \phi_0), \alpha\phi \right)$$
$$+ \left(\frac{1}{2} R_0 \cos\phi_0, \frac{1}{2} R_0 \sin -\phi_0, 0 \right). \tag{21}$$

We also advance here the information that the values of curvature and torsion are twice the values corresponding to the helix of eq. (19), or

$$k = \frac{2R_0}{R_0^2 + \alpha^2}, \ \tau = \frac{2\alpha}{R_0^2 + \alpha^2}. \tag{22}$$

This also means that the radius and pitch of this solution will be half of these values.

3. Elementary Theory of Geodesic Curves in $\mathbb{R}^3$

From the constant metric requirement as given by eqs. (7) and (3), we write the Frenet-Serret orthonormal and local basis as[4]

$$\hat{t} = a^{-1}\frac{\partial\vec{r}}{\partial\phi} \ ; \ \hat{n} = \frac{a^{-2}}{k}\frac{\partial^2\vec{r}}{\partial\phi^2} \ ; \ \hat{b} = \hat{t} \times \hat{n} = \frac{a^{-3}}{k}\frac{\partial\vec{r}}{\partial\phi} \times \frac{\partial^2\vec{r}}{\partial\phi^2} \tag{23}$$

where $k(\phi)$ stands for curvature function,

$$k^2(\phi) = a^{-4}\frac{\partial^2\vec{r}}{\partial\phi^2} \cdot \frac{\partial^2\vec{r}}{\partial\phi^2}. \tag{24}$$

The Frenet-Serret equations can be now written,

$$\frac{\partial\hat{t}}{\partial\phi} = ak\hat{n} \ ; \ \frac{\partial\hat{n}}{\partial\phi} = a(-k\hat{t} + \tau\hat{b}) \ ; \ \frac{\partial\hat{b}}{\partial\phi} = -a\tau\hat{n} \tag{25}$$

where $\tau(\phi)$ is the torsion to be obtained from eq. (24) and

$$k^2(k^2 + r^2) = a^{-6}\frac{\partial^3\vec{r}}{\partial\phi^3} \cdot \frac{\partial^3\vec{r}}{\partial\phi^3} - a^{-2}\left(\frac{\partial k}{\partial\phi}\right)^2. \tag{26}$$

Eqs. (23) to (26) lead to the identities

$$\frac{1}{k}\frac{\partial^3\vec{r}}{\partial\phi^3} - \frac{\tau}{k}\frac{\partial\vec{r}}{\partial\phi} \times \frac{\partial^2\vec{r}}{\partial\phi^2} - \frac{1}{k^2}\frac{\partial k}{\partial\phi}\frac{\partial^2\vec{r}}{\partial\phi^2} + ka^2\frac{\partial\vec{r}}{\partial\phi} = 0 \tag{27}$$

$$\frac{1}{k}\frac{\partial \vec{r}}{\partial \phi} \times \frac{\partial^3 \vec{r}}{\partial \phi^3} - \frac{1}{k^2}\frac{\partial k}{\partial \phi}\frac{\partial \vec{r}}{\partial \phi} \times \frac{\partial^2 \vec{r}}{\partial \phi^2} + \frac{\tau}{k}a^2\frac{\partial^2 \vec{r}}{\partial \phi^2} = 0. \tag{28}$$

We notice that eq. (28) is the vector product of eq. (27) by $\frac{\partial \vec{r}}{\partial \phi}$. The internal product of eq. (27) by the vectors $\vec{m} = \frac{\partial^2 \vec{r}}{\partial \phi^2}$, $\vec{n} = \frac{\partial \vec{r}}{\partial \phi}$, $\vec{p} = \vec{m} \times \vec{n}$ leads to the non-trivial identity

$$\frac{\partial^3 \vec{r}}{\partial \phi^3} \cdot \frac{\partial^2 \vec{r}}{\partial \phi^2} \times \frac{\partial \vec{r}}{\partial \phi} + \tau k^2 a^6 = 0, \quad \frac{\partial k}{\partial \phi} \neq 0 \tag{29}$$

where $k(\phi)$ and $\tau(\phi)$ are given by eqs. (24) and (26) above.

If eq. (29) is given as a definition of the torsion function, ir is easy to proof that it is an identity by observing the relations

$$n^2 = a^2\,;\ m^2 = a^4 k^2\,;\vec{m} \cdot \vec{n} = 0\,;\ m^2 + \vec{n} \cdot \vec{p} = 0 \tag{30}$$

and

$$(\vec{p} \times \vec{m}) \times \vec{n} = \pm m^2 \vec{m} \tag{31}$$

Eq. (29) has a suggestive appeal for defining the torsion as a measure of chirality of molecular configurations[5], due to pseudoscalar properties of the triple product.

4. Techniques of Solution

The fundamental equation to be solved for obtaining the curves with the properties shown at fig. 1 is eq. (18). We need more information to be provided by a careful analysis of the biomolecular structure in order to have physical insights for the function $z(\phi)$. However some generic techniques to solve eq. (18) can be advanced in this section. In the next section, we introduce some important information on the mathematical modelling of the structure.

We write eq. (18) in the form

$$R'^2 + R^2 = G^2, \quad (') = \frac{d}{d\phi} \tag{32}$$

where

$$G(\phi) = (a^2 - z'^2)^{1/2}. \tag{33}$$

If $z(\phi) = a\phi$, $G = (a^2 - a^2)^{1/2}$ is a constant and we can have the solution given by eq. (20). A generic class of solutions can be obtained from a transformation of dependent variables $R(\phi) \rightarrow u(\phi)$ such that

$$R' = R \cot u(\phi), \quad R = G(\phi) \sin u(\phi). \tag{34}$$

After eliminating R from eqs. (34), we get for eq. (32),

$$u' + \frac{G'}{G} \tan u = 1 \,. \tag{35}$$

Another transformation of dependent variables $u(\phi) \to v(\phi)$ with

$$v(\phi) = \tan u(\phi); \quad R(\phi) = G(\phi) \frac{v(\phi)}{\sqrt{1 + v^2(\phi)}} \tag{36}$$

will change eq. (35) into

$$v' = (1 + v^2) \left(1 - \frac{G'}{G} v \right) \,. \tag{37}$$

This is an Abel differential equation[6] and its associated theory is one of the most beautiful parts of mathematics. However the special form of the right hand side of eq. (37) can lead us to try the solution of the equation for the inverse function $\phi(v)$:

$$\frac{d\phi}{dv} = \frac{1 + \frac{G'}{G} v}{(1 + v^2)\left(1 + \frac{G'^2}{G^2}\right)} + \frac{G'^2/G^2}{\left(1 - \frac{G'}{G} v\right)\left(1 + \frac{G'^2}{G^2}\right)} \,. \tag{38}$$

Elementary cases will correspond to the possibility of an equation which is linear. These could be given by $G'/G = k = \text{const.}$, say. The solution can be written by taking v as a parameter.

$$R = \frac{v}{\sqrt{1 + v^2}} e^{k(\phi - \phi_0)} \tag{39}$$

$$\phi - \phi_0 = \frac{1}{1 + k^2} \left(\arctan v + k \ln \frac{1 + kv}{\sqrt{1 + v^2}} \right) \,. \tag{40}$$

Some results related to the treatment of Abel's equations can give us also other solutions[6]. As an example, we could apply the fact that if the quantity

$$I(\phi) = \frac{1}{3} \left(\frac{G''}{G} + \frac{G'^2}{G^2} + \frac{2}{9} \right) \tag{41}$$

is identically zero, then eq. (37) has the special solution

$$v(\phi) = \frac{1}{3G'/G} \,. \tag{42}$$

From eq. $I(\phi) = 0$ and after substituting into eq. (36), we get

$$G = R_0 \left[\cos(2(\phi - \phi_0)/3) \right]^{1/2} ; \quad v = -\cot(2(\phi - \phi_0)/3)$$

where R_0 is an arbitrary constant.

The solution to eq. (32), will be obtained from eq. (37) as

$$R(\phi) = R_0 \left[\cos(2(\phi - \phi_0)/3)\right]^{3/2} . \tag{43}$$

We can also notice that there are generic solutions of the form

$$R(\phi) = R_0[\cos 2(\phi - \phi_0)/c]^{c/2} , \quad G(\phi) = R_0[\cos 2(\phi - \phi_0)/c]^{c/2-1} \tag{44}$$

where c is a constant, to eqs. (32).

There is a subclass of solutions for

$$G(\phi) = R_0 \sec^2(\phi - \phi_0) \tag{45}$$

and we observe that in this subclass there is a special solution

$$R(\phi) = R_0 \sec(\phi - \phi_0) . \tag{46}$$

We can also find other solutions corresponding to eq. (45) by going back to eq. (35). This equation can be now written

$$u' + 2 \tan \phi \tan u = 1 . \tag{47}$$

Instead of a mere transformation of dependent variable, we now transform the integration domain: $u(\phi) \to v(\varsigma)$, such that

$$v(\varsigma) = \tan u , \quad \varsigma = \tan \phi . \tag{48}$$

From eqs. (47) and (48), we can write,

$$(1 + \varsigma^2) \frac{dv}{d\varsigma} = (1 + v^2)(1 - 2\varsigma v) . \tag{49}$$

We now use the transformation:

$$T(y)f(y) = \frac{y + f(y)}{y f(y) - 1} . \tag{50}$$

We notice that

$$T^2(y)f(y) = \frac{y + T(y)f(y)}{y T(y)f(y) - 1} = f(y) . \tag{51}$$

This transformation could be extended to an algebraic structure with interesting properties. It could also be very useful in the kind of modelling we are proposing in this work.

The following development aims to show the usefulness of the transformation (50). After transforming eq. (49) according to

$$v(\varsigma) = T(\varsigma)N(\varsigma) = \frac{\varsigma + N(\varsigma)}{\varsigma N(\varsigma) - 1} , \tag{52}$$

we get,

$$\frac{dN}{d\varsigma} = \frac{2(1 + N^2)}{\varsigma N - 1}. \tag{53}$$

The inverse function $\varsigma(N)$ satisfies a linear differential equation:

$$\frac{d\varsigma}{dN} = \frac{\varsigma N}{2(1 + N^2)} - \frac{1}{2(1 + N^2)}. \tag{54}$$

The elementary quadrature with an integrating factor will lead to a particular solution

$$\varsigma(N) = -\frac{1}{2}(1 + N^2)^{1/4} \int (1 + N^2)^{-5/4} \, dN. \tag{55}$$

The integral is an eliptic integral of the 2nd kind and we have to proceed through numerical steps in order to obtain the $R(\phi)$ function, namely $\varsigma(N) \to N(\varsigma) \to v(\varsigma) \to R(\phi) = \sec^2 \phi \frac{v}{\sqrt{1+v^2}}$.

5. Geodesic Curves Along Steiner Points

Many arguments have been posed in recent works[7,8] for considering the position sites of carbon and nitrogen atoms as candidates for Steiner points. As we have emphasized in the 2nd section, the conformation of evenly spaced atom sites is essential for modelling. We could think that the consideration of Steiner points which also fulfil this regularity criterion is a fundamental part of the information which is missing for a feasible modelling. In this contribution, we do not develop the theory of Steiner points and Steiner trees and its usefulness for understanding the biomolecular architecture. Interested readers are recommended to study the recent literature. The only information about Steiner trees which will be used here is the condition of meeting edges at angles $2\pi/3$ on each Steiner point. This is known to lead to the relation

$$z(\phi) = R(\phi)\sqrt{A(A + 1)} \tag{56}$$

where

$$A = 1 - 2\cos(\phi - \phi_0). \tag{57}$$

Eq. (32) will be written as

$$(B + 1)R'^2 + B'RR' + \left(1 + \frac{B'^2}{4B}\right)R^2 = a^2 \tag{58}$$

where

$$B(\phi) = A(A + 1). \tag{59}$$

Eq. (58) cannot be written as a product of two linear differential equations of the form:

$$(a_1 R' + a_2 R + a_3)(b_1 R' + b_2 R + b_3) = c \qquad (60)$$

where $a_1 = a_1(x)$, $a_2 = a_2(x)$, $\ldots$, $c = c(x)$.

Since this will lead to

$$2\frac{a_2}{a_1} = \frac{B'}{B+1} = 2\frac{b_2}{b_1}$$

$$2\frac{a_1}{a_2} = \frac{4BB'}{4B+B'^2} = 2\frac{b_1}{b_2} \qquad (61)$$

$$\frac{a_1}{a_2} = \frac{b_1}{b_2}$$

or

$$B'^2 + 4B(B+1) = 0. \qquad (62)$$

This last equation cannot be satisfied by $B = A(A+1)$, $A = 1 - 2\cos\phi$ and instead of the last technique, we shall use the transformation

$$R(\phi) = \eta(\xi)(B+1)^{-1/2} ; \; \xi = \int \frac{d\phi}{H} \qquad (63)$$

where

$$H^2 = \frac{4B(B+1)^2}{B'^2 + 4B(B+1)} \qquad (64)$$

and we get

$$\eta'^2 + \eta^2 = H^2 . \qquad (65)$$

We have solved an analogous problem before. We make the transformations:

$$\eta' = \eta \cot\omega(\xi) ; \; \eta(\xi) = \pm H \sin\omega(\xi) \qquad (66)$$

and

$$M(\xi) = \tan\omega(\xi) . \qquad (67)$$

We have:

$$M' = \frac{a}{H}(1 + M^2)\left(1 - \frac{H'}{a}M\right) . \qquad (68)$$

This is also an Abel differential equation which solution is given in the general case by elliptic functions. The original function $R(\phi)$ can be obtained from eqs. (63), (59), (65) and (67). We can write

$$R(\phi) = \pm H\left[(B+1)(1+M^2)\right]^{-1/2} . \qquad (69)$$

We do not solve these equations here. Our aim was to present the possibility of alternatives to helices to model biomolecular structures. Actually, Nature uses many of these alternatives as can be observed from the molecular conformation of natural polymers.

6. Concluding Remarks

The summary of the ideas advanced in this work can be given in the following way: Researchers do not look for helices at first. They look for regularity of distribution of atom sites in the biomolecular structure. They look for a special law of formation of structure to be derived from their observations. They use their primitive notions of Euclidean distance and length to be measured along spatial curves. The first trial for identifying regularity is the assumption of evenly spaced points according to the self-assumed Euclidean norm. The additional requirement of evenly spaced points according the metric defined on the curve is always possible to be made, by restricting the search to geodesic curves. This is due to the fact that the values of the chosen parameter can be always made to be evenly spaced in its 1-dimensional domain (a straight line for cartesian coordinates, a circumference for cylindrical coordinates and so on). If the two assumptions come together in the mind of the researcher, then the geodesic curves are restricted to geodesics of the right circular cylinder and these are helices or straight lines. The set of two conditions is then necessary and sufficient for the existence of helices as the unique possible non-trivial solution. However, if we keep the freedom of looking for regularity only along a hypothetical curve which is the candidate to model the sequence of atom sites, then we will be able to find many interesting spatial curves for the modelling of atom distribution in natural polymers. Some of the solutions to the equations derived in the previous sections can be also provide examples of successful modelling. A similar history should be told for the requirement of regularity leading to surface modelling on biomolecules. This is now in progress and will be published elsewhere.

References

1. J. D. Watson and F. H. C. Crick, *Molecular Structure of Nucleic Acids. A Structure for Deoxyribose Nucleic Acid. Nature* **171**, 737 (1953).
2. L. Pauling, R. B. Corey and H. R. Branson, *The Structure of Proteins: Two Hydrogen-bonded Helical Configurations of the Polypeptide Chain. Proc. Natl. Acad. Sci. USA* **37**, 729 (1951).

3. E. Schroedinger, *What is Life? Mind and Matter. Cambridge University Press, Cambridge* (1967).

4. Z. Nakayama, H. Segur and M. Wadoti, *Integrability and the Motion of Curves. Phys. Rev. Lett.* **69**, 2603 (1992).

5. R. P. Mondaini, *The Steiner Tree Problem and Its Application to the Modelling of Biomolecular Structures. In Mathematical Modelling of Biosystems, Springer Verlag, Heidelberg* (2008).

6. E. Kamke, *Differentialgleichungen - Lösungsmethoden und Lösungen. B. G. Teubner Stuttgart, Band I* (1983).

7. R. P. Mondaini, *Euclidean Full Steiner Trees and the Modelling of Biomolecular Structures. In BIOMAT 2006 - World Scient. Co. Pte. Ltd., Singapore, London* (2007).

8. R. P. Mondaini, *An Analytical for Derivation of the Steiner Ratio of 3D Euclidean Steiner Trees. Journal of Global Optimization, June 2007, Springer Verlag, in press, available online.*

AGENT-BASED MODELS
OR DIFFERENTIAL EQUATIONS:
TWO WAYS TO LEARN ABOUT SELECTION
MECHANISMS IN GERMINAL CENTRES[*]

MICHAEL MEYER-HERMANN[†]

Frankfurt Institute for Advanced Studies (FIAS)
Ruth-Moufang-Str. 1
60438 Frankfurt/Main
Germany

The germinal centre reaction selects B cells from a large diversity of clones in order to optimise the efficiency of an immune response. We discuss two rather different approaches to tackle the puzzle of selection mechanisms in germinal centres with mathematical methods. A space-averaged differential equation approach is compared to a space-resolved agent-based approach. The same two novel selection mechanisms could be localised with both methods which increases the predictive power of the result. In addition, the comparison of both methods allows interesting conclusions about the suitability of diverse approaches in Theoretical Biology.

1. Introduction

Theoretical Biology has entered a computational phase. For the first time complex systems can be considered with detailed models concerning the constituents of the system. Differential equations have been applied to biological systems with great success for some decades and with today's computational power, the question arises whether they should be complemented by involved computational agent-based models. A critical benchmark for this question is the predictive power that is achieved by differential equation-based versus agent-based modelling approaches. The discussion is controversial: On one hand it is believed that only most realistic simulations are able to capture the complexity of biological systems. On the other hand it is claimed that only drastic simplifications of the biological

---

[*]This work is supported by the EU-NEST project MAMOCELL.
[†]FIAS is supported by the ALTANA AG.

system have the potential to unravel essential interactions and to give novel insights.

The present article will contribute to this discussion by comparing different modelling attempts for germinal centre reactions. Germinal centre reactions are a central part of adaptive immunity during which antibodies are optimised for invading pathogens. This process involves a quasi-evolutionary process of mutation and selection. The basis of selection is not fully understood, and the mathematical models are used to shed light on which kind of selection mechanism is efficient while maintaining experimental constraints.

It will be shown that, indeed, agent-based models have the potential to be physically realistic and richer in properties than corresponding differential equation models. However, the general behaviour is also well captured by the latter models. A discrimination between both models is presently not possible because the experimental data concerning the properties of the agents are not sufficiently known to determine the agent-based model on the quantitative level. Once these data are available, the richer agent-based models have the potential to make quantitative predictions of cellular behaviour. In contrast the differential equation model may not go far beyond the insights it can already provide today.

2. Selection in Germinal Centres

The germinal centre is a central part of humoral immune responses[1,2]. B cells develop in this environment to antibody producing plasma cells and are, thus, central to an efficient immune response. While all cells of the body have the property that they conserve the inherited genetical information, this is different for B cells. By recombination of modular genetic information a large diversity of antibodies is already achieved with the germline B cells stemming from the bone marrow. However, in the periphery these cells have the remarkable property to change their genetic information by a process called somatic hypermutation. It is exactly this process that is located in the germinal centre environment.

At first B cells have to be activated by encounter of some antigenic stimulus. The encoded antibody in a primary reaction stems from the germline repertoire and will generally be of low or moderate affinity[3]. The first encounter of antigen happens in follicles situated in secondary lymphoid tissue. Under normal conditions B cells will need T cell help to get fully activated. To this end B cells migrate to the border of the B cell follicle where the T zone is located. After successful interaction with T cells the B cells initiate the germinal centre reaction in the follicle.

In a first phase the B cells expand with an extremely high proliferation rate of about 6 hours[4,5]. After three days of expansion a population of about 15000 cells is reached. The process of somatic hypermutation[6,7] generates a large diversity of clones and the mutations are believed to be random[8,9]. However, the mutations are focussed on the specific variable region encoding the antigen specific part of the antibody[10,11]. The question which will be addressed in this article is how the high affinity clones in this diverse pool of antibodies are localised and further expanded to induce a high affinity immune response. This process is denoted as affinity maturation[12].

Within the germinal centre two types of B cells can be distinguished by corresponding surface markers. The first is proliferating and somatically hypermutating, the second is not but instead apoptosis is activated. We will denote these two classes of B cells as centroblasts and centrocytes, respectively. While in centroblasts the antibody expression is downregulated, it is upregulated in centrocytes. This makes centrocytes potential targets of the selection mechanism.

As centrocytes are in a state of activated apoptosis it is plausible that they have to compete for survival signals[13,14]. It is generally believed that centrocytes compete for limited amount of antigen which is presented on follicular dendritic cells (FDC). However, this conviction was also subject to intense discussions[15,1,16,2]. In this article we will provide evidence that antigen is not the limiting factor of competition for survival signals. Instead, the mathematical models presented below show that the efficiency of affinity maturation is optimised if other mechanisms are assumed, which are further characterised.

In earlier mathematical models of the germinal centre reaction several insights were already possible. A fundamental result was that positively selected centrocytes will not necessarily differentiate to plasma cells right away. A successful germinal centre reaction was shown to be possible only under the assumption of selected B cells to re-proliferate and further mutate[17,18,19] — a process denoted as recycling. By this a step-wise optimisation process of antibodies becomes possible. The probability of high affinity clones is to low without this process such that only a few high affinity plasma cells would be produced per reaction. On the basis of experimental data it was estimated that 80% of the selected cells are recycled[20]. The mechanisms assumed in mathematical models so far are competition for antigen or for antigen presenting sites[21,18,22,23,19,20,24,25,26,27,28]. In the following we propose new selection mechanisms[29,30].

Antibodies are represented in both subsequent models in the shape space[31]. This is a phenomenological space in which each mutations leads to a displacement to a nearest neighbour. The affinity between antibody and antigen is defined as a Gaussian

$$\rho(\phi) = \exp\left\{-\frac{||\phi - \phi^*||_1^2}{\Gamma^2}\right\} \tag{1}$$

in dependence on the squared Hamming-distance of a clone ϕ to the optimal clone ϕ^* (which has affinity 1) and with a width of $\Gamma = 2.8$[32]. The considered shape space is assumed four dimensional[20,33] and the number of clones per dimension is 13 giving rise to 28561 clones.

3. Differential Equations for Germinal Centres

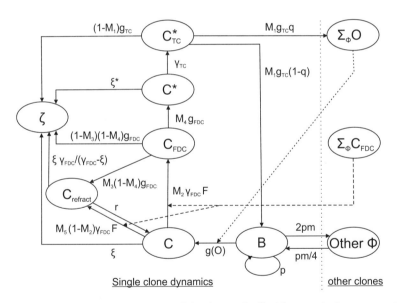

Figure 1. **The germinal centre model schematically** The germinal centre models as defined by Eqs. (2) – (9) is illustrated schematically for a single clone (left to the vertical dotted line). The connection to other clones (right of the vertical dotted line) is only given structurally. The differentiation states of the B cells are shown in the ellipses. Each arrow denotes a reaction and the corresponding rates are attributed to the arrows. The quantities and mechanisms are defined after Eqs. (2) – (9). Depending on the values of M_i different selection mechanisms are represented. This figure was previously published[30].

A set of coupled equations describes the dynamics of the germinal centre where different selection mechanisms are included as shown in Fig. 1. The set of coupled ordinary differential equations is solved for each of the 28561 clones[30]:

$$\frac{dB(\phi)}{dt} = (p - 2pm - g(O))B(\phi) + \frac{pm}{4} \sum_{||\Delta\phi||=1} B(\phi + \Delta\phi)$$

$$+ g_{TC}(1 - q)M_1 C_{TC}^*(\phi) \quad (2)$$

$$\frac{dC(\phi)}{dt} = g(O)B(\phi) + r C_{\text{refract}}$$

$$- (\gamma_{\text{FDC}}(M_2 + M_5(1 - M_2)) F(A, C_{\text{FDC}}) + \xi) C(\phi) \quad (3)$$

$$\frac{dC_{\text{FDC}}(\phi)}{dt} = \gamma_{\text{FDC}} M_2 F(A, C_{\text{FDC}}) C(\phi) - g_{\text{FDC}} C_{\text{FDC}}(\phi) \quad (4)$$

$$\frac{dC_{\text{refract}}(\phi)}{dt} = M_3 g_{\text{FDC}}(1 - M_4) C_{\text{FDC}}(\phi) - \left(r + \frac{\gamma_{\text{FDC}}\xi}{\gamma_{\text{FDC}} - \xi}\right) C_{\text{refract}}$$

$$+ M_5 \gamma_{\text{FDC}}(1 - M_2) F(A, C_{\text{FDC}}) C(\phi) \quad (5)$$

$$\frac{dC^*(\phi)}{dt} = g_{\text{FDC}} M_4 C_{\text{FDC}}(\phi) - (\gamma_{TC} + \xi^*) C^*(\phi) \quad (6)$$

$$\frac{dC_{TC}^*(\phi)}{dt} = \gamma_{TC} C^*(\phi) - g_{TC} C_{TC}^*(\phi) \quad (7)$$

$$\frac{dO}{dt} = \sum_{\phi} g_{TC} q M_1 C_{TC}^*(\phi) \quad (8)$$

$$\frac{dA}{dt} = -M_6 g_{\text{FDC}} M_4 \sum_{\phi} C_{\text{FDC}}(\phi) \quad (9)$$

with

$$F(A, C_{\text{FDC}}) = \frac{A}{A + M_6 K_A} \left(1 - \frac{M_0 \left(\sum_{\phi'} C_{\text{FDC}}(\phi')\right)}{\left(\sum_{\phi} C_{\text{FDC}}(\phi')\right) + K_{\text{FDC}}}\right). \quad (10)$$

Centroblasts B proliferate with rate $p = 1/6hr$ and mutate with probability $m = 0.5$. They differentiate to centrocytes C with rate $g(O)$

$$g(O) = \frac{g_0}{1 - f_g \frac{O^2}{O^2 + K_O^2}} \quad (11)$$

(with $K_O = 1000$, $f_g = 0.4$ and g_0 adapted to fit realistic population kinetics for each selection mechanism) providing a feedback on the achieved success of the reaction reflected by the total produced output O. Centrocytes die with rate $\xi = 1/6hr$ unless they bind to FDC with rate $\gamma_{\text{FDC}} F(A, C_{\text{FDC}})$. Here the function F induces a reduction of the binding probability depending on the consumption of antigen A with half reduction at K_A. The successful C_{FDC} release the FDC with rate $g_{\text{FDC}} = 1/2hr$. Depending on the success of the interaction they go into a refractory state C_{refract}, form which they return to the initial state C with rate r or die, or they got survival signals and turn into the state $C^\star$. These cells find T cell help with rate $\gamma_{\text{TC}} = 2/hr$ or die with the small rate $\xi^\star = 1/50hr$. The signalling process and further differentiation lasts $1/g_{\text{TC}} = 6hr$. The bound centrocytes $C_{\text{TC}}^\star$ differentiate to plasma or memory cells O with probability $q = 0.2$ ($= 0$ for the first 52 hours) or recycle back to the centroblast state with probability $1 - q = 0.8$. All values are based on experiments[30]. If no value is given the parameter is subject to a considered selection mechanism.

The M_i (with $i = 0, \ldots, 6$) distinguish the different selection mechanisms:

M_i		M_0	M_1	M_2	M_3	M_4	M_5	M_6
CC-FDC binding affinity	A	0	1	$\rho(\phi)$	0	1	0	0
Antigen competition	B	0	1	$\rho(\phi)$	0	1	0	1
Negative FDC-selection	C	0	1	1	0	$\rho(\phi)$	0	0
Two-step FDC-selection	D	0	1	$\rho(\phi)$	0	$\rho(\phi)$	0	0
Multiple CC-FDC	E	0	1	$\rho(\phi)$	1	$\rho(\phi)$	0	0
CC refractory time	F	0	1	$\rho(\phi)$	0	1	1	0
CC refractory time 2	F2	0	1	$\rho(\phi)$	1	$\rho(\phi)$	1	0
Site competition	G	1	1	$\rho(\phi)$	0	1	0	0
CC-TC selection	H	0	$\rho(\phi)$	$\rho(\phi)$	0	1	0	0
Realistic germinal centre	I	0	$\rho(\phi)$	$\rho(\phi)$	0	$\rho(\phi)$	1	1

where CC and TC stand for centrocytes and T cells, respectively. The capital letters serve as acronym for the different models. Model A assumes that the binding process of centrocytes to FDC depends on the affinity of antibody to presented antigen. This is in contrast to model C which assumes that binding occurs anyway but FDC signalling to the centrocyte depends on the affinity, which corresponds to a kind of negative selection. Both affinity-dependent mechanisms are combined in model D. This is extended in model E which allows centrocytes that did not get survival signals in interaction with FDC to return to state C and get a next chance. The frequency of centrocyte-FDC binding attempts might be limited by a refractory time which is done in model F. Such a refractory time might also apply to failed centrocyte-FDC signalling (not binding) which extends model E to F2. Competition for antigen is realised in model B. This is to be distinguished from site competition where the number of sites and not the amount of antigen is limiting (model G). Finally, a new mechanism is proposed in model H which allows T cells to sense the affinity of the antibody of B cells after a successful interaction with antigen on FDC. Model I combines these mechanisms to get a realistic and robust affinity maturation model for the germinal centre.

The success of affinity maturation is shown in Fig. 2 for the different selection models.

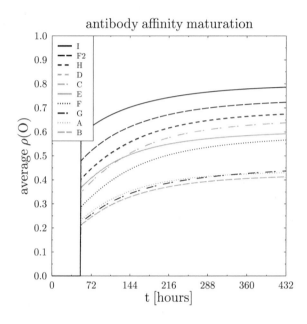

Figure 2. **Plasma cell affinity maturation in the differential equation model:** Affinity maturation is measured by the average affinity of antibody producing output cells $\overline{\rho}(O)$. The inset depicts the mechanisms in the sequence of the final antibody affinity (note that model D is hidden behind the line of model H). The asymptotic average antibody affinity of the models A, B, and G (CC-FDC-binding affinity, antigen competition, and site competition, respectively) is clearly lower than for all other selection mechanisms. This figure was previously published[30].

It is striking that the most discussed selection mechanisms, namely antigen competition (B) and site competition (G) together with simple affinity-dependent centrocyte-FDC binding (A) exhibit the weakest result. All other selection mechanisms show a similar amount of affinity maturation. Note that these results are based on comparable results in terms of population kinetics. This result turns out to be robust against variation of the parameters[30].

4. Agent-based Model for Germinal Centres

While the approach presented in Sec. 2 averages all spatial quantities the germinal centre was also modeled by space-resolved techniques[20,24,25,27,28,29]. The method employed here is a two-dimensional cellular automaton (CA). Each cell is represented by occupation of a node on a regular grid. The motility of the agents is adopted to fit the recent two-photon intravital experiments[34,35,36,37,38]. Each cell carries properties like the type of encoded antibody, its presentation state, the proliferation potential and the cell's differentiation state. It can interact with neighbouring objects (von Neumann neighbourhood) in order to change its intrinsic state. The system is complemented by a second lattice for solubles on which the reaction-diffusion-equation is solved. This soluble-grid contains chemokines and differentiation signals.

All parameters of the model are directly related to quantitative measurements[29]. This implies that the CA is, more specifically, an hybrid dimer stochastic scaled CA with asynchronous update. *Hybrid* denotes the complementation with the soluble grid. *Dimer* because the cell motility involves changing not only the state of one node but also of its neighbour node where the cell is eventually going to[39]. *Stochastic* because all interactions are formulated as reaction rates which are equivalent to probabilities of interaction. Thus, at each time step the object will not make an deterministic change but will perform possible interactions with a specified probability only. *Scaled* means that the units of the CA are all gauged to real world quantities. Thus, the time is gauged with a well-measured duration of a process. Space is given in units of micrometer, and all speeds and reaction rates are provided relative to these quantities. This allows to use physiological measurements in order to determine the parameter values rather than performing a qualitative simulation only. *Asynchronous* defines the way of how the objects on the grid are updated: In each time step each object is touched once and instantly updated. The objects are updated in a randomized sequence which is chosen in every time step in order to avoid any anisotropic effects.

The model for the germinal centre is almost identical to the one introduced in Sec. 2. Therefore, only deviations of this model are considered here, while the full model is described elsewhere[29] Fig. 1 well represents the differentiation states of the individual cells on the CA. Just the reaction rates have to be adopted: The rate g_{TC} in Eq. (7) effectively incorporates the interaction of centrocytes with T cells and its further differentiation.

In the CA this is naturally separated into one duration of interaction and one rate of differentiation. There is no point of introducing an additional corresponding equation in the differential equation model because there is no branching in the intermediate state.

The rates for centrocytes to find an FDC or a T cell are fully deleted in the CA. Finding an interaction partner is now a result of the spatial distribution of the cells. Thus, these two parameters are simply ignored and replaced by the spatial organisation of the CA. The interaction with T cells lasts for 2.1 hours where the TC has to remain polarised to the B cell with highest affinity for 2 hours in order to provide positive differentiation signals. All B cells that do not fulfill this requirement die by apoptosis.

It turns out that the feedback of the produced output, which was introduced for stability reasons in the differential equation model, is not necessary in the CA. In contrast, the centroblast differentiation rate is complemented by a soluble differentiation signal provided by FDC and consumed by centroblasts. Only when this signal was encountered these cells start the differentiation process with a rate that guarantees comparable population kinetics (fit parameter).

The feedback of the total (clone integrated) population of centrocytes bound to FDC is replaced by a limited number of FDC sites in space and by a finite amount of antigen which is presented and consumed by centrocytes.

An additional mechanism is considered here for the purpose of rescuing competition for antigen as a relevant mechanism. To this end masking of antigen on FDC by soluble antibodies which are produced by already differentiated plasma cells is included. Antibodies b are produced by plasma cells at some rate r_{ab}, diffuse on the lattice, and locally bind to antigen a on FDCs to form immune complexes c according to the chemical rate equation

$$\frac{dc}{dt} = k_+ a(t)b(t) - k_- c(t). \tag{12}$$

While the affinity of the secreted antibodies will increase during the course of the immune response, we simplified the simulation by using binding rates characteristic of high affinity antibodies: $k_+ = 10^6/(Ms)$ and $k_- = 10^{-3}/s$[40,41]. This simplification is appropriate because even under such conditions antibody production at a physiological rate does not sufficiently increase affinity maturation.

The results of affinity maturation are shown in Fig. 3.

Again it is striking that the two previously not discussed mechanisms lead to the best results, namely the refractory time between two trials of a centrocyte to get into contact with FDC and the affinity-dependent selec-

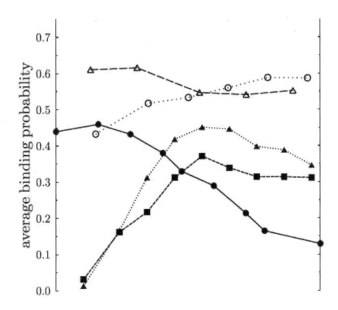

Figure 3. **Plasma cell affinity maturation in the agent-based model:** As in the differential equation model affinity maturation is measured by the average affinity of antibody producing output cells $\bar{p}(O)$. The respective limiting quantity is varied on the horizontal axis in the physiological relevant regime on a logarithmic scale[29]. Symbols denote the assumed selection mechanism: Antigen competition (filled squares), antigen competition with antigen masking (filled triangles), site competition (filled circles), refractory time (open circles), affinity-dependent selection by T cells (open triangles).

tion of centrocytes by T cells. Competition for antigen is too weak when antigen is abundant in the beginning of the reaction and gets selective when affinity maturation in real germinal centres has already occurred. This is not repaired by masking of antigen by soluble antibodies. More importantly, the mechanism is not robust against the antigenic load, while germinal centres develop equally well over a wide range of antigen concentrations[42]. Site competition does not provide a sufficiently strong selective pressure alone but might well contribute to the overall affinity maturation process.

5. Biological Conclusions

A refractory time of centrocytes after an unsuccessful encounter of FDC seems to be a rather intuitive mechanism. At least refractory times of 10 minutes or so may already be induced by cellular and intracellular reorganisations associated with a polarised contact between centrocytes and FDC. Even though unspecific contacts are dissolved in the range of minutes,[35] experiments with lipid bilayers have also shown that especially at low to medium affinity antigen binding may fail to induce the formation of an immunological synapse but can still result in a firm integrin-dependent attachment of B cells[43]. Such short refractory times already have an effect and lead to better affinity maturation than the best case scenarios with site or antigen competition. Mechanisms for longer refractory times further improve affinity maturation but are less plausible.

There is general agreement that centrocyte have to get signals by T cells in order to survive in the germinal centre,[44] and it was seen that affinity maturation depends on the help by T cells[45]. Also the number of T cells in the germinal centre area is quite limited to $5 - 20\%$[46] which makes it plausible that T cell help is a limiting factor. Also it was observed that T cells can interact with several B cells at a time[47] and polarise towards one of these cells. This involves a reorientation of the microtubule organizing center (MTOC) and the Golgi apparatus[47,48,49]. The direction is determined by the location of T cell receptor signalling[50]. When the T cell interacts with several B cells it polarises towards the B cell which is characterised by the highest intensity of peptide MHC presentation[51]. In view of these experimental facts, it was proposed that the intensity of peptide-MHC presentation by B cells reflects the affinity of the encoded antibody to the antigen present on FDC[29]. This idea was recently included into a selection model within the framework of intravital observations of lymphocyte motility[52]. One might consider that B cells encounter antigen on FDC and digest more or less of the immune complexes they take up depending on their B cell receptor affinity to the antigen. The processed immune complexes are then presented in correspondingly higher or lower affinity and, indeed, reflect the affinity of the antibody. It is worth to consider this hypothesis, which challenges the presently held view of selection mechanisms in germinal centres, in further experimental studies.

6. Methodological Conclusions

The two methods used to model selection mechanisms of germinal centres are fundamentally different. One is space-averaging and continuous, while the other is space-resolved and discrete. The agent-based models consider where an individual cell of specific affinity is while the differential equations ignore that information. Each B cell type with a different antibody is represented by an additional population in the differential equation model, while the agent-based approach only considers a single object but with dynamic internal properties. The agent-based model is also more constraint because it has not only to exhibit the same population dynamics as the differential equation model but also to respect the specific morphology of the germinal centre – namely the development of dark and light zone[12,53].

Despite these fundamental differences the results are surprisingly coherent. It is the same selection mechanisms that fail and the same that exhibit the most efficient affinity maturation. This strongly supports the general conclusions and increase the predictive power of the results.

Note, however, that the ordinary differential equation approach is less robust than the agent-based approach. In particular, a feedback of output production (plasma and memory cells) had to be introduced (see Eq. (11)) in order to guarantee the stability of the population kinetics. This effectively captures some effects of the space-resolved representation of the cells and their migration[35,36,37,38] in the agent-based model, which does not need that additional feedback loop. Also the lack of individual-agents inferred that the corresponding looping mechanisms are not well-represented in the differential equation approach. This, in particular, applies to the multiple trial models that allow the same centrocyte to bind and/or interact with an FDC several times within their life time. It is exactly this life time of an individual which is lost as information in the differential equation. As a consequence the efficiency of the selection mechanism relying on a refractory time between two consecutive interaction trials is reduced in the differential equation approach.

The future of Theoretical Biology is a close correlation between well-defined experimental systems and development of quantitative and computationally expensive models with predictive power. It is only in such optimal conditions that it is worth to invest in agent-based models. However, the present discussion has demonstrated that even today, still relying on rather uncertain data, the agent-based models more intuitively captures essential properties of complex systems which are difficult to be represented in differential equations. This, in particular, includes the details of the T cell-related selection mechanism Sec. 5 which can only be effectively modeled in the differential equation model and can be tested in a realistic framework in the agent-based approach.

Acknowledgments

This work is supported by the EU-NEST project MAMOCELL within FP6. I thank Marc-Thilo Figge, Alexandre Garin, Matthias Gunzer, Marie Kosco-Vilbois, and Kai-Michael Toellner for regular, intense and fruitful discussions. Parts of the presented results stem from a collaboration with Dagmar Iber and Philip K. Maini within an EU-Marie-Curie Fellowship in FP6 for a project at Oxford University. FIAS is supported by the ALTANA AG.

References

1. I. C. M. MacLennan. *Annu. Rev. Immunol.* **12**, 117–139 (1994).
2. T. Manser. *J. Immunol.* **172**(6), 3369–3375, Mar (2004).
3. F. G. Kroese, A. S. Wubbena, H. G. Seijen, and P. Nieuwenhuis. *Eur. J. Immunol.* **17**, 1069–1072 (1987).
4. M. G. Hanna. *Lab. Invest.* **13**, 95–104 (1964).
5. Y. J. Liu, J. Zhang, P. J. Lane, E. Y. Chan, and I. C. M. MacLennan. *Eur. J. Immunol.* **21**, 2951–2962 (1991).
6. C. Berek and C. Milstein. *Immunol. Rev.* **96**, 23–41 (1987).
7. K. M. Toellner, W. E. Jenkinson, D. R. Taylor, M. Khan, D. M. Y. Sze, D. M. Sansom, C. G. Vinuesa, and I. C. M. MacLennan. *J. Exp. Med.* **195**, 383–389 (2002).
8. M. Weigert, I. Cesari, S. Yonkovich, and M. Cohn. *Nature* **228**, 1045–1047 (1970).
9. M. D. Radmacher, G. Kelsoe, and T. B. Kepler. *Immunol. Cell Biol.* **76**, 373–381 (1998).
10. J. Jacob, J. Przylepa, C. Miller, and G. Kelsoe. *J. Exp. Med.* **178**, 1293–1307 (1993).
11. I. MacLennan, C. G. de Vinuesa, and M. Casamayor-Palleja. *Philos. Trans. Roy. Soc. Lond. B Biol. Sci.* **355**, 345–350 (2000).
12. G. Nossal. *Cell* **68**, 1–2 (1991).
13. J. M. Dal Porto, A. M. Haberman, G. Kelsoe, and M. J. Shlomchik. *J. Exp. Med.* **195**(9), 1215–1221, May (2002).
14. T.-A. Y. Shih, M. Roederer, and M. C. Nussenzweig. *Nat. Immunol.* **3**(4), 399–406, Apr (2002).
15. G. W. Siskind and B. Benacerraf. *Adv. Immunol.* **10**, 1–50 (1969).
16. M. H. Kosco-Vilbois. *Nat. Rev. Immunol.* **3**(9), 764–769, Sep (2003).
17. T. B. Kepler and A. S. Perelson. *Immunol. Today* **14**, 412–415 (1993).

18. M. Oprea and A. S. Perelson. *J. Immunol.* **158**, 5155–5162 (1997).

19. M. Oprea, E. v. Nimwegen, and A. S. Perelson. *Bull. Math. Biol.* **62**, 121–153 (2000).

20. M. Meyer-Hermann, A. Deutsch, and M. Or-Guil. *J. Theor. Biol.* **210**, 265–285 (2001).

21. M. Oprea and A. S. Perelson. *J. Theor. Biol.* **181**, 215–236 (1996).

22. A. Rundell, R. Decarlo, H. Hogenesch, and P. Doerschuk. *J. Theor. Biol.* **194**, 341–381 (1998).

23. C. Kesmir and R. J. d. Boer. *J. Immunol.* **163**, 2463–2469 (1999).

24. M. Meyer-Hermann. *J. Theor. Biol.* **216**, 273–300 (2002).

25. T. Beyer, M. Meyer-Hermann, and G. Soff. *Int. Immunol.* **14**, 1369–1381 (2002).

26. D. Iber and P. Maini. *J. Theor. Biol.* **219**, 153–175 (2002).

27. C. Kesmir and R. J. De Boer. *J. Theor. Biol.* **222**(1), 9–22, May (2003).

28. M. Meyer-Hermann and P. K. Maini. *J. Immunol.* **174**, 2489–2493 (2005).

29. M. Meyer-Hermann, P. K. Maini, and D. Iber. *Math. Med. Biol.* **23**, 255–277 (2006).

30. M. Meyer-Hermann. *Adv. Compl. Sys.* **10(3)** (2007).

31. A. S. Perelson and G. F. Oster. *J. Theor. Biol.* **81**, 645–670 (1979).

32. M. Meyer-Hermann and T. Beyer. *Bull. Math. Biol.* **66**, 125–141 (2004).

33. A. Lapedes and R. Farber. *J. Theor. Biol.* **212**(1), 57–69, Sep (2001).

34. S. H. Wei, I. Parker, M. J. Miller, and M. D. Cahalan. *Immunol. Rev.* **195**, 136–159, Oct (2003).

35. M. Gunzer, C. Weishaupt, A. Hillmer, Y. Basoglu, P. Friedl, K. E. Dittmar, W. Kolanus, G. Varga, and S. Grabbe. *Blood* **104**, 2801–2809 (2004).

36. C. D. Allen, T. Okada, H. L. Tang, and J. G. Cyster. *Science* **315**, 528–531 (2007).

37. T. A. Schwickert, R. L. Lindquist, G. Schakhar, G. Livshits, D. Skokos, M. H. Kosco-Vilbois, M. L. Dustin, and M. C. Nussenzweig. *Nature* **446**, 83–87 (2007).

38. A. E. Hauser, T. Junt, T. R. Mempel, M. W. Sneddon, S. H. Keinstein, S. E. Henrickson, U. H. von Andrian, M. J. Shlomchik, and A. M. Haberman. *Immunity* **26**, 655–667 (2007).

39. B. Schoenfisch and K. P. Hadeler. *Physica D* **94**, 188–204 (1996).

40. F. Batista and M. Neuberger. *Immunity* **8**, 751–759 (1998).

41. A. Fersht. *Structure and Mechanism in Protein Science.* W.H. Freeman and Company, (1998).

42. L. G. Hannum, A. M. Haberman, S. M. Anderson, and M. J. Shlomchik. *J. Exp. Med.* **192**, 931–942 (2000).

43. Y. R. Carrasco, S. J. Fleire, T. Cameron, M. L. Dustin, and F. D. Batista. *Immunity* **20**(5), 589–599, May (2004).

44. C. de Vinuesa, M. Cook, J. Ball, M. Drew, Y. Sunners, M. Cascalho, M. Wabl, G. Klaus, and I. MacLennan. *J. Exp. Med.* **191**, 485–493 (2000).

45. Y. Aydar, S. Sukumar, A. Szakal, and J. Tew. *J. Immunol.* **174**, 5358–5366 (2005).

46. G. Kelsoe. *Semin. Immunol.* **8**(3), 179–184, Jun (1996).

47. H. Kupfer, C. R. Monks, and A. Kupfer. *J. Exp. Med.* **179**(5), 1507–1515, May (1994).
48. W. J. Poo, L. Conrad, and C. A. J. Janeway. *Nature* **332**(6162), 378–380, Mar (1988).
49. A. Kupfer, T. R. Mosmann, and H. Kupfer. *Proc. Natl. Acad. Sci. USA* **88**(3), 775–779, Feb (1991).
50. C. Sedwick, M. Morgan, L. Jusino, J. Cannon, J. Miller, and J. Burkhardt. *J. Immunol.* **162**, 1367–1375 (1991).
51. D. Depoil, R. Zaru, M. Guiraud, A. Chauveau, J. Harriague, G. Bismuth, C. Utzny, S. Muller, and S. Valitutti. *Immunity* **22**, 185–194 (2005).
52. C. D. C. Allen, T. Okada, and J. G. Cyster. *Immunity* **27**, 190–202 (2007).
53. S. Camacho, M. Kosco-Vilbois, and C. Berek. *Immunol. Today* **19**, 511–4 (1998).

SYNCHRONIZATION PHENOMENA IN PROTOCELL MODELS

ALESSANDRO FILISETTI

European Center for Living Technology
Ca'Minich, Calle del Clero, S.Marco 2940
30124 Venice, Italy
E-mail: alessandro.filisetti@unimore.it

ROBERTO SERRA

Dipartimento di Scienze Sociali, Cognitive e Quantitative
Università di Modena e Reggio Emilia
via Allegri 9, 42100 Reggio Emilia, Italy
E-mail: rserra@unimore.it

TIMOTEO CARLETTI

Département de Mathématique, Facultés Universitaires Notre Dame de la Paix
8 rempart de la Vierge, B5000 Namur, Belgium
E-mail: timoteo.carletti@fundp.ac.be

IRENE POLI

Dipartimento di Statistica, Università Ca' Foscari
San Giobbe - Cannareggio 873, 30121 Venezia, Italy
E-mail: poli@unive.it

MARCO VILLANI

Dipartimento di Scienze Sociali, Cognitive e Quantitative
Università di Modena e Reggio Emilia
via Allegri 9, 42100 Reggio Emilia, Italy
E-mail: marco.villani@unimore.it

1. Introduction

Almost all life forms known today, are composed by cells, fundamental constituting units able to *self–replicate* and *evolve* through changes in genetic information; it is generally believed that this was not the case when first life–forms emerged on Earth almost 4 billion years ago. These *protocells* were much simpler, probably exhibiting only few simplified functionalities, that required a primitive embodiment structure, a protometabolism and a rudimentary genetics, so to guarantee that offsprings were similar to their parents[1,2,3].

Artificial protocells have not yet been reproduced and intense research programs are being established aiming at developing reference models[4,5] to capture the essence of the first protocells appeared on earth and enable to monitor their subsequent evolution. The interest for these researches is motivated either by the quest to understand which are the minimal requirements for a life form to exist and evolve, or by the search for indications about the way in which primitive life might have developed on earth. Moreover besides from their interest for the origin–of–life problem, protocells may be of practical interest in applications[2]: obtain populations of protocells that grow and reproduce, specialized for useful tasks, like drug synthesis and reduce pollution.

Because protocells didn't yet exist, in order to study how they can develop researchers have considered simplified models able to capture general behaviors, without carefully adding complicating details[6]. It is widely accepted that a protocell should comprises at least one kind of "container"(typically an amphiphile) and one kind of replicator molecule, typically a linear polymer able to self–replicate or a system of two or more kinds of polymers which catalyze each others synthesis. One can thus emphasize the existence of two kinds of *key reactions* which are crucial for the good functioning of the protocell: those which synthesize the container molecules and those which synthesize the replicators.

Let us observe that these key reactions may take place at different rates, however, to achieve sustained protocell growth and avoiding death by dilution[7], it is necessary that the two are proceed at equal rate, i.e. that the genetic material has to double when the protocell splits into two offsprings, a condition referred to as *synchronization*. Of course the requirement of duplication of the genetic material at duplication time refers to the average behavior, while each single event is affected by noise and fluctuations.

Let us finally observe that synchronization leads to *exponential growth* of the population of protocells (a straightforward consequence of constant doubling time) and therefore to strictly *Darwinian selection* among protocells, even if the kinetic equations for the replicators have sub linear growth terms[8]. The introduction of a mesoscopic structure (the container) and emergent synchronization change the features of the process. Selection is moved from replicators level to protocells level.

In this paper we consider protocell models where the key reactions take place on the surface [a] of the protocell membrane, that's way they are called *Surface Reaction Models*[9,11]. Our models are inspired by the so–called "Los Alamos bug"[10,5], however due to their abstraction they can cover a larger class of protocell models. In the Los Alamos bug replicators are PNAs which can be found in the membrane, either in its interior or on its surface, but replication takes place when a single–stranded PNA on the surface ligates (via Watson–Crick pairing) the corresponding nucleotides which are supposed to be freely available. The PNA's on the membrane surface also catalyze the formation, from lipid precursors, of amphiphiles which are then incorporated in the membrane.

We have been able to prove under general assumptions[9,11] that synchronization is an emergent property of our models, in contrast to earlier models, like the well–known Chemoton[14] where synchronization was achieved by ad hoc hypotheses concerning the form of kinetic equations. In these papers we considered models with one or two non–interacting replicators in each protocell, here we extend our analysis by considering an arbitrary, but finite, number of replicators inside each protocell, and moreover we allow the replicators to interact between them: by catalyzing or inhibiting the replication of others.

The analytical tools we set up[9] to study such models, are improved to cover these more general cases to answer to the synchronization question; moreover to complete this analysis, dedicated numerical simulations have been performed to deal with cases where the theoretical tools weren't able to provide an answer[15,12].

Our numerical simulations show that synchronization is obtained in a class of models more general than the ones studied previously, but now it is a *fragile phenomenom*: by introducing small changes one can destroy it. Cell duplication time and replicators quantities show an oscillatory behavior in relations to the initial values of parameters. In some cases we observed a time transient where synchronization seems to be achieved, but then an oscillatory behavior enters.

---

[a]Of course this is not the only possibility, and models where the key reactions take place in the interior of the protocell have been proposed[4,7].

2. Surface Reaction Models of Protocells

Let us start by briefly recalling the main basic model with some related results, and referring the interested reader to[9] for a more detailed discussion and analysis. In the second part we then generalize this model to the case of N self–replicating molecules interacting each other.

In the case of a single replicator in the protocell lipid phase [b], let its quantity (mass) be denoted by X and let also C be the total quantity of "container"(e.g. lipid membrane in vesicles or bulk of the micelle). Let V be its volume, which is equal to C/ρ (where ρ is the density, which will be assumed constant) and with S we will denote the surface area, which is a function of V (S is approximately proportional to V for a large vesicle with a very thin surface, a condition which will be referred to as the "thin vesicle case", and to $V^{2/3}$ for a micelle).

We assume, according to the Los Alamos bug hypothesis, that X favors the formation of amphiphiles, and that only the fraction which is near the external surface is effective in doing so. That is because precursors are found outside the protocell. We also assume that the replication of X takes place near the external surface, too.

Let us further assume that

- spontaneous amphiphile formation is negligible, so that only the catalyzed term matters;
- the precursors (both of amphiphiles and templates) are buffered;
- S is proportional to V^{β}, and therefore also to C^{β} (β ranging between $2/3$ for a micelle and 1 for a very thin vesicle);
- diffusion is very fast within the protocell, so concentrations can be assumed to be constant everywhere in the lipid phase;
- the protocell breaks into two identical daughter units when its container size reaches a certain threshold;
- the rate limiting step which may appear in the replicator kinetic equations does not play a significant role when the protocell is smaller than the division threshold.

Under these hypotheses, as discussed in detail in[9,11], one obtains the following approximate equation which describes the growth of a protocell between

---

[b]This model is invariant with respect to the way in which either C or X are measured; for example, if they were measured as number of molecules the equations would retain exactly the same form (of course, the units of the kinetic constants would be different).

two successive divisions:

$$\frac{dX}{dt} = \eta C^{\beta-1} X \quad \text{and} \quad \frac{dC}{dt} = \alpha C^{\beta-1} X, \tag{1}$$

moreover we assume that once C reaches a critical value, here named θ, the protocell breaks into two equal offsprings (halving hypothesis), hence at the beginning of each duplication cycle the initial amount of X equals one half of the value attained at the end of the previous cycle. In between two successive divisions the system is again ruled by Eq. 1.

The generalization to the case where there are more replicators is straightforward. Let:

$$\vec{X} = (X_1, X_2, \dots, X_N), \tag{2}$$

denote the total quantity (mass) of N different types of replicating molecules in the protocell lipid phase. Obviously, all the X_i's must be real and non negative. The N–dimensional generalization of Eq. 1 is then

$$\frac{d\vec{X}}{dt} = C^{\beta-1} M \vec{X} \quad \text{and} \quad \frac{dC}{dt} = C^{\beta-1} \vec{\alpha} \cdot \vec{X}, \tag{3}$$

where $\alpha = (\alpha_1, \dots, \alpha_N)$ and the (constant, real) matrix element M_{ij} represents the contribution of X_j to the growth of X_i. Without loss of generality we will consider the case where det $M \neq 0$, if this were not the case, some of the differential equations for the X_i's would be redundant (i.e. their values at time t could be expressed as a function of the values of the other variables at t) and they could therefore be removed from the set of differential equations under consideration.

An important simplification can now be considered: as it was demonstrated in[9], in order to determine whether there is synchronization in the asymptotic time limit, one can limit himself to consider the $\beta = 1$ case (the final result does not depend on β, while of course this parameter affects the speed with which it is approached). With this simplification, the basic equations (which are valid between two successive divisions) are then

$$\frac{d\vec{X}}{dt} = M \vec{X} \quad \text{and} \quad \frac{dC}{dt} = \vec{\alpha} \cdot X. \tag{4}$$

As outlined above, we assume that division takes place when the mass (or equivalently the volume, since density is assumed constant) of the protocell reaches a certain critical size. Without loss of generality we may then assume that the initial size is one half of the final value (indeed, if the size of the very first protocell were different then it would suffice to consider the evolution from the following generation).

So, starting with an initial quantity of container C at time T_0 equal to $\theta/2$, we assume that once C reaches the critical value θ it will divide into two equal protocells of size $\theta/2$. Let ΔT_0 be the time interval needed to double C from this initial condition, and let $T_1 = T_0 + \Delta T_0$ be the time when the critical mass θ is reached. Since the initial value for C is fixed, ΔT_0 is a function of the initial quantity of replicators, $\vec{X}_0$. The final value of $\vec{X}$, just before the division is denoted by $\vec{X}(T_1)$. Because we assume perfect halving at the division, each offspring will start with an initial concentration of replicators equal to $\vec{X}_1 = \vec{X}(T_1)/2$. The successive doubling time will be denoted by $T_2 = T_1 + \Delta T_1$, and the third generation will start with an initial value $\vec{X}_2 = \vec{X}(T_2)/2$, a.s.o.

We generalize the preceding discussion with the following equations, which refer to the k-th cell division cycle that starts at time T_k and ends at time T_{k+1}:

$$\frac{\theta}{2} = \int_{T_k}^{T_{k+1}} \frac{dC}{dt}(t)\, dt \quad \text{and} \quad \vec{X}_{k+1} = \frac{1}{2}\vec{X}(T_{k+1})\,. \tag{5}$$

Note that in general $\vec{X}(T_{k+1}) \neq 2\vec{X}(T_k)$ and $\Delta T_{k+1} \neq \Delta T_k$, however we will prove in the next section that these conditions can be asymptotically approached.

3. Synchronization in Linear Surface–reaction Models

We will now consider under which conditions the system described in the previous section displays synchronization, in the sense that $\lim_{k\to\infty} \vec{X}(T_k) = \vec{X}_\infty$, for some finite positive value $\vec{X}_\infty$, so that, after several cell divisions, the initial quantity of all inner chemicals between successive duplications approaches a constant value. This requires that

$$\lim_{k\to\infty} \left(\vec{X}(T_{k+1}) - \vec{X}(T_k) \right) = 0\,. \tag{6}$$

As observed above, this implies that, as k grows, also the division time approaches a constant value, so that

$$\lim_{k\to\infty} \Delta T_k = \Delta T_\infty\,. \tag{7}$$

Let us therefore consider the behavior of the system in the continuous growth phase between two successive generation, ruled by Eq. 4. From the linearity of this equation one immediately infers that, during the first replication (i.e. when $0 \leq t \leq T_0$)

$$\vec{X}(t) = e^{M(t-T_0)}\vec{X}_0\,, \tag{8}$$

so that

$$\vec{X}(T_1) = e^{M\Delta T_0}\vec{X}_0 \quad \text{and} \quad \vec{X}_1 = \frac{1}{2}e^{M\Delta T_0}\vec{X}_0\,. \tag{9}$$

The same reasoning applies to all generations, so

$$\vec{X}(T_{k+1}) = e^{M\Delta T_k}\vec{X}_k \quad \text{and} \quad \vec{X}_{k+1} = \frac{1}{2}e^{M\Delta T_k}\vec{X}_k\,. \tag{10}$$

From these last equations one derives a necessary and sufficient condition to ensure synchronization

$$\vec{X}_\infty = \frac{1}{2}e^{M\Delta T_\infty}\vec{X}_\infty\,. \tag{11}$$

Namely $\vec{X}_\infty$ must be an eigenvector of the matrix $e^{M\Delta T_\infty}$, belonging to the eigenvalue 2, i.e. it must be an eigenvector of M belonging to the eigenvalue $\log 2/\Delta T_\infty$:

$$M\vec{X}_\infty = \lambda\vec{X}_\infty \quad \text{and} \quad \lambda = \frac{\log 2}{\Delta T_\infty}\,. \tag{12}$$

Remember that the X_i's are the quantities of the different replicators, therefore they must be real and non negative, so in order for synchronization to take place in a linear system the (real) matrix M must have a real positive eigenvalue λ with such a real, nonnegative eigenvector. The conditions under which these requirements are satisfied are discussed in the next section 4, where we also discuss which eigenvalue has to be chosen among those of the matrix M. In the rest of this section we will assume that λ is a simple positive eigenvalue of the coefficient matrix M associated with a positive eigenvector. Observe that since eigenvectors are determined up to a multiplicative constant, Eqs. 12 do not suffice to determine a unique solution, and we will now provide the formula which determines the actual values of the X_i's.

Since its determinant is not null, the matrix M is invertible, so from Eqs. 4 we get:

$$\frac{dC}{dT} = \vec{\alpha}\cdot M^{-1}\frac{d\vec{X}}{dt}\,, \tag{13}$$

hence the quantity $Q(t) = C(t) - \vec{\alpha}\cdot M^{-1}\vec{X}(t)$, is a first integral, i.e. a quantity constant during each division cycle (the proof is straightforward, $dQ/dt = 0$ derives from Eq. 13). Evaluating $Q(t)$ at the beginning and the end of the k–th division we obtain

$$C(T_k) - \vec{\alpha}\cdot M^{-1}\vec{X}(T_k) = C(T_{k+1}) - \vec{\alpha}\cdot M^{-1}\vec{X}(T_{k+1})\,, \tag{14}$$

recalling that C takes an initial value equal to $\theta/2$ and a final value equal to θ and using the definition of $\vec{X}_k$ (see Eq. 5) we finally get:

$$\frac{\theta}{2} = \vec{\alpha} \cdot M^{-1} \left(\vec{X}_{k+1} - \vec{X}_k \right) , \tag{15}$$

in the limit of large k, calling $\vec{X}_k \to \vec{X}_\infty$, we get:

$$\frac{\theta}{2} = \vec{\alpha} \cdot M^{-1} \vec{X}_\infty . \tag{16}$$

By multiplying the first relation of Eqs. 12 by M^{-1} and then taking the scalar product with α, from Eq. 16 we get:

$$\Delta T_\infty = \frac{\theta \log 2}{2\vec{\alpha} \cdot \vec{X}_\infty} . \tag{17}$$

which is the required relationship. The general approach is now clear: from the matrix of the coefficients M one computes the eigenvalues, λ, which in turn determine the asymptotic interval between two successive divisions ΔT_∞ (Eq. 12b). The components of the eigenvector $\vec{X}_\infty$ are determined except for a constant, which can be determined from Eq. 16, a proof of this statement can be found in[13].

4. Eigenvalues and Eigenvectors

Since the matrix M may have different eigenvalues, it is necessary to find which one should be used in Eq. 12. From Eqs. 9 and 10 one obtains

$$\vec{X}(T_2) = e^{M\Delta T_1} \vec{X}_1 = e^{M\Delta T_1} \frac{\vec{X}(T_1)}{2} = e^{M\Delta T_1} e^{M\Delta T_0} \frac{\vec{X}_0}{2} = e^{M(T_2 - T_0)} \frac{\vec{X}_0}{2} , \tag{18}$$

which can be iterated to yield

$$\vec{X}(T_k) = e^{M(T_k - T_0)} \frac{\vec{X}_0}{2^{k-1}} . \tag{19}$$

Note that, although $2^k \to \infty$, the r.h.s does not vanish as $k \to \infty$ since, at every generation, the numerator is multiplied by a new term. Recall that T_k measures the total time elapsed from the origin of time to the end of the k–th generation. Indeed as $k \to \infty$, $\vec{X}(T_k)$ tends to $\vec{X}_\infty$ and at each generation the r.h.s of Eq. 19 is multiplied by $e^{M\Delta T_\infty}/2$.

We will now suppose that M is diagonalizable, i.e. it has N independent eigenvectors. In this case there exists a nonsingular matrix A such that $A^{-1}MA = \Lambda$, where Λ is a diagonal matrix whose diagonal elements are the eigenvalues of M. The columns of A are the corresponding eigenvectors.

Recalling that $M = A\Lambda A^{-1}$, and that $e^{AMA^{-1}} = Ae^{M}A^{-1}$, from Eq. 19 one gets:

$$\vec{X}(T_k) = Ae^{\Lambda(T_k-T_0)}A^{-1}\frac{\vec{X}_0}{2^{k-1}} \quad \text{and} \quad A^{-1}\vec{X}(T_k) = \frac{1}{2^{k-1}}e^{\Lambda(T_k-T_0)}A^{-1}\vec{X}_0 \,. \tag{20}$$

By introducing a new variable $\vec{Y}(T_k) = A^{-1}\vec{X}(T_k)$, one obtains:

$$\vec{Y}(T_k) = e^{\Lambda(T_k-T_0)}A^{-1}\frac{\vec{Y}_0}{2^{k-1}} \quad \text{and} \quad Y_i(T_k) = \frac{1}{2^{k-1}}e^{\lambda_i(T_k-T_0)}Y_{0i} \,, \tag{21}$$

where Y_i denotes the i-th component of the vector $\vec{Y}$.

If, for every i, $\Re\lambda_i < 0$, then $\vec{Y}$ asymptotically tends to 0 and so does $\vec{X} = A\vec{Y}$ (recall that $\det A \neq 0$) . The same holds, due to the growing denominators in Eq. 21, if $\Re\lambda_i = 0$ for every i. In all these cases the quantities of replicators asymptotically vanish.

Let us then consider the case where, for some i, $\Re\lambda_i > 0$. Let us also suppose that there is a *single* eigenvalue with *largest real part*, without loss of generality we can suppose that this eigenvalue is the first one, λ_1. We will also suppose that λ_1 is a *simple eigenvalue* and we will denote its eigenvector as $\vec{v}_1$. So $\Re\lambda_1 > \Re\lambda_j$ for every $j \neq 1$. As k increases, T_k goes to infinity so does the ratio Y_1/Y_j for all $j > 1$, see Eq. 21b, hence Y_j becomes negligible with respect to Y_1. Therefore $\vec{Y}_\infty$ is proportional, up to a multiplicative constant, to $(1, 0, \ldots, 0)$. But since A diagonalizes M, its columns are the eigenvectors of M, hence $\vec{Y}_\infty$ is proportional to the first column of A, i.e to the eigenvector $\vec{v}_1$.

By definition we get $\vec{X}_\infty = A\vec{Y}_\infty$ and we come therefore to the conclusion that the long term behavior of the system is ruled by the eigenvalue with the largest real part, and by the corresponding eigenvector. Let us call for brevity *ELRP* the eigenvalue with the largest real part.

As we have seen, if the real part of the ELRP is null or negative, the system dies out, and the asymptotic quantities X_i's vanish in successive generations. We may have sustained growth and synchronization only if the real part of the ELRP is positive.

Let us now analyze the physical conditions ensuring that the matrix M has a single eigenvalue with largest real part and a corresponding positive eigenvector. Let us first discuss the important case where all the matrix elements are non negative, i.e $M_{ij} \geq 0$, for all $i, j = 1 \ldots, N$. This implies that there is *no negative interference* between different replicators i and j, the only possible alternatives being that either i favors (e.g. catalyzes) the formation of j or that it does not influence it in any way. Moreover, we must also require that at least one of the entries M_{ij} does not vanish, since otherwise there would be no replication at all.

We can therefore apply the Perron–Frobenius theorem[17,16], which states that if the matrix M is non-negative and non–null then the eigenvalue with the largest module is real, positive and unique, and that there is a non–negative eigenvector belonging to that eigenvalue. Since it is real, the eigenvalue with the largest module is also the ELRP, which thus rules the long time behavior of the system.

In Fig. 1 a simulation of a system of this type is reported: note that the cell division time converges, in successive generations, to the value given by Eq. 12. Moreover, one also observes that the quantity of genetic material at the beginning of the protocell growth cycle tends to a constant value as generations follow generations. Similar results are obtained with non–negative matrices of arbitrary size.

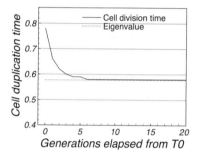 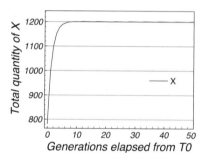

Figure 1.   Data refer to simulations of the system described by Eq. 3 with matrix $M_{11} = M_{22} = 0$, $M_{12} = M_{21} = K$, whose eigenvalues are $\pm K$. On both panel, we report on the x axis the number of generations elapsed from T_0, while on the y axis, on the left panel, cell division time si shown (the level predicted by the theory is also shown), on the right one, the initial amount of X_k at generation k is reported (numerical values of parameters used for this simulation can be found in Appendix B).

Let us now consider the case where some entries of the real matrix M can be negative, while it still possesses N independent eigenvectors. In this case M can still be diagonalized and therefore the eigenvalue(s) with the largest real part determine the long term behavior of the system, recall that the results following Eq. 21 have been obtained by supposing only that M is diagonalizable.

If it happens that the eigenvalue of M with the largest real part is real and positive, and that its eigenvector is non negative, then the behavior of the system is exactly the same as described above (as confirmed by numerical simulations). However now in general: i) the eigenvector of the ELRP may have negative components and ii) the ELRP may be complex, so the previous equations loose their physical meaning.

Let us first consider the case of a real eigenvalue whose eigenvector has positive and negative components [c]. A possible answer could be to try to extend the theory to deal with these cases by assuming that, whenever one of X_i's becomes negative, it has to be interpreted as being actually equal to zero (the non–physical negative value indicating some limitation of the model used). The rationale is that if X_i, starting from a positive value, "becomes negative", it must have passed through the value 0: in this case there is no more replicator in the system, and it is justified to set its value equal to 0. The value of X_i may become positive again at a later time if it is produced by reactions involving other replicators .

Since the analytical theory is not applicable we resort to simulations which show that in this case it often happens that some components get permanently extinguished. If we drop from the matrix M those components which the simulation shows go to extinction, we obtain a reduced matrix M'. If its ELRP is positive and its eigenvector non–negative then the previous analytical theory applies and correctly predicts asymptotic duplication time and quantities of replicators, see Fig. 2.

It may however happen that this latter condition is not satisfied. While several simulations show synchronization, we have indeed also found some different behaviors, where the duplication time does not reach a constant value but seems to oscillate periodically in time, see Fig. 3.

Let us now consider an example where the long time behavior is ruled by a complex conjugate pair of eigenvalues. A simple 2×2 example is given by the following

$$\begin{cases} \frac{dX}{dt} & = aX - qY \\ \frac{dY}{dt} & = qX + aY \\ \frac{dC}{dt} & = \alpha X + \alpha' Y \,. \end{cases} \tag{22}$$

---

[c]Note that, since the components of eigenvectors are determined up to a multiplicative constant, if $\vec{v}$ is an eigenvector so it is also $-\vec{v}$: there is no absolute sign attached to the components, saying that the eigenvector is non–negative means that all its components have the same sign. Therefore the case we are considering now is indeed that of components of both signs. Nonetheless, for brevity, we will sometimes refer to it as the case with "negative components".

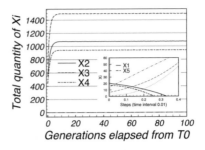

Figure 2. An example of a 5×5 matrix M with negative entries: the replicators which survive are those which might have been predicted by inspection of $\vec{v}_1$ (numerical values of parameters used for this simulation can be found in Appendix B).

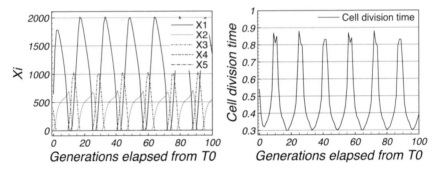

Figure 3. An example of a 5×5 matrix M with negative entries where simple synchronization is not achieved: the graphs show (left panel) the time behavior of the values of the components of $\vec{X}_\infty$ and (right panel) the duplication time in function of the generation number (numerical values of parameters used for this simulation can be found in Appendix B).

The eigenvalues and eigenvectors are

$$\lambda_1 = a + iq \Rightarrow (1, -i) \quad \bar{\lambda}_1 = a - iq \Rightarrow (1, i) , \tag{23}$$

and the continuous time solution (between two successive divisions) is

$$(X(t), Y(t)) = 2e^{at} \left(d \cos qt - b \sin qt, b \cos qt + d \sin qt \right) , \tag{24}$$

where d and b are real coefficients determined by the initial conditions

$$(X(0), Y(0)) = 2 \left(d, b \right) . \tag{25}$$

The system described by Eq. 24 oscillates in time, and it is impossible to guarantee that both X and Y remain positive. It is possible to simulate the behavior of this system and to prove that Y survives while X gets extinct, even if the quantity of X is greater at the beginning (see Fig. 4).

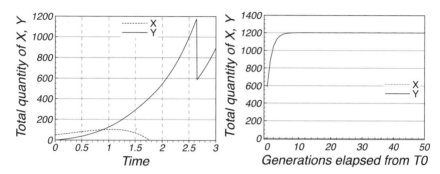

Figure 4.   The behavior of the system described by Eq. 22: left panel, variation in time of the quantities of the two replicators in the first steps of the simulation; right panel, the value of the quantity of X at duplication time approaches a constant value (numerical values of parameters used for this simulation can be found in Appendix B).

Briefly, one can conclude that the analytical method precisely describes the system behavior when the ELRP is real and positive and its eigenvector non–negative. In different cases one has to resort to simulations, however the analytical theory may still help in understanding the system's behavior (like in the case where the reduced matrix M' has the properties required to apply it).

The above analysis could be extended also to non-diagonalizable matrices, which can be reduced to Jordan normal form. The idea being to compute the exponential of the matrix M as we did in Eq. 21 by using some standard linear algebra computation, and to observe that the first component of $\vec{Y}_k$, associated to the single eigenvalue with largest real part grows faster that all the other, and thus in the long run it prevails over the remaining components.

Let us now consider in detail the case where M is not diagonalizable. We will consider here the case where there is a single Jordan block and we left to theAppendix A the analysis of the more general situation involving several Jordan blocks. In the present case there is one real eigenvalue with algebraic multiplicity n and then there exists a non-singular matrix A such that:

$$AMA^{-1} = M_j = \begin{pmatrix} \lambda & 1 & 0 & 0 \\ 0 & \ddots & \ddots & 1 \\ \vdots & \ddots & \ddots & 1 \\ 0 & \cdots & 0 & \lambda \end{pmatrix} \tag{26}$$

where M_j is the standard Jordan form. For all division event k, let us introduce the auxiliary variables, Y_k such that:

$$\vec{Y}_k = A\vec{X}_k, \tag{27}$$

hence eq. 10 can be rewritten as:

$$\vec{Y}_{k+1} = \frac{1}{2}Ae^{M\Delta T_k}A^{-1}\vec{Y}_k = \frac{1}{2}Ae^{AMA^{-1}\Delta T_k}\vec{Y}_k = \frac{1}{2}e^{M_j\Delta T_k}\vec{Y}_k. \tag{28}$$

It is a standard result of linear algebra the computation of the exponential of a $n \times n$ matrix in standard Jordan form:

$$e^{M_j\Delta T_k} = \begin{pmatrix} e^{\lambda\Delta T_k} & \Delta T e_k^{\lambda\Delta T_k} & \frac{(\Delta T_k)^2}{2}e^{\lambda\Delta T_k} & \cdots & \frac{(\Delta T_k)^{n-1}}{(n-1)!}e^{\lambda\Delta T_k} \\ 0 & \ddots & \ddots & \ddots & \vdots \\ \vdots & \ddots & \ddots & \ddots & \frac{(\Delta T_k)^2}{2}e^{\lambda\Delta T_k} \\ \vdots & \ddots & \ddots & \ddots & \Delta_k e^{\lambda\Delta T_k} \\ 0 & \cdots & & 0 & e^{\lambda\Delta T_k} \end{pmatrix}. \tag{29}$$

Let us observe also that for all k and m the matrices $M_j\Delta T_k$ and $M_j\Delta T_\infty$ do commute thus:

$$e^{M_j\Delta T_k}e^{M_j\Delta T_m} = e^{M_j(\Delta T_k+\Delta T_m)} = e^{M_j\Delta T_m}e^{M_j\Delta T_k}. \tag{30}$$

Eq. 28 can be iterated back in such a way we can express $\vec{Y}_{k+1}$ in terms of $\vec{Y}_0$, it follows thus from Eq. 30 that:

$$\vec{Y}_{k+1} = \frac{1}{2^{k+1}}e^{M_j\sum_{m=0}^{k}\Delta T_m}\vec{Y}_0, \tag{31}$$

This relation is the key point to conclude that also in this case, the long-term behavior can be explicitly determined. In fact let us call for short $S_k = \sum_{m=0}^{k}\Delta T_m$, then from Eq. 28, Eq. 29 and Eq. 31 we get:

$$Y_{k+1}^{(i)} = \frac{1}{2^{k+1}}e^{\lambda S_k}\left(Y_0^i + S_k Y_0^{(i+1)} + \frac{(S_k)^2}{2}Y_0^{(i+2)} + \dots + \frac{(S_k)^{n-1}}{(n-1)!}Y_0^{(n)}\right), \tag{32}$$

for all $i = 1, \ldots, n$. Let us now compute the following ratios for all $i = 2, \ldots, n$:

$$\frac{Y_{k+1}^{(i)}}{Y_{k+1}^{(1)}} = \frac{Y_0^{(i)} + S_k Y_0^{(i+1)} + \frac{(S_k)^2}{2} Y_0^{(i+2)} + \cdots + \frac{(S_k)^{n-1}}{(n-1)!} Y_0^{(n)}}{Y_0^{(i)} + S_k Y_0^{(2)} + \frac{(S_k)^2}{2} Y_0^{(3)} + \cdots + \frac{(S_k)^{n-1}}{(n-1)!} Y_0^{(n)}}$$

$$= \frac{1}{(S_k)^{i-1}} \frac{\frac{Y_0^{(i)}}{(S_k)^{n-i}} + \frac{Y_0^{(i+1)}}{(S_k)^{n-i-1}} + \frac{1}{2}\frac{Y_0^{(i+2)}}{(S_k)^{n-i-2}} + \cdots + \frac{1}{(n-i)} Y_0^{(n)}}{\frac{Y_0^{(1)}}{(S_k)^{n-1}} + \frac{Y_0^{(2)}}{(S_k)^{n-2}} + \frac{1}{2}\frac{Y_0^{(3)}}{(S_k)^{n-3}} + \cdots + \frac{1}{(n-1)} Y_0^{(n)}} \tag{33}$$

hence, observing that $S_k \to \infty$, being [d] $\Delta T_k \to \Delta T_\infty > 0$, we can conclude that for all $i = 2, \ldots, n$:

$$\frac{Y_{k+1}^{(i)}}{Y_{k+1}^{(1)}} \xrightarrow[k \to \infty]{} 0, \tag{34}$$

and excluding the unboundedness of $Y_\infty^{(1)}$, we can conclude that:

$$Y_{k+1}^{(1)} \to Y_\infty^{(1)} \text{ and } Y_{k+1}^{(i)} \to 0 \; \forall i = 2, \ldots, n, \tag{35}$$

from which we can drawn the same conclusions as in the case where M was diagonalizable: the long–term behavior is driven by the eigenvalue with largest real part (trivially in this case, because we suppose to have only one eigenvalue with algebraic multiplicity n), while the asymptotic amounts of SRMs are described by the first eigenvector. This result can be straightforwardly generalized as to include the general case where the matrix M has p eigenvalues, λ_i, each one with algebraic multiplicity m_i, and moreover λ_1 is real and has the largest real part of all the remaining eigenvalues. The corresponding treatment is given in the Appendix A.

5. Conclusions

Let us first comment on a simplification which has been used throughout this work, namely that of assuming that the surface is proportional to a power of the volume. This is certainly the case for a spherical micelle (with exponent $2/3$), but in the case of a vesicle it holds (with exponent 1) only in the limit of a very large size.

It can be shown that the finite size effects can be taken into account without modifying our results: synchronization is still obtained. In fact

---

[d]Let us remark in fact that here we don't need to assume the existence of the limit $\Delta T_k \to \Delta T_\infty$, we only need ΔT_k to be definitely strictly positive.

assuming a generic relation between the volume, and thus the container size, and the surface, $S = f(C)$, for some positive increasing function f, then Eqs. 3 have to be modified into:

$$\frac{d\vec{X}}{dt} = \frac{f(C)}{C} M\vec{X} \quad \text{and} \quad \frac{dC}{dt} = \frac{f(C)}{C} \vec{\alpha} \cdot X. \tag{36}$$

But then we can observe that the function given by Eq. 13 is still a first integral and thus the same analysis follows. Another explanation of this result is that we can "rescale" the time [e] by the positive function $C/f(C)$ and thus identifying Eqs. 36 and Eq. 13. This result is supported by a dedicated numerical simulation of a linear system with a single self–replicating molecule X (remember that $C = \rho V$) in a "realistic" vesicle with a thick membrane, which is reported in Fig. 5. Thus we can conclude that synchronization is robust with respect to the finite size and the details of the geometry of the protocell.

In the present paper we address some relevant questions about the synchronization phenomenon for systems where the kinetic equations are linear, while of course non–linear terms may play a key role. While the analysis of non–linear kinetics lies beyond the scope of the present work, let us briefly mention that there are indeed some cases where unbounded growth of the replicator can be observed, as it may happen (depending upon the values of some parameters) when there are two replicators X and Y whose growth rate is proportional to XY.

We are also considering a model where the growth of each replicator is proportional to its quantity multiplied times a sigmoid function which depends upon the presence of other replicators, i.e. a system of the kind:

$$\begin{cases} \frac{dC}{dt} &= \vec{\alpha} \cdot \vec{X} \\ \frac{d\vec{X}}{dt} &= \vec{X} \cdot \vec{\sigma}(W\vec{X}), \end{cases} \tag{37}$$

where $\sigma_i(W\vec{X}) = \tanh\left(\sum_k W_{ik} X_k\right)$ play the role of an activating function.

In several cases synchronization is achieved but, depending upon the values of the entries of the matrix W, a more intriguing phenomenon can sometimes be observed, where the system seems to approach synchronization, but at a certain point there is a sudden drop of one replicator, with a dramatic increase of the replications time. This is followed by a recovery,

---

[e]More precisely let us introduce a new non–linear time $\tau = \int^t C^{-1}(s)f(C(s))\,ds$ and let us denote the quantities C and $\vec{X}$ using this new time, respectively by $c(\tau)$ and $\vec{x}(\tau)$, then Eq. 36 is formally equivalent to Eq.3.

which may be followed by a further "crisis", a.s.o. Contrary to what has been observed in some systems with linear kinetics, the crises do not seem to be periodic in time. Further studies are necessary to give a comprehensive account of the behavior of these non–linear systems.

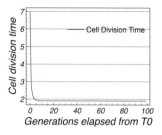 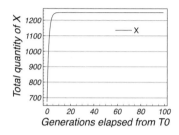

Figure 5. Synchronization in a thick vesicle (we suppose a spherical cell and that the volume of spherical shell increases with the same thickness). On the left the cell division time, while on the right the initial amount of X self replicating molecule in function of the time (numerical values of parameters used for this simulation can be found in Appendix B).

Appendix A. Several Jordan Blocks

In this section we briefly show how the synchronization result of the previous sections can be extended as to include the case where the matrix M has p eigenvalues, each one with algebraic multiplicity m_i, and moreover λ_1 is real and has the largest real part of all the remaining eigenvalues.

In fact in this case Eq. 26 can be generalized by stating the existence of a non–singular matrix A such that:

$$AMA^{-1} = M_j = diag\left(J_{m_1}(\lambda_1), ..., J_{m_p}(\lambda_p)\right),\qquad\text{(A.1)}$$

where $J_{m_i}(\lambda_i)$ is a standard Jordan matrix $m_i \times m_i$ with eigenvalue λ_i. Then we can introduce once again for all k auxiliary variables, $\vec{Y}_k$, $\vec{Y}_k = A\vec{X}_k$. Hence Eq. 10b can be rewritten as

$$\vec{Y}_{k+1} = \frac{1}{2}Ae^{M\Delta T_k}A^{-1}\vec{Y}_k = \frac{1}{2}e^{AMA^{-1}\Delta T_k}\vec{Y}_k = \frac{1}{2}e^{M_j\Delta T_k}\vec{Y}_k.\qquad\text{(A.2)}$$

The remarkable fact is that one can write a relation similar to Eq. 29 and

Eq. 31:

$$\vec{Y}_{k+1} = \frac{1}{2^{k+1}} e^{M_j \sum \Delta T_m} \vec{Y}_0$$

$$= \frac{1}{2^{k+1}} diag \left(e^{J_{m_1}(\lambda_1) \sum\limits_{m=0}^{k} \Delta T_{m_1}}, \ldots, e^{J_{m_p}(\lambda_p) \sum\limits_{m=0}^{k} \Delta T_{m_p}} \right) \vec{Y}_0 \,. (A.3)$$

The algebraic structure is such that the previous analysis performed on the unique Jordan block, is still applicable to each Jordan blocks, so we can conclude that in the long–time behavior we have:

$$Y_{k+1}^{(1)} \to Y_\infty^1, Y_{k+1}^{(m_1+1)} \to Y_\infty^{(m_1+1)} \tag{A.4}$$
$$Y_{k+1}^{(m_1+m_2+1)} \to Y_\infty^{(m_1+m_2+1)}, \ldots, Y_{k+1}^{(m_1+\cdots+m_{p-1}+1)} \to Y_\infty^{(m_1+\cdots+m_{p-1}+1)},$$

and $Y_{k+1}^{(i)} \to 0$ otherwise. But we can say something more about the remaining components, in fact by Eq. 32 we have:

$$Y_{k+1}^{(1)} = \frac{1}{2^{k+1}} e^{\lambda_1 S_k} \left(Y_0^1 + S_k Y_0^2 + \frac{(S_k)^2}{2} Y_0^3 + \cdots + \frac{(S_k)^{m_1-1}}{(m_1-1)!} Y_0^{m_1} \right)$$

$$Y_{k+1}^{(m_1+1)} = \frac{1}{2^{k+1}} e^{\lambda_1 S_k} \left(Y_0^{m_1+1} + S_k Y_0^{m_1+2} + \frac{(S_k)^2}{2} Y_0^{m_1+3} + \cdots \right.$$
$$\left. + \frac{(S_k)^{m_1+m_2-1}}{(m_1+m_2-1)!} Y_0^{m_1+m_2} \right), \tag{A.5}$$

till

$$Y_{k+1}^{(m_1+\cdots+m_{p-1}+1)} = \frac{1}{2^{k+1}} e^{\lambda_1 S_k} \left(Y_0^{(m_1+\cdots+m_{p-1}+1)} \right. \tag{A.6}$$

$$+ S_k Y_0^{(m_1+\cdots+m_{p-1}+2)} + + \frac{(S_k)^2}{2} Y_0^{(m_1+\cdots+m_{p-1}+3)}$$

$$\left. + \cdots + \frac{(S_k)^{m_1+\cdots+m_p-1}}{(m_1+\cdots+m_p-1)!} Y_0^{(m_1+\cdots+m_p)} \right).$$

By assumption λ_1 has the largest real part, hence recalling that $S_k \to \infty$, we easily obtain:

$$\frac{Y_{k+1}^{(m_1+1)}}{Y_{k+1}^{(1)}} \to 0, \frac{Y_{k+1}^{(m_1+m_2+1)}}{Y_{k+1}^{(1)}} \to 0, \ldots, \frac{Y_{k+1}^{(m_1+\cdots+m_{p-1}+1)}}{Y_{k+1}^{(1)}} \to 0, \tag{A.7}$$

thus in the long–therm behavior the only positive component is $Y_{k+1}^{(1)} \to Y_\infty^{(1)} > 0$ which hence determine the asymptotic amount of SRMs.

Appendix B. Parameters and Initial Values Used to Perform the Numerical Simulations

In this last section we show all the parameters and initial values used to perform the numerical simulations and some indication about the simulator. The simulator is developed using Matlab's standard solver for ordinary differential equations (ODE) and in particular function $ode45$[f] with param $nonNegative$ active. The integration is stopped when $C = \theta C(0)$ (in our simulations $\theta = 2$).

Figure 1

$C(0) = 1000, X(0) = 200, Y(0) = 400, K = 1.2$

Figure 2

$C(0) = 1000, \vec{X}(0) = [19.8746, 1.4148, 6.7402, 15.747, 17.054]$

$$M = \begin{pmatrix} 2.1312 & -0.3739 & -0.8514 & -1.2244 & -2.6129 \\ 0.0004 & 1.5765 & -0.3896 & 2.8089 & -1.9983 \\ 2.2535 & 1.2555 & 2.0946 & 0.7982 & -1.1329 \\ 2.9557 & -1.3967 & 2.3441 & 1.3712 & 1.7104 \\ 1.3183 & 2.3212 & -2.0992 & -2.4288 & 0.5195 \end{pmatrix}$$

Figure 3

$C(0) = 1000, \vec{X}(0) = [16.9072, 35.5262, 24.3672, 5.20082, 8.73393]$

$$M = \begin{pmatrix} 2.7144 & -2.9588 & 0.3313 & 2.6009 & 0.3950 \\ 0.4427 & 0.9361 & -1.3434 & -2.0490 & 0.3880 \\ -0.6124 & 0.2124 & 1.6742 & 1.9529 & 0.6036 \\ -1.4740 & -2.3815 & -0.8131 & 1.9732 & 1.0030 \\ -1.9342 & -0.3103 & -0.8019 & 0.4851 & 1.9618 \end{pmatrix}$$

Figure 4

$C(0) = 1000, X(0) = 50, Y(0) = 0, \alpha = 0.9, \alpha\prime = 1, q = 1.2$

---

[f]This function implements a Runge-Kutta method with a variable time step for efficient computation, for details see Matlab website (http://www.mathworks.com) and in particular *Matlab Function Reference* section.

Figure 5

$$\begin{cases} \frac{dC}{dt} = \alpha X \frac{S}{C} \\ \frac{dX}{dt} = \eta X^{\gamma} \frac{S}{C^{\gamma}} \end{cases}$$
$$C(0) = 2000, X(0) = 100, \eta = 1, \alpha = 1.6, \gamma = 1$$

Acknowledgments

Support from the EU FET–PACE project within the 6th Framework Program under contract FP6–002035 (Programmable Artificial Cell Evolution) is gratefully acknowledged.

References

1. Alberts B. et al. : *Molecular Biology of the Cell*, Garland, New York, 2002.
2. Rasmussen S. et al. : *Science*, **303**, 2004, pp 963.
3. Szostak D., Bartel P. B. and Luisi P. L. : *Nature*, **409**, 2001, pp 387.
4. Luisi P. L., Ferri F. and Stano P : *Naturwiss.*, **93**, 2006, pp 1.
5. Rasmussen S., Chen L., Stadler B. and Stadler P. : *Origins of Life and Evol. Biosph.*, **34**, 2004, pp 171.
6. Kaneko K. : *Life: an introduction to complex system biology*, Springer-Verlag, Berlin, 2006.
7. Oberholzer T. et al. : *Biochemical and biophysical Research Communications*, **207**, 1, 1995, pp 250.
8. Maynard-Smith J. and Szathmary E. : *Major transitions in evolution*, Oxford University Press, New York, 1997.
9. Serra R., Carletti T. and Poli I. : *Artificial Life*, **13**, 2007, pp 1.
10. Rasmussen S. et al : *Artificial Life*, **9**, 2003, pp 269.
11. Serra R., Carletti T. and Poli I. : *proceedings BIOMAT2006, World Scientific*, 2007.
12. Serra R. et al. : *Accepted proceedings ECCS–07: European Conference on Complex Systems*, 2007.
13. Serra R., Carletti T., Poli I., Villani M., Filisetti A. : *Sufficient conditions for emergent synchronization in protocell Models*, Submitted, 2008.
14. Ganti T. : *Chemoton Theory, Vol. I: Theory of fluyd machineries: Vol. II: Theory of livin system*, Kluwer Academic/Plenum, New York, 2003.
15. Filisetti A. : *M. Sc thesis, Dept. of Social, Cognitive and Quantitative Sciences, Modena and Reggio Emilia University*, 2007.
16. Minc H. : *Nonnegative matrices*, John Wiley, New York, 1988.
17. Lutkepohl H. : *Handbook of matrices*, John Wiley, New York, 1996.

INDEX